Introductory PLANT SCIENCE

Introductory
PLANT SCIENCE

Cynthia Campbell
Joyce Newton

Kruger Brentt
Publishers
2024

Kruger Brentt Publishers UK. LTD.
Company Number 9728962

Regd. Office: 68 St Margarets Road, Edgware, Middlesex HA8 9UU

ISBN: 978-1-78715-190-1

For information on all our publications visit our website at http://krugerbrentt.com/

PREFACE

Plant science is the study of plants and their interactions with the environment. It encompasses a range of scientific disciplines, including botany, ecology, and agronomy. Plant science is an important field of study, as plants are essential to human life, providing us with food, fiber, medicine, and oxygen. Plant anatomy is concerned with the study of the structure of plants. This includes understanding the different parts of the plant, such as the roots, stems, leaves, flowers, and seeds. Plant physiology is concerned with the study of the function of plants. This includes understanding how plants grow, reproduce, and respond to their environment. Photosynthesis is the process by which plants use sunlight, water, and carbon dioxide to produce glucose and oxygen. Respiration is the process by which plants use glucose and oxygen to produce energy for growth and reproduction. Understanding photosynthesis and respiration is important for understanding how plants interact with the environment and for managing plant growth in agriculture.

Plant ecology is concerned with the study of how plants interact with their environment. This includes understanding the role of plants in ecosystems, such as in nutrient cycling and carbon storage. It also includes studying the impact of environmental factors, such as climate change and pollution, on plant growth and distribution.

Plant biotechnology is concerned with the use of genetic engineering and other biotechnological techniques to improve plant growth and productivity. This includes developing crops that are resistant to pests and diseases, that require less water and fertilizer, and that produce higher yields. Plant science is an important field of study, as it helps us to understand the role of plants in the environment and their importance to human society. By studying plant anatomy and physiology, photosynthesis and respiration, plant ecology, and plant biotechnology, plant scientists can develop new crops and management techniques that improve food security and environmental sustainability. As the world's population continues to grow, the study of plant science will become increasingly important for meeting the challenges of feeding a growing population while preserving the Earth's natural resources.

The present book contains 15 chapters namely plants in our lives; taxonomy and seed germination; how plants grow, part 1; how plants grow, part 2; inside plants; cells, tissues, and woody growth; meristems and flowers; fruit; seeds; grafting; water and light; soils, fertility, and plant growth; sexual reproduction; variation and plant breeding & invasive plants and gmos. It describes every part of each plant in detail, starting with a section on roots, stems and branches, leaves, flowers, followed by seeds. In order to have better understanding of the topic figures have been incorporated. The subject matter is presented in a lucid and succinct style so that even those who do not have a prerequisite knowledge of the subject can easily understand it. A must-have volume for all plant lovers, including naturalists, budding botanists, ecologists as well as gardeners.

We are grateful to all those persons as well as various books, manuals, periodicals, magazines, journals etc. that helped in the preparation of this book. In spite of the best efforts, it is possible that some errors may have occurred into the compilation and editing of the book. Further queries, constructive suggestions and criticisms for the improvement of the book are always welcomed and shall be thankfully acknowledged.

Cynthia Campbell

Joyce Newton

CONTENTS

CHAPTER 1
PLANTS IN OUR LIVES

In this, you'll discover what horticulture is and how it relates to other disciplines that involve the cultivation of plants, and take a deep dive into the different types of scientific experimentation. Then you'll explore some of the plant parts that you eat, so you can start thinking about the plants that are all around us and how we use them in our daily lives.

Plants contribute to our lives in countless ways: from the foods we eat to the clothes we wear, from the structures we build to the flowers we grow in our gardens. David Mark. Pixabay license

Learning objectives

- Define horticulture and describe its disciplines and sub-specialities.

- Apply the principles of experimental design to your own experiments in this course and in daily life.
- Use biological language to describe the parts of the above-and below-ground plants parts that contribute to your diet.

1.1 WHAT IS HORTICULTURE?

Learning Objectives

- Define the term horticulture.
- Describe disciplines related to horticulture.
- Describe some of the specialties in the field of horticulture.

Lance Cheung. Public domain

Horticulture and related disciplines

Horticulture

Horticulture is the art and science of the development, sustainable production, marketing, and use of high-value, intensively cultivated food and ornamental plants. The word is derived from the Latin words hortus (garden plant) and cultura (tilling the soil). Horticulture includes ornamental and food plants that are grown with intensive and individualized care, and often in a small space rather than in an expansive field.

Horticultural plants overview	
Ornamental plants	**Food plants**
Flowers, ornamental shrubs, ornamental trees, turfgrass, native grasses, and forbs are all horticultural plants.	The plants producing the vegetables and fruits we eat are all horticultural plants.
They all have a fairly high value per acre.	They have a high value per acre and, like the ornamental plants, require intensive management.

Agronomy

Agronomy is another term commonly used in reference to food production, and refers to the management of plants grown over large areas with less intensive management than that normally provided to garden plants. Its etymology is from the Greek agros (= field) and nomos (~management). Agronomy fields are larger than gardens, so the plants grown in these fields are less intensively and individually managed than those in most gardens. It is estimated that a single agronomy farm produces food for over 150 people.

Agronomy refers to large scale production of commodity crops like grains.
Photo by huntz. CC BY-NC-ND 2.0

Extensive agronomic crop production requires fewer person-hours of management per acre than intensive horticultural production, which requires more person-hours of management. In contrast, agronomy refers to management of field crops such as cereals (e.g. corn, wheat, rice, barley) and legumes (e.g. soybeans, common beans, peanuts, alfalfa) and a few other high-acreage crops, like cotton. These are typically plants that have a low dollar value per acre, and in many cases the crops are used for animal feed rather than for direct human consumption.

These are grown over extensive areas with less intensive management, or at least with fewer people per acre involved in managing the crop than would be typical of horticultural crops.

Forestry

Forestry is the science or practice of propagating, planting, managing, and caring for forests, and of harvesting products from them. Forestry, which focuses on trees for building materials, pulp, and paper, is a third type of plant-production system, considered separately from horticulture and agronomy, and is not covered in this course.

Agriculture

Agriculture encompasses all farming practices, including this large-scale strawberry operation in Argentina. Photo from World Bank Photo Collection. CC BY-NC-ND 2.0.

Agriculture is the science or practice of farming, including cultivating soil for growing crops and rearing animals to provide food, fiber, and other products. The term is derived from the Latin ager (field) and cultura (tilling the soil). While the Latin root means "field" and implies a larger land area than "garden," "agriculture" typically encompasses both horticulture and agronomy. For instance, the University of Minnesota College of Food, Agricultural and Natural Resource Sciences (UMN CFANS) includes both the Department of Horticultural Science and the Department of Agronomy and Plant Genetics. There is, however, no hard, distinct line separating horticulture and agriculture. While horticulture deals with plants you might find in a garden, it's common to find those same plants (like vegetables and fruits) grown in large fields and harvested in volumes sufficient to supply grocery stores. Other ornamental garden plants, such as annual and perennial flowers, ornamental shrubs, and trees, are planted in extensive, designed landscapes. Field corn used for animal feed is considered an agricultural crop, while sweet corn is considered a horticultural crop, yet they are the same species of plant.

Here is a summary of terms:

- Horticulture: Requires intensive management on fewer acres and higher human input per acre, and produces a higher value per acre. Includes ornamental plants and whole foods (like those found in the produce aisle).
- Agronomy: Requires extensive production on more acres with lower human input per acre, and produces a lower value per acre. Includes animal feed and processed food ingredients, (such as oil, protein, sugar, and starch).
- Agriculture encompasses both horticulture and agronomy.

Domesticated plants and wild plants

The plants grown in horticulture and agronomy are usually domesticated rather than wild, meaning that humans have selected them, intentionally or unintentionally, for particular characteristics such as adaptation to cultivation in a garden, large showy flowers, or large, sweet fruits. You will learn about the science of plant improvement and domestication in the section on plant breeding.

Because garden plants are grown in modest-sized spaces, the gardener can provide intensive management such as a complex garden design, special care for the soil and plant health, and regular weed control.

In general, then, "Horticulture" refers to domesticated ornamental and food plants that humans grow in modest-sized spaces where they provide intensive management.

Horticulture and Plant Propagation

Science of plants

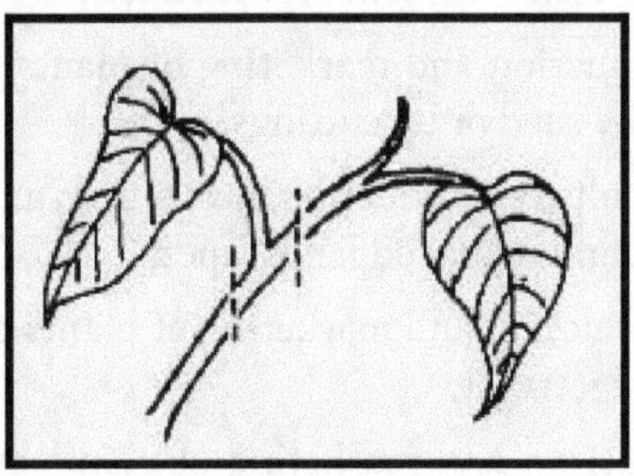

Rooting stem cuttings is a common propagation technique. Photo by Laura Irish. Plant science explores how a plant is put together and how its parts work together during a plant's life cycle — from seed to seed.

Throughout this course, you will study plant structure, growth, and reproduction, applying what you learn to plant propagation practices in the lab portion of this course.

Science in our lives

For many of you, this course might be the only science course you take. The course therefore goes beyond the subject of plants to help you to see the world as a scientist might see it. Science is a systematic enterprise that builds and organizes knowledge in the form of testable explanations based on observations and predictions.

You will learn how to propagate plants, and learn about plant structure and function. But perhaps more importantly, you will learn about science as a way of understanding and appreciating the world around you — in this case, the horticultural world around you.

Review Questions

- What is the difference in meaning between the Latin words *hortus* **and** *agros*?
- Differentiate among horticulture, agronomy, and agriculture.
- In general, which would you expect to provide the highest value per hour of human management: horticultural plants like vegetables, or agronomic commodity crops like corn?

Horticulture Specialties

Within the industry, and also within universities, horticulture is often subdivided into specialties according to the use of the plant or plant part that is produced. Here are six of these specialties:

- Breeding and genetics: development of new cultivars (cultivated varieties) of plants for production via sexual reproduction.
- Floriculture: production and marketing of plants valued for their flowers and propagated by seed or by cuttings.
- Landscape horticulture: production, marketing, and maintenance of plants used in designed and managed landscapes.
- Olericulture: production and marketing of plants or plant parts valued for culinary use as vegetables.
- Pomology: production and marketing of plants or plant parts valued for their culinary use as fruits including nuts.
- Post-harvest management: development of practices that maintain quality and prevent spoilage of harvested horticultural plants or plant parts during storage and transportation.

Review and looking ahead

Plant propagation refers to plant multiplication, or making many plants from just a few. Two broad categories of multiplication will be addressed in this course:

- Asexual reproduction: ausing new plants to arise from plant parts like leaves, stems, or roots, or from storage organs like tubers or rhizomes.
- Sexual reproduction: making new plants from spores or seeds.

1.2 SCIENCE AND EXPERIMENTATION

Learning objectives

- Describe why science is considered a discipline of philosophy.
- Summarize the four basic types of experiments.
- Apply the principles of experimental design in this course and in your daily life.

Thinking about science

The primary goal of this section is to help you think about the nature of science. You might be taking this course to fulfill an undergraduate requirement for a biology course with a lab. This course fulfills that requirement because we investigate the process behind using science as a way of learning about the natural world around us. If you're starting down the path to becoming a plant scientist, understanding the nature of science will be essential for you in your career

Regardless of whether you're going to pursue a career as a scientist, now is a good time reflect on the nature of science, and to understand how scientific thinking can become a strategy for resolving many issues that you confront during daily life.

Scientific inquiry

While "science" is a word commonly used in our culture, in popular use it is rarely spoken of as a philosophy. By identifying science as a philosophy we are taking an epistemic view, one focusing on how knowledge is acquired.

At its core, science is a mode of inquiry: a way of acquiring new knowledge about the world around us and a strategy for understanding the inner workings of elements in that world. Scientists believe that if we follow the principles of this philosophy we will continue to expand our knowledge about how things work in the world around us. This systematic approach is called the "scientific method."

There are two key steps in the scientific method:

- Hypothesis building through reflective observation.
- Hypothesis testing through experimentation.

A "hypothesis" is a question or proposed explanation made on the basis of limited evidence and used as a starting point for experimentation. Experimentation is commonly equated with science—rightly so, because hypotheses evaluated on the basis of evidence generated through experiments. Experimentation, however, isn't the whole story. Science—including the development and testing of new hypotheses—is also a creative endeavor.

Watch this video about scientific inquiry:

Scientific inquiry has generated a vast body of knowledge about the world around us. Your school science classes might have required you to memorize facts and relationships, and pay attention to detail. Sometimes such memorization leads students to believe that science is just an accumulation of facts rather than the process behind discovering all of that information.

Scientific discovery builds on what is already known. Even the most accomplished scientists initially approach a problem by learning what is already known. Armed with that information, they then apply their own creativity to form new hypotheses about something they have observed, and design experiments to test those hypotheses. They also communicate their results publicly so that others can benefit from their work and have the opportunity to challenge conclusions. In this way, science builds on itself.

The foundational knowledge you learn in science classes prepares you to develop and test hypotheses and to make new discoveries of your own. While a good memory may help you pass a science classes, you will absorb a body of knowledge more effectively when you learn how facts fit and work together in systems rather than learning through the brute force of memorization.

In this section we work from the point of view that science is a way of acquiring knowledge—a mode of inquiry—and that this mode of inquiry follows a process called the scientific method. Those who follow the philosophy of science:

- Use it to understand how the natural world works.
- Start by learning what is already known.
- Carefully observe the subjects of their scientific inquiry and look for details about form, function, and interaction with the environment.
- Develop hypotheses about the inner workings of natural phenomena not yet understood.

- Test their hypotheses by making observations, conducting experiments and collecting and evaluating evidence.
- Communicate with others about their hypotheses, experiments, and the outcomes of their studies so that others can repeat, validate, and build upon their work.

Although science is typically used to understand how the natural world works, it is also regularly applied to the development of new technologies that are based on these natural phenomena and to the solving of problems associated with the natural world.

Putting the scientific method to work

As noted, the scientific method relies on building hypotheses and then testing them through experimentation. In the lab section of this course you will develop hypotheses about the effects of various treatments on propagation success and then conduct experiments to test those hypotheses. Because experimentation is such a key component of the scientific method, we'll spend time characterizing and examining four types of experimentation and explore whether they are part of the scientific method. While each is valuable when applied in the right circumstances, only one clearly follows each step of the scientific method to uncover new knowledge about the natural world.

Types of experiments

The types of experimentation we will cover are:

- Demonstration
- Evaluation
- Exploration
- Discovery

Demonstration experiments

Demonstration experiments are a classic method used in educational settings to help students learn and understand known relationships already discovered by others. Learners will usually have had prior exposure to the relationships through preliminary observations, lectures, reading, and discussions, and will have some sense of what the experimental outcome might be.

Good demonstration experiments actively involve the learner, who manipulates the experimental materials, applies the treatments, and observes the outcomes, then gathers, analyzes, and interprets the resulting data. Poor demonstration experiments, in contrast, make learners only passive witnesses to something done by an expert at the front of the classroom.

In the plant propagation labs for this course, you will be actively engaged in demonstration experiments. Although you won't be creating new knowledge, the knowledge will likely be new to you. The hands-on experience of conducting the experiments will help you to learn the concepts more effectively than if you only read a textbook or listened to a lecture. The techniques you learn and use in demonstration experiments often contribute to the learning experience as much as the relationships revealed at the experiment's conclusion. Employing these techniques will help you gain an understanding of many biological functions, such as the production of adventitious roots and mechanisms for seed dispersal.

Many experiments conducted in lab courses are demonstration experiments. Photo by Salish Sea Expeditions. CC BY-NC-ND 2.0

While demonstration experiments are valuable for actively learning a body of scientific knowledge previously discovered and communicated by others, the experience is specifically orchestrated for teaching and learning, not for the discovery of new information. Yet since the knowledge is new to the learner, it can still bring the joy of personal discovery and a sense of accomplishment.

In summary, demonstration experiments:

- Are designed for teaching and learning.
- Address relationships that may be new to you, but are otherwise known.
- In their best forms, actively involve the learner.
- May emphasize experimental techniques, in addition to outcomes, as part of the learning experience.
- Are not the types of experiments that are at the core of practicing science as a way to uncover new knowledge.

Evaluation Experiments

Evaluation experiments are designed to help us make decisions, and to choose from a number of options. They might, for instance, help us determine the efficacy of a new treatment relative to a known treatment, or decide on further experimentation. An evaluation experiment will highlight a compound, a technique, a piece of equipment, or an organism, and will include a control and/or other alternatives.

Evaluation experiments are common in horticultural and agronomic research, where the purpose of the experiment is to identify, for example, the best cultivar, production method, pest control, fertility regime, or light intensity for growing a crop. Correct experimental design is crucial for assuring that conclusions from the experiment are meaningful and credible.

This field experiment is testing different living mulches between rows of strawberries. Photo by University of Minnesota West Central Research and Outreach Center.

These experiments are typically used in the development of new technologies to identify the best method for the desired purpose (e.g., which pesticides are effective against the target insect, but not harmful to non-target insects). They are not used to discover new knowledge about how the world works, as they typically don't advance our understanding of the natural world. The information from an evaluation experiment might, however, point the way to additional experimentation that does help us discover new knowledge. This is particularly true if the outcome of an evaluation experiment is unexpected or novel.

In summary, evaluation experiments:

- Are used to help in decision-making.
- Help users choose a winner or determine efficacy relative to other alternatives.
- Are commonly used when evaluating and recommending horticultural production methods.
- Can be useful in solving problems and developing technologies.
- Require proper experimental design (e.g., comparison to a control) for credibility and meaningfulness.

Exploration Experiments

Some scientists specialize in observing and cataloging nature, and some, like members of the group that discovered a new species of hominid, Homo naledi in a South African cave, aggressively search for previously unknown phenomena. In the botanical realm, such scientists study the diversity of organisms within habitats, discover new species, or are in other ways very skilled in "seeing" nature. Explorer-scientists recognize and appreciate detail and can identify the enormous diversity among plants by comparing characteristics that might be overlooked by others. They may also have the capacity to recognize possible interrelationships among organisms and with habitats, making their work particularly important to science. They might notice, for instance, that a particular species of plant is commonly found in wet areas but not in dry, or that a particular vegetable tastes sweeter when grown at higher altitudes than when grown closer to sea level. They don't confirm the cause of these relationships, but are the first to notice them.

This scientist is collecting plants in Ecuador to identify unknown species and to determine relatedness to other plants. Photo by Dr. Eric Tepe, University of Cincinnati.

Explorers' observations are essential to stimulating the development of sound, testable hypotheses. The possible relationships they propose must be tested to determine whether those relationships actually exist, or are artifacts of other effects. Explorers help develop hypotheses, but the work of exploration, cataloging, and seeing possible relationships don't prove or disprove the hypotheses or necessarily generate new knowledge about relationships. The work does, however, result in new information about the existence of the object or phenomenon itself. An exception is exploration done to test a hypothesis, such as a mission to test the hypothesis that a particular type of ecosystem is required for reproduction of a particular plant species.

Scientists must resist jumping to conclusions based on exploration and observation alone. If you see two people together many times, for example, you might conclude that they are a romantic couple, when in fact they are brother and sister. Relationships hypothesized as a result of exploration and observation must be experimentally tested before they are accepted or rejected.

Exploration experiments uncover new things, many of which can be exciting and eventually change our view of the world. While one of their greatest values is that they lead to the development of new and stronger hypotheses about how the world works, they go so far as to test those hypotheses or fully engage in the cycle of knowledge generation associated with the scientific method. Additional experiments based on this new information are required to put this new information in context and to advance our understanding of how the natural world works.

In summary then, exploration experiments:

- Focus on detailed observation of organisms and habitats.
- Increase our knowledge of the natural world.
- Identify potential relationships that need to be tested.
- Are essential to sound and testable hypothesis-building.

Discovery Experiments

Discovery experiments are central to the use of the scientific method in tasks ranging from problem solving to the discovery of new knowledge. They focus on uncovering new relationships and solving problems, follow the scientific method, test hypotheses and their predicted outcomes, and utilize a careful design in order to maintain meaningfulness and credibility.

The similarity between the scientific method and Kolb's Experiential Learning Cycle is not an accident. The scientific method is a practical strategy based on how we sense and experience the world around us and used to solve problems encountered during those experiences.

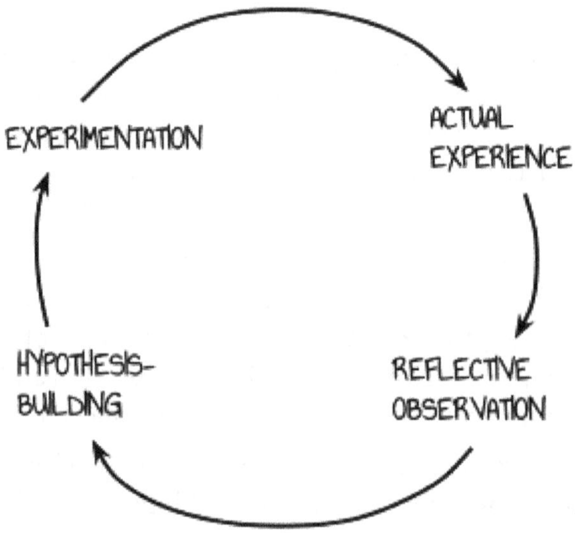

The scientific method is a great example of the experiential learning theory. Illustration by Emily Tepe.

The diagram above illustrates a combination of the scientific method and Kolb's four-step experiential learning, describing a cyclic process for solving problems that can be applied to disciplines as diverse as molecular biology, global warming, and even appliance repair. While you might initially think that appliance repair doesn't belong in that list, the difference is one of application, not method. Though far removed from the esoteric scientific discoveries we associate with scientific method, appliance repair follows the same steps. Appliances are often, and quite literally, boxes, where you don't know what is going on inside. But what's going on inside is knowable, and through that knowledge comes repair.

The learning/problem solving/scientific process could theoretically start anywhere in Kolb's cycle. But it will likely start with a problem that needs to be solved, something you don't understand but would like to know more about. You become aware that there is a problem or that you lack understanding because you have an experience where you observe something and then step back and said, "I wonder how that works," or perhaps, "why is that broken?" Through observation you develop a sufficiently adequate description of the problem to start doing some research on what is already known.

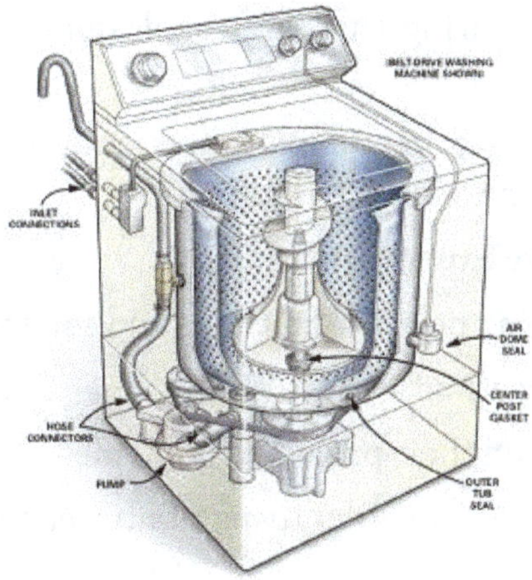

We can apply the scientific method to everyday problems like figuring out why a washing machine isn't working. Image used with permission © HomeTips.com.

With a good description of the problem in hand, you can begin to review what is known through the work of others, and think about what might be going on in your situation and how your new understanding can be applied to the problem. This is "reflective observation." It isn't just sitting back and thinking in a vacuum. You need raw material for your mind to work on, and that only comes through the tough task of gathering and engaging with the background information. There is a very important quiet phase in this process when you let your mind assemble and sort through ideas until alternatives begin to emerge that might lead to a solution. Talking with others and sharing ideas is an important part of this quiet phase.

Sometimes the alternatives are no- brainers (blown fuse?), and sometimes they're more creative (residue from the wrong detergent gunking up the water level sensor?). Regardless of their simplicity or complexity, these become hypotheses that need to be tested. The hypothesis-building stage includes both a statement of how something works or why it isn't working, and predictions about what might happen if the hypothesis is true. In appliance repair, for example, the prediction will likely be that the appliance will function normally. In horticultural molecular biology, it might be that you will see accumulation of a particular type of fatty acid in the cotyledons.

You put the hypothesis to the test by designing an experiment that assesses whether your predictions were right. If the outcome doesn't match your prediction, you reject the hypothesis (the fuse was ok, so that wasn't the problem). If the outcome does match your prediction, you tentatively accept the hypothesis pending further observation (when the fuse was replaced the washing machine worked again, so it might have been a blown fuse, but on the other hand maybe it was just because the motor had time to cool down). As with evaluation experimentation, experimental design is important in assuring that the conclusions from the experiment are meaningful and credible.

Experimentation leads to new experiences and an incremental increase in knowledge, and then the cycle begins again.

In summary, discovery experiments:

- Focus on uncovering new relationships and solving problems.
- Follow scientific method.
- Test hypotheses and their predicted outcomes.
- Utilize a careful design in order to maintain meaningfulness and credibility.

Summary

Of the four types of experiments, only the discovery experiments are core to the process of science in the narrow sense of being a way of acquiring new knowledge. The other three types of experimentation are still important; demonstration and evaluation experiments are valuable for learning and decision-making and for technology development, and exploration experiments are essential for developing testable hypotheses. But discovery experiments are core to science.

Remember: the methodology of effective washing machine repair, when applied to what is unknown about the physical world, is the methodology of science. It's not esoteric; it's good appliance repair.

You might argue that, when applied to a broken washing machine, a discovery experiment results in knowledge that is probably already known by those skilled in appliance repair, so it isn't really new knowledge about how the world works. That's a fair criticism. Use of the scientific method can result in new knowledge about how the world works, but whether it uncovers new knowledge depends on the object of experimentation.

Experimental Design

The methods for designing experiments are carefully studied and often discipline-specific. Methods used in molecular biology, for instance, will be somewhat different from those used in chemistry or in field evaluations of horticultural plants. There are, however, some generalizations we can make about good experimental designs.

Emphasize Comparisons

Experiments include more than just one treatment. "Treatment" refers to the factor that you are varying in your experiment—for example, different cultivars of tomato, different fertilizers, or different amounts of light. Experimental designs incorporate comparison of treatments. You usually compare the treatments to one another and often to a control, which is either the application of no treatment or the application of a customary or standard level of treatment.

If you grow a particular type of tomato in your garden, and find that it produces tasty fruit, would you declare it to be the best tomato variety you could grow? Certainly not. You couldn't even say with certainty that it was the best tomato variety you have ever grown (unless it is the only one you have grown). Next year, however, you could grow that tomato as your control, and grow two other varieties that your neighbors like, and compare fruit quality (appearance, flavor, yield, sugar content). You could then say something definitive about the three tomato varieties because you have compared them to each other after growing them next to each other in the same year and environment.

Replicate Treatments

The same treatment is applied to more than one "experimental unit"—the object that receives the treatment. In the example above, the tomato plant is the experimental unit, and you would perhaps plant two or three seedlings of each tomato variety rather than just one. Think of a treatment as something like a fertilizer spread on a patch of land. The patch of land is the experimental unit, while the fertilizer is the treatment.

By applying the treatment to more than one experimental unit you can estimate the variation you get when two experimental units are treated the same, and compare this to the variation when experimental units are given different treatments. If the treatments actually differ in their effectiveness, you would expect the variation between experimental units given different treatments to be much greater than the variation between those given the same treatment. This is one of the fundamental ways in which experiments are statistically analyzed and treatments declared significantly different or not.

Randomize Treatments

Once you know how many treatments you are going to apply, and how many replications you want, the product of these two quantities (# treatments × # replications) equals the number of experimental units you need. For instance, if you have three fertilizers you want to test, plus a control, you have four treatments. If you want three replications of each treatment, then you 4 treatments x 3 replications = 12 experimental units or patches of land where you will apply the fertilizers. The treatments will be randomly assigned to each experimental unit (patch of land). This is done using a random number table and is not just haphazard picking. Randomization helps minimize any bias you haven't recognized in advance and controlled for in other ways.

1.3 PLANT PARTS WE EAT

Learning Objectives

- Summarize the various above- and below-ground plant parts that contribute to your diet.
- Use the correct language of biology when identifying parts of plants.
- Appreciate the diversity of edible plant parts.

Above-ground plant parts we eat

Edible leaves and petioles

In this image of an iceberg lettuce cut in half, you can see how the leaf blades are packed and folded together tightly in the lettuce head. Lettuce is an example of a plant shoot with very short internodes on the stem. This results in a compact but leafy plant. Iceberg lettuce is a type of heading lettuce where older leaves envelop newer leaves forming a solid or semi-solid ball or head of lettuce leaves.

"Iceberg Lettuce" by Кулинарно, CC BY-NC-SA 2.0

Romaine and leaf lettuces exhibit a more open architecture, with the leaves forming a looser head with upright leaves. Romaine lettuce has elongated leaves. There may be some tendency of older leaves to enclose newer leaves, but it is much less pronounced than in iceberg lettuce, and may be absent altogether in some of the garden types. Leaf lettuce lacks the tendency to form heads.

Image by Peter Drache. Pixabay license

Lettuce leaves generally lack a petiole. The blade narrows a bit, but attaches directly to the node. A leaf lacking a petiole is called a "sessile" leaf. The point of attachment of the leaf to the stem is at a node. If you tear the leaves from a lettuce plant you are left with a short stem made up of many nodes and short internodes.

You can see in the romaine lettuce that its morphology is similar to that of the iceberg lettuce and that it has some tendency to wrap newer leaves within older. However, the nodes are a bit longer than what you see in iceberg lettuce, and that makes the node locations more apparent.

A few examples of leaf parts we eat

Modified petioles

Celery is an example of a leaf with a petiole. The parts that you eat are the petioles, while the leaf blades are often not present in the bunch of celery you purchase. If you buy a bunch of celery and pull off the large, outside petioles, inside you will find shorter petioles with the leaf blades still attached.

Celery is a geophyte (covered in a later lesson). Some of the celery petiole — the pale part at the bottom where it attaches to a node on the stem — grows underground. This part is pale because it lacks chlorophyll; the petioles were not exposed to sunlight and chlorophyll failed to develop.

Intentionally covering the petioles to discourage chlorophyll and encourage white, tender stems is called blanching. Blanched celery is more attractive to some cooks and consumers, although it may not be as nutritious.

Edible stems

The photo below shows an asparagus shoot. You can tell it is a shoot by the regular node/internode construction of the stem. Most of the shoot is stem tissue. The triangular growth at each node is colloquially called a "bract" by asparagus growers, but it is actually a very small, scale-like leaf. If the shoots are left unharvested, branches grow from the nodes and then repeatedly branch into soft, feathery green foliage, as shown in the next photo.

Image by Mark's Postcards from Beloit, CC BY-NC-ND 2.0.

Asparagus shoots that are not harvested grow taller and produce branches and feathery foliage.

Edible inflorescences

Broccoli and cauliflower are eaten as immature inflorescences. The dark green exterior of the broccoli inflorescence is made up of tight flower (or floret) buds that have not yet opened. The term "floret" is often used for the name of a flower born on a complex inflorescence.

Photo by Fir0002/Flagstaffotos, CC BY-NC.

Inside the inflorescence, the flower buds are supported by short, thin pedicels. The pedicel is attached to a series of increasingly thick internal stalks which make up the rachis structure of the inflorescence. The rachi all connect to the main stem of the inflorescence, which is the peduncle, and which then attaches to a node on the stem. The peduncle is the bit of the stalk that extends from the node where the inflorescence is attached to where the first rachis branches off. Above that, the central axis of the inflorescence is also called the rachis. An inflorescence can contain many rachi.

Nasturtium flowers and leaves are edible. Photo by henna lion. CC BY-NC 2.0.

These plant parts may look similar, but they're in different positions in the inflorescence. The terminology is important for distinguishing between parts, making observations, and describing different aspects of the plant, especially for data collection in experiments. Nasturtiums, shown above, are also an inflorescence. They can be used as a colorful addition to a salad, and have a pleasant spicy flavor. In this case we are eating a single open flower, instead of a mass of immature rachi, pedicels, and florets as we are with broccoli.

It's important to remember that not all flowers are edible, and some are even poisonous. If you're interested in edible flowers, you must learn which species are safe to eat and how to identify and prepare them.

Review questions

Cauliflower has very tight flower clusters, but otherwise has a very similar morphology to broccoli. Can you identify the pedicel, rachis, peduncle, and florets?

Cauliflower. alsen. Pixabay license

This Chinese cabbage has a morphology similar to that of romaine lettuce. Can you identify the stem, nodes, and leaf blades?

Chinese cabbage. misskursovie2013. Pixabay license

Fruits

"Apple half" EverydayTrish. CC BY-NC 2.0

We will deal with fruits in detail later in the course. Just a bit of introductory information: a fruit is a mature ovary that was part of a flower

Sometimes the botanical and culinary definitions conflict with one another. Botanically speaking, for instance, a nut is a fruit, as are a corn kernel, a pumpkin, a tomato, and **an** orange. As we'll discuss, you could make a botanical argument that an apple isn't a true fruit because the juicy part we eat isn't ovary tissue; the ovary tissue is the core that we throw in the compost.

Below-ground plant parts we eat: Geophytes

Geophytes are plants with underground organs where the plant stores energy or water. Geophytes are often called bulbs, but they are far more diverse than that. Many of these plants protect buds using structures other than bulbs, such as rhizomes or enlarged roots. These modified parts include:

- Bulbs
- Tubers
- Rhizomes
- Roots, including storage and enlarged tap roots

Underground shoots

Bulb – onion

Half of a purple onion, close-up, by Marco Verch. CC BY 2.0.

A bulb is a specialized, underground organ with a short, fleshy basal stem enclosed by thick, fleshy scales modified for storage. A true bulb consists of both leaf and stem tissue. The compressed stem, or basal plate, has attached to it a set of modified leaves called scales. These scales serve as the primary storage tissue for carbohydrates, nutrients, and water.

The main stem and apical meristem are protected by the layers of leaves. Axillary buds are born at the junction of the scale and basal plate (leaf and stem). Bulbs with a papery outer covering, like onions, are called tunicated. Plants that produce bulbs without this covering, like lilies, are non-tunicated.

Underground Stems

Tuber – Potato

Russet potato cultivar with sprouts. ZooFari. Public domain

A tuber is a thickened, enlarged underground stem typically produced from a swelling of a stolon or rhizome. The stem tissue serves as the primary storage tissue for carbohydrates, nutrients, and water. The potato tuber is a typical example. Potato tubers are born on stolons that emerge from nodes near the soil surface.

It is common garden and agricultural practice to "hill" potatoes (mound loose soil around the base of the main stem of the plant) so the stolon grows into the mound of soil, where the tip swells into a tuber. Unlike a corm, which is also stem tissue, a tuber has no basal plate, but rather is fleshy throughout. The "eyes" of the potato are the meristems or buds from which new, above-ground growth initiates when conditions are favorable. These eyes are found at nodes on the tuber, which indicates that the tuber is shoot tissue rather than root.

Rhizome – Ginger

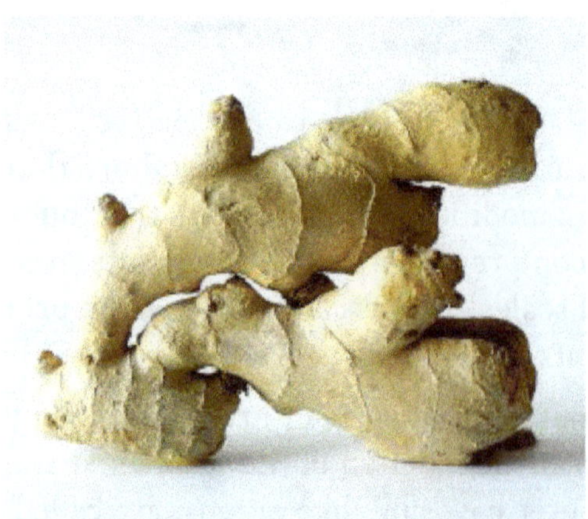

Organic ginger root, by artizone. CC BY-NC-ND 2.0

Rhizomes are horizontal-growing underground stems that arise from nodes at or below the soil surface. In plants with fleshy rhizomes, these underground stems store nutrients and swell a bit. The stems are not usually as enlarged as a potato tuber, and the node/internode structure typical of stems is usually more obvious than on a potato. The stem tissue itself is the primary storage tissue, and it grows horizontally in the soil.

Storage roots - sweet potato

Storage roots are enlarged fleshy portions of root tissue, and are are the primary storage tissue. There may be a bit of stem — the crown — attached to these roots, where you will find the buds from which new above-ground growth will initiate.

Photo by HarvestPlus, CC BY-NC 2.0.

In plants with storage and fleshy roots, including dahlias and daylilies, it is important to protect the buds on the crown of the root because new shoots **originate** from there. Roots of some plants can produce shoots directly from the root tissue, Sweet potatoes are propagated this way; roots are cut, new shoots emerge from the cut roots, and these shoots are transplanted into the sweet potato field. Not all plants can produce new shoots from roots.

Tap root – radish, carrot, parsnip, beet, turnip

"Parsnip, Farmers Market" by See-ming Lee (SML), CC BY-SA 2.0

The swollen primary root is the storage organ. New growth initiates from buds at the crown, which is a small area of stem tissue sitting atop the tap root.

Hypocotyl – radish

The below-ground organ of the Raphinus sativa (radish) is not a root or shoot, but the continued growth of the hypocotyl — the part of the embryo arising from the cotyledonary node (where the cotyledons attach to the beginning of the root), and evident when a seed germinates.

Image credit: Matt Clark

TERMS

These are the important terms from this chapter to be sure to know. You might also find these in later chapters.

Flashcards

Agriculture	The science or practice of farming, including cultivation of the soil for the growing of crops and the rearing of animals to provide food, wool, and other products.
Agronomy	The science and technology of producing and using plants for food, fuel, fiber, and land restoration on an extensive scale. The value per acre is lower than for a typical horticultural crop.
Asexual propagation	A form of propagation that results in plants with genetics identical to those of the parent plant.
Bract	A modified leaf or scale, usually small, with a flower or flower cluster in its axil.
Bulb	A specialized, underground organ with a short, fleshy stem axis (basal plate) enclosed by thick, fleshy scales modified for storage.
Control (in an experiment)	Used to verify or regulate a scientific experiment by conducting a parallel experiment or by comparing with another standard.
Demonstration experiment	A method for actively learning the body of scientific knowledge that has been previously discovered and communicated by others; specifically orchestrated for teaching and learning, not for the discovery of new information about the world around us.
Discovery experiment	A method focused on uncovering new relationships and solving problems, following the scientific method, testing hypotheses and their predicted outcomes, and utilizing a careful design in order to maintain meaningfulness and credibility.
Evaluation experiment	A method typically used during the development of new technologies to identify the best products for a desired purpose (e. g., which pesticides are effective against a target insect, but not harmful to non-target insects), but not used to discover new knowledge about how the world works, and thus not typically advancing our understanding of the natural world. Used to pick a winner from among a number of options.
Experimental design	The process of planning an experiment to test a hypothesis.
Experimental unit	The entity to which a specific treatment combination is applied.
Exploration experiment	A method focused on detailed observation of organisms and habitats, used to increase our information about the natural world and to identify potential relationships that need to be tested, and essential to the building of a sound and testable hypothesis.
Floriculture	Discipline of horticulture concerned with the production and marketing of plants valued for their flowers.
Forestry	The science or practice of propagating, planting, managing, and caring for forests; includes harvesting.
Fruit	Ripened ovary together with the seeds within the ovary.

Geophytes	Plants with underground organs in which the plant stores energy or water. New growth begins underground, and the function of this growth is the storage of food, nutrients, and water during adverse environmental conditions.
Horticulture	The art and science of the development, sustainable production, marketing, and use of high-value, intensively cultivated food and ornamental plants.
Hypothesis	Scientific means of forming a question or proposed explanation made on the basis of limited evidence as a starting point for experimentation. In science, a testable statement.
Inflorescence	Complete flower structure of a plant; includes the flower, pedicle, rachis, and peduncle.
Internode	Stem regions between nodes in plants.
Leaf	A usually green, flattened, lateral structure attached to a stem and functioning as a principal organ of photosynthesis and transpiration in most plants.
Leaf blade	Broad portion of a leaf; does not include the petiole.
Monocotyledon	Seed plant that produces an embryo with a single cotyledon and parallel-veined leaves; includes grasses, lilies, palms, and orchids.
Node	Stem region of a plant where one or more leaves attach; location of lateral buds.
Olericulture	Discipline of horticulture concerned with the production and marketing of plants or plant parts valued for culinary use as vegetables.
Pedicel	Short stalk that holds up the flower.
Peduncle	Large, central stalk that attaches the rachi to the stem of the plant.
Petiole	Stalk by which most leaves are attached to a stem; part of the leaf structure, not the stem.
Pomology	Production and marketing of plants or plant parts valued for their culinary use as fruits, including nuts; propagated by cuttings and grafting (asexual propagation).
Rachis	Stalk of a flower that is situated between the peduncle and the pedicel.
Randomization	Act of randomly assigning treatments to experimental units using a random number table or computer-generated randomization to help minimize any bias that has not been recognized in advance and controlled for in other ways.
Replication	Application of the same treatment to more than one experimental unit.
Rhizome	Horizontal stem growing just below the soil surface.
Science	Systematic study of the structure and behavior of the physical and natural world through observation and experiment.
Scientific discovery	Process of scientific inquiry; builds on what is known by testing hypotheses.
Sessile	A leaf that lacks a petiole; called a sessile leaf.
Sexual propagation	Form of propagation that results in plants with genetics that differ from those of the parent plants; also called seed propagation.

Stem	Supporting and conducting organ, usually developed initially from the epicotyl and growing upward; consists of nodes and internodes.
Treatments	Administration or application of agents to a plant to prevent disease or facilitate growth.
Tuber	Swollen, underground, modified stems that store food.

CHAPTER-2

TAXONOMY AND SEED GERMINATION

This chapter addresses the importance of **binomial names**, sometimes referred to as **scientific names**. At a garden center, a plant labeled a bluebill could be *Scilla non-scripta*, from the monocotyledones, or *Mertensia virginica*, from the dicotyledones. Two different plants, both from division Anthophyta, but from different Classes. Common names have their place, but they can be ambiguous. Binomial nomenclature is more precise; its use ensures that you'll get the correct plant and the correct information on how to grow and propagate it.

The second section in this chapter addresses seed germination, arguably the most common and important method of propagating plants.

Garden centers and nurseries tend to use common plant names, but knowing the scientific names will eliminate confusion. "Langley Circle Farm Tour: Cedar Rim Nursery Plants Galore" by Queenie C, CC BY-NC 2.0

Learning objectives

- List the seven levels of the plant classification system.
- Recognize the taxonomic diversity of common foods in your diet.
- Summarize the contribution of *Linneaus* **to binomial nomenclature.**
- Understand how the two-part scientific naming system works and its applications.
- Describe the differences between epigeal and hypogeal seedling emergence.

2.1 PLANT TAXONOMY

Learning objectives

- List the seven levels of the plant classification system.
- Recognize the taxonomic diversity of common foods in your diet.
- Summarize the contribution of *Linneaus* **to binomial nomenclature.**
- Understand how the two-part scientific naming system works and its applications.

Plant taxonomy

Here are some introductory definitions:

- Taxonomy (or systematics): The science of classifying organisms.
- Classification: A grouping of plants according to shared qualities or characteristics.
- Plant taxonomy: A hierarchical classification system based on morphological (see below) and phylogenetic (see below) similarities among plants.
- Nomenclature: A formal system of names attached to taxonomic groupings.
- Hierarchy: A system of grouping in which each classification is a subset of a superior grouping, and may contain subordinate categories. As an example: The landmass of the United States (used here as the highest or most inclusive level of classification) is partitioned into states (a middle level of classification). States, in turn, are partitioned into counties (the lowest level in this hierarchy). Counties are subsets of states, which are in turn subsets of the nation. This hierarchical type of grouping system is used in plant taxonomy.
    - Morphology: The appearance (shape and structure) of a plant. Plant taxonomy is a hierarchy primarily based on grouping together plants that exhibit structural (phenotypic) similarities.

- Phylogeny: Ancestral, evolutionary relationships among plants. While plant taxonomy has historically been based on plant morphology, these relationships are currently being verified and expanded using new molecular genetic technologies that uncover genetic similarities through comparisons of shared DNA sequences. In general, plants sharing more DNA are considered more similar from an evolutionary standpoint, and considered to have diverged from each other more recently in evolutionary time than plants that share less DNA.

Taxonomy in the pantry: Classification exercise

"Food on shelf" by Jaranda, CC BY-SA 2.0

To start becoming familiar with taxonomic categories, go to your fridge, cupboard, or pantry and choose a variety of fruits, vegetables, and grains. These might include cans of mushrooms, green peas, black-eyed peas, chickpeas, butter beans, and sweet corn; bags of pine nuts and coconut, and perhaps a banana that was getting a bit too brown.

Think about the many ways in which you could group these foods. You might, for instance, categorize a food by whether it is canned or fresh, by size, manufacturer, or **color**, or by the meal in which you would typically eat it.

Or you could apply a biological, hierarchical classification system, categorizing them by the morphology and phylogeny of the plant on which they grew. The plant systematics hierarchy we will use in this course is as follows, from highest (most inclusive) to lowest level:

- Kingdom
- Division (or Phylum, although Phylum is more commonly associated with animal taxonomy)
- Class
- Order
- Family
- Genus
- Specific epithet (usually a species name)

Memorize this hierarchy, so it rolls off your tongue like a multiplication table.

Now apply this taxonomic system to your foods. An easy way to do this is to search for each food on the U.S. Department of Agriculture's site USDA Plants Database. If you type "tomato" into the search bar, select "Common Name" from the dropdown menu, and click "go," you'll see all the plants with "tomato" in their common name. Click on *Solanum lycopersicum* **L. (garden tomato) and you'll get** this entry, the description for the common garden tomato. Scroll down to see the "Classification" section, which lists the taxonomic classification and includes Kingdom, Division, Class, Order, Family, Genus, and Specific epithet. Notice that this database has finer divisions of hierarchy than you are required to know, including Subkingdom, Superdivision, and Subclass.

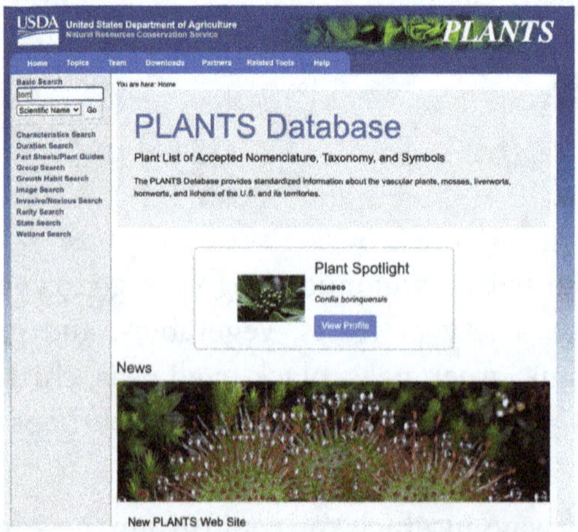

Here's a look at how you can use the USDA Plants Database to find the full classification of plants, like tomato. Try it out!

You can use the information in the database to classify your foods. For products with several ingredients, pick one from the ingredient list, such as wheat in crackers or tomato in spaghetti sauce.

While you can sometimes find this info on Wikipedia, be aware that Wikipedia is not always reliable and you'll want to cross-reference with other sources. If you enter "tomato" into the Wikipedia search bar you'll get this page. The right sidebar includes the taxonomic classification. "Unranked" is used instead of Division and Class, which means there is some disagreement on whether those names are the **correct** Division or Class names. You might also see several hierarchical terms listed as "Clade," rather than the proper terms. If you can't find complete information on Wikipedia, use the USDA site.

For the foods in our hypothetical pantry — mushrooms, green peas, black-eyed peas, chickpeas, butter beans, sweet corn, pine nuts, coconut, and banana — we can divide them into the following Kingdoms: Plantae and Fungi.

We can separate the products within the Plantae kingdom into two Divisions:

- Pinophyta: the pine nuts, which come from a conifer
- Magnoliophyta: everything else in this kingdom, which come from flowering plants

"Public Flower Garden in downtown Seattle" by FallenPegasus, CC BY-NC 2.0

The cans and bags in the Magnoliophyta division can be separated into these Classes:

- Liliopsida (Monocotyledons— one embryonic leaf in the seed, parallel leaf veins, and petals and sepals in multiples of three): corn, coconut, and banana
- Magnoliopsida (Dicotyledons —two embryonic leaves in seed, and branched leaf veins): green peas, black-eyed peas, chickpeas, and butter beans

In comparison to the other levels, Order is a relatively arbitrary set of classifications that were created in part to make subsequent classifications more manageable. Order will be addressed in the section on phylogeny.

Next, the products can be subdivided by Family.

Depending on details of the particular plant classification system used, there are approximately 230 plant families. Families are often based on types and organization of flower parts and fruit type, including the number of petals, sepals, stamens, and pistils, and the location of the ovary relative to petals. In this website from the University of California Cooperative Extension **(optional) the authors identify many of the characteristics used to group plants into families.**

Among our food examples, the Family hierarchy includes:

- Arecaceae (coconut, which comes from a palm tree)
- Poaceae (corn, which is a grass)
- Musaceae (banana)
- Fabaceae (the three peas and the butter beans, which are legumes)

Genus and Specific Epithet are the last two classifications. The pairing of genus and specific epithet to name a plant is called binomial nomenclature. Tthe first letter of the genus is capitalized, and the entire binomial is either underlined or written in italics.

Watch this video for an explanation of plant taxonomy.

Review questions

Think about these questions and be able to discuss the answers or know how to find the them using the resources provided:

- Where are blue spruce trees (a conifer) taxonomically separated from lilies (flowering plants) at the Kingdom or Division level? (Hint: Use the online resources that were provided in this section.)
- Are lilies (monocots) separated from beans (dicots) at the Class or Order level?
- At which taxonomic level are flowering plants separated into different classifications based on flower and fruit characteristics?

Example taxonomy tree

Now that you have the names for each of your plants, you can organize them into a taxonomic tree that more clearly shows their relationships to one another. Below is an example tree based on some of the foods found in our hypothetical pantry:

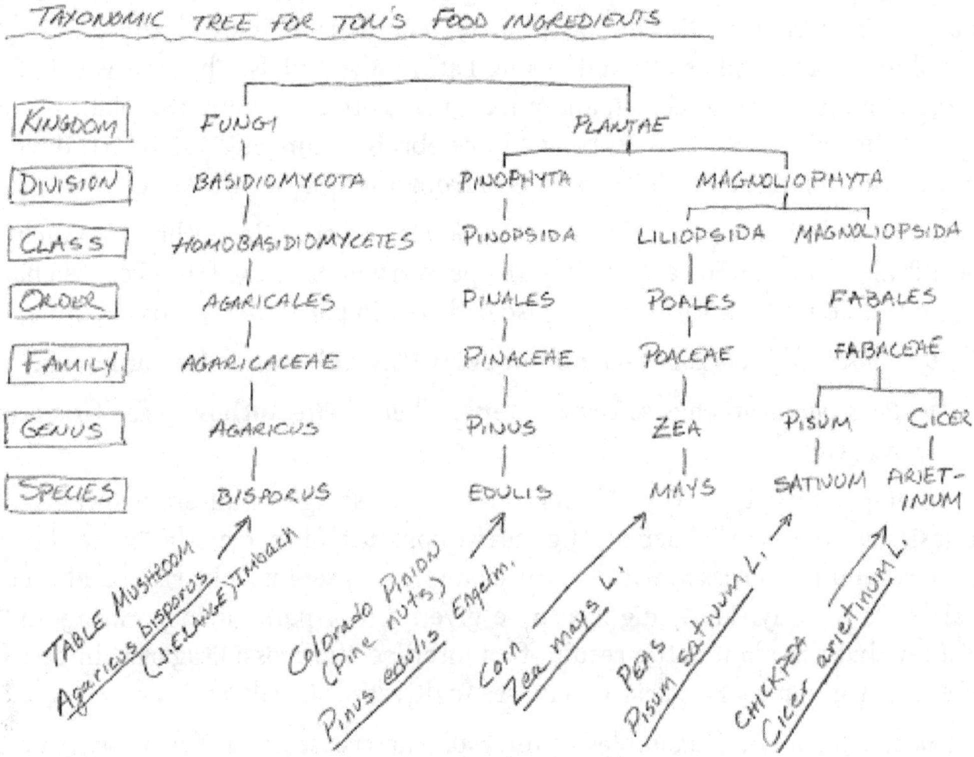

Image credit: Tom Michaels

Now, try it yourself. Determine which plants you want to use, look them up on the USDA Plants Database, write down the Kingdom, Division, Class, and so on for each, and begin constructing the tree to show relationships and points of divergence. In this example, the mushrooms diverge from everything else at the Kingdom level, pine nuts diverge from the other three at the Division level, corn is in a different Class than pea and chickpea, and pea and chickpea diverge at Genus.

The point of this exercise is for you to understand that relationships among plants are known, and are categorized in a sophisticated taxonomic system. Some of the plants we commonly eat have close relationships, like the various plants in the Solenaceae family (tomato, eggplant, potato), but others are much more distant.

Linneaus and plant taxonomy

Binomial nomenclature

Carolus Linnaeus (1707–1778), a Swedish professor, is widely recognized for developing the binomial nomenclature for plants. Binomial nomenclature is a scientific classification in which each organism is given two names. In his 1753 book *Species Plantarum* **(kinds of plants)**, Linnaeus employed this system to describe a great number of plants using Latin polynomials. The first word of the polynomial became the genus, and a marginal note describing the plant became the specific epithet. Several years ago we celebrated Linnaeus' 300th birthday, and you can find a long set of links about him from a simple Google search.

A proper binomial, in addition to the Genus and specific epithet, also includes the initials of the naming authority — the person who proposed the accepted name. Previous naming authorities might also be listed in parentheses. For example:

- *Phaseolus vulgaris* **L. — common bean. The "L" stands for Linnaeus.**
- *Phaseolus acutifolius* **A. Gray — tepary bean. The authority for this one is A. Gray.**

Interspecific hybrids (hybrids formed from crossing two different species) may be designated with an "x" separating the two constituent species; the "x" can be read as shorthand for "crossed with" — for example, Phaseolus vulgaris L. x Phaseolus acutifolius A. Gray. They might also be given a new name incorporating an "x" to show that the plant is the result of an interspecific cross: Fragaria chiloensis x Fragaria virginiana = Fragaria x ananassa (cultivated strawberry)

Notice, from these examples of interspecific crosses, that the ability to cross and to have fertile offspring isn't a firm definition of species. It is generally true that breeding is restricted to within-species boundaries, but there are exceptions.

While some plant names have been updated to reflect the most recent knowledge about their morphology and phylogeny, their older names might still be in common use in some settings. Coleus, for example, has **the** following binomials, all for the same plant:

- *Ocimum scutellarioides* **L.**
- Plectranthus scutellarioides (L.) R. Br. (Notice that "L" is now in parentheses, showing that Linneaus was the earliest naming authority, but that his original name for the plant has now been superseded.)
- *Coleus scutellarioides* **(L.) Benth.**
- *Coleus blumei* **Benth.**

Important notes about binomial naming conventions:

- The Genus is always capitalized and either italicized or underlined.
- The specific epithet is lowercase and either italicized or underlined.
- The naming authority is capitalized and often abbreviated; if the species has been renamed, the first authority is in parentheses.
- An "x" between the Genus and specific epithet denotes an interspecific cross.
- A "x" before the Genus denotes an intergeneric cross.

Future of plant taxonomy and systematics

Taxonomy might first seem an old and dull science, sorting plants into a database using a system developed by someone born more than 300 years ago. But plant exploration experiments and the discovery of previously unknown species can take researchers to the far corners of the world, and taxonomy is important in classifying and naming these new discoveries. Also, for already discovered species, there is continual discussion about the real relationships among these plants and others and whether currently classified plants should be reclassified based on new information. With advances in molecular genetics through techniques that reveal a plant's DNA sequence, for example, taxonomy is moving more and more toward a phylogenetic basis, based on evolutionary relationships established through DNA similarities and differences instead of solely on morphological characteristics (features about the plant that you can see).

- Traditional taxonomy relies on morphological phenotype (the appearance of the plant).
- Molecular taxonomy relies on genotype (the particular combination of alleles of each gene in the organism).

The Angiosperm Phylogeny Group (APG) is a group of taxonomists who are working together to modify flowering plant taxonomy using molecular systematics. The APG's work is focused at the taxonomic level of Order and, to some extent, Family. While Order has long been a fairly arbitrary categorization, it may now be based more on molecular relationships.

The utility of classification goes beyond the satisfaction of good organization. Classification can inform us of new or lesser-studied plants that share valuable characteristics with plants already familiar to us. Now we have tools and knowledge that give us increasing control over the transfer of DNA among plants. Plant breeders can use insights from taxonomists to identify DNA sequences in related plants that might provide new sources of resistance to disease and insects, or new quality attributes, if transferred to food crops.

2.2 INTRODUCTION TO SEED GERMINATION

Learning objectives

- Describe the differences between epigeal and hypogeal seedling emergence.
- Understand the terms that are used to describe different parts of the seedling as it emerges.

Seeds and their importance

A seed, in botanical terms, is an embryonic plant enclosed inside its seed coat. Typically, the seed also has stored energy (proteins and carbohydrates) that are used by the seed during germination to establish itself when environmental conditions are favorable for growth. The stored energy is what makes seeds valuable for humans, too. Seeds are important in our daily lives because they feed us (food), feed livestock (aka, feed), and provide us with fuel and fiber for personal, home, and industrial purposes.

"Germinating bean seed" by Jose Bañuelos, CC BY-NC 2.0

Seeds are by far the most common mode by which plants reproduce, and most people are familiar with their role in plant propagation and reproduction. The evolutionary advantage of reproduction by seed is the mixing of genetic material through meiotic recombination and the transfer of gametes (pollen) from one parent to another. This mixing of male and female parent genetics results in seeds, and thus seedlings, that are unique from one another and from the parents. The seeds may be dispersed locally or distributed far away through many mechanisms, such as wind, animals, insects, and water. Seedlings will germinate and grow, and those that are most fit in the environment will reproduce and pass on their genes to the next generation. This ability of plants to adapt to local environments and to pass on their genes is evolution in action, as new variations and even new species emerge and disappear from the landscape.

One advantage of seed production is that plants generally produce copious amounts of seed. Each seed may have slightly different germination requirements, a reflection of the diversity resulting from sexual recombination and an evolutionary strategy that allows seeds to germinate at different times. Seeds are able to remain dormant until the conditions are suitable for plant growth and survival, and have mechanisms that prevent germination before winter, during droughts, or in low-light environments. Some weedy species are excellent at interpreting these signals and may lie dormant for years in the soil "seed bank," only germinating when the seed has been exposed.

Seedling Emergence

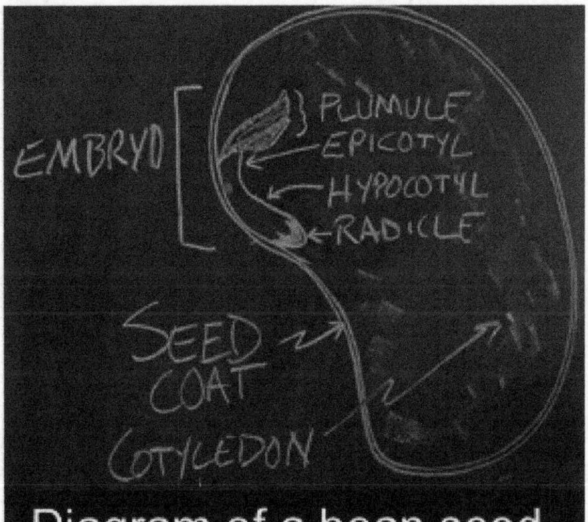

Diagram of a bean seed.

Image credit: Tom Michaels

Most seeds have a very slow metabolism when they are mature, which puts them in a state of quiescence: alive, but not growing and not physiologically active. At germination, the seed's metabolic pathways are activated, leading to embryo growth and emergence of a new seedling. Germination begins with activation by water uptake. We call this imbibition, and sometimes the seed or fruit requires special treatment for water to get into the seed and start this process. We often use the emergence of the radicle (the embryonic root) from the seed coat as a measure of successful **germination.** Water uptake alone is not an indication that the seed is alive and growing, despite the expansion of seed tissues. Cell division is taking place in the epicotyl, and the hypocotyl and the shoot and root are beginning to break through the seed coat. The new plant is beginning to grow and emerge from the soil.

Two types of seedling emergence

Epigeal and hypogeal

Epigeal and hypogeal are terms used to describe the position of the cotyledonary node during germination, indicating whether the node is above or below ground once the seedling has become established.

Epi **means** above **while** Hypo **means** below. The location of the cotyledonary node following seedling emergence is a characteristic used as a first step to differentiate plant species. The position of the cotyledon is affected by the rapidity of cell division in the hypocotyl region of the plant during germination and early seedling growth. The epicotyl is the embryonic shoot region above the attachment point of the cotyledons, and the hypocotyl the embryonic region below the cotyledon attachment point and extending down to where the root begins.

Epigeal

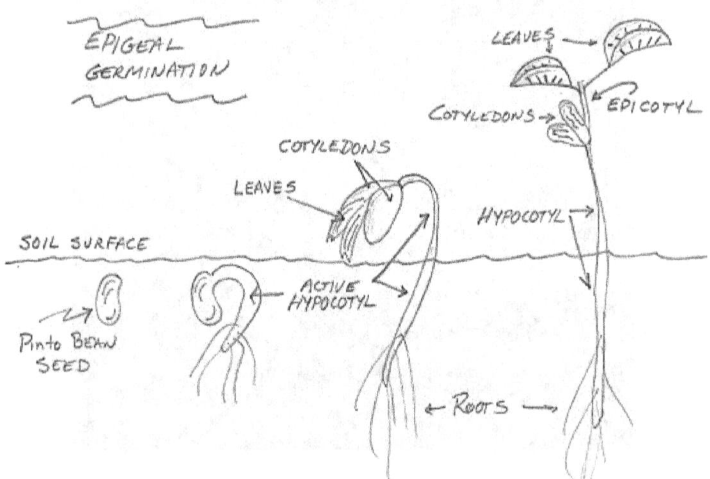

Epigeal germination. Image credit: Tom Michaels

In this type of seedling emergence, cell division in the hypocotyl is initially more active and rapid than cell division in epicotyl. The actively dividing meristem in the hypocotyl causes cell growth and elongation that pushes some of the hypocotyl, as well as the cotyledonary node and epicotyl, above the soil surface. The cotyledonary node is above the ground – epigeal. The drawing shows four stages in the emergence of a pinto bean (*Phaseolus vulgaris* L.) which exhibits epigeal germination.

Hypogeal

In this type of seedling emergence, the apical meristem at the tip of the epicotyl is more active than the hypocotyl. This cell division and elongation pushes the epicotyl above the soil while the cotyledons and all of the hypocotyl remain below the soil surface. The cotyledonary node is below the ground — it is hypogeal. The example above is a pea (*Pisum sativum* L.) **which exhibits hypogeal germination.**

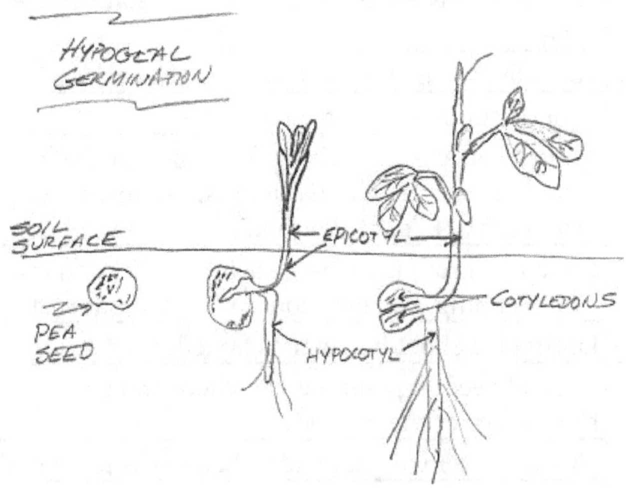

Hypogeal germination. Image credit: Tom Michaels

TERMS

Here are the terms from this week's lessons that you will need to be familiar with for your assignments and for the quiz.

Flashcards

Binomial nomenclature	System of naming in which two terms are used to denote a species of living organism, the first indicating the Genus and the second the Specific Epithet.
Class	Taxonomic rank below Division and above Order.
Cotyledonary node	Food storage structure used in germination.
Dicotyledon	Seed plant that produces an embryo with paired cotyledons, floral organs arranged in cycles of four or five, and leaves with net-like veins.
Division	Highest taxonomic category, consisting of one or more related classes, and corresponding approximately to a Phylum in zoological classification.

Emergence	Germination, when the embryo becomes active and the radicle grows through the seed coat.
Epicotyl	Portion of the stem of a seedling or embryo located between the cotyledons and the first true leaves.
Epigeal	Type of seedling emergence where cell division in the hypocotyl is initially more active and rapid than cell division in the epicotyl. Cotyledons are brought above the soil surface as the hypocotyl expands.
Family	Taxonomic rank below Order and above Genus.
Genotype	Genetic composition of an organism.
Genus	Group of species possessing fundamental traits in common but differing in other lesser characteristics; taxonomic rank below Family and above Specific Epithet.
Hierarchy	System of grouping where each classification is a subset of a superior grouping, and may contain subordinate categories.
Hypocotyl	Embryonic shoot below the cotyledons.
Hypogeal	Type of seedling emergence where the cotyledons remain below the surface of the ground.
Nomenclature	Formal system of names attached to the taxonomic groupings.
Order	Taxonomic rank below Class and above Family.
Phenotype	Physical appearance of an organism.
Phylum	Taxonomic rank below Kingdom and above Class; used in zoological classification.
Radicle	Embryonic root that breaks through the seed coat during germination and develops into the seedling's root system.
Seed	Ripened ovule containing a seed covering, food storage, and an embryo.
Seed coat	Outer layer of the seed.
Seed germination	Activation of metabolic pathways of the embryo leading to the emergence of a new seedling.
Specific epithet	Uncapitalized Latin adjective or noun that follows a capitalized Genus name in binomial nomenclature and serves to distinguish a species from others in the same genus, as **saccharum** in **Acer saccharum** (sugar maple).
Taxonomy	Science of classifying organisms.

CHAPTER 3
HOW PLANTS GROW, PART 1

The organization of the plant is not unlike that of our own bodies. At the simplest level, cells are organized into tissues; these form organs that make up the plant body. At each level of this organization are specializations for the specific functions that occur during the plant's life cycle. In this chapter, you'll explore the structures and functions of leaves, shoots, and roots.

Each part of a plant — from tiny root hairs to the concealed buds — contributes uniquely to the overall growth of the plant. Pixabay. Pixabay license

Learning objectives

- Identify the unique features that distinguish shoots, leaves, and roots.
- Describe ways in which stolons and rhizomes are modified stems.
- Identify the types and parts of the major shoots, leaves, and roots.

3.1 LEAVES

Learning objectives

- Identify the parts of the angiosperm leaf.
- Describe some of the ways in which leaf parts differ from plant to plant.
- Recognize the basic patterns of leaf shape and orientation of the veins in the leaves.

Leaves

Leaves are shoot structures that attach to stems and branches at nodes. Leaves are made up of cells that usually contain a high concentration of chloroplasts (cell organelles unique to plants) and are specialized sites for photosynthesis. We will explore photosynthesis in greater detail later; for now, remember that photosynthesis is the process of capturing light energy and converting it into chemical energy that can be stored in plants (like starch and sugar). In some plants, leaves may be modified for nutrient storage (as with onions, where the bulb is made up of fleshy leaves), or for support (as with peas, where some leaves are modified into tendrils that wrap around a trellis).

Photo by yooperann, CC BY-NC-ND 2.0.

Leaves are also the surface where water that has moved from the soil into the roots and up through the plant finally evaporates back into the atmosphere in a process called transpiration.

Angiosperms, which are flowering plants whose seeds develop inside an ovary, tend to have flattened leaves. Many perennial angiosperms (flowering plants that can grow for many years) have leaves that senesce, or die, at the end of each growing season and are replaced at the beginning of the next growing season. Gymnosperms, plants whose seeds are produced without the protection of an ovary, tend to have needle-like leaves. Perennial gymnosperms tend to hang on to their leaves for a number of years. This saves energy, since the plant doesn't need to grow a whole set of new leaves every year. The needle-like form helps retain moisture in harsh, dry climates, including those that are very cold and dry.

Leaf parts and venation

Angiosperm leaves typically have a blade or lamina, a flattened part with high chloroplast concentration. They may also have a petiole, the stalk that attaches the blade to the stem at a node. Stipules, small leaf-like bracts at the point of attachment of the petiole to the stem, may also be present. Some leaves have no petiole at all, and are termed sessile.

Photo by Matt Lavin, CC BY-SA 2.0.

In contrast to the blade-petiole structure, grasses have a sheath-type structure in which the blade attaches to an envelope of leaf tissue that wraps around the shoot of the plant and then attaches to a lower node on the stem.

Leaf blades also have characteristic patterns of venation. In grasses, the veins lie parallel to each other and to the long edges of the leaf. We call this parallel venation, and it is typical of monocots. Most other angiosperms have a strong major midrib with veins branching from the midrib, smaller veins branching from those, and so on to form a netted venation throughout the leaf. This type of venation is typical of dicots.

Leaves may also have a palmate venation where several veins radiate from the point where the petiole attaches to the blade. Ginkgo tree leaves have palmate venation. The veins fork, then travel a bit, then fork again, travel, fork, and so on until the veins reach the margin (edge) of the leaf. Sugar maple leaves have a classic palmate venation with five lobes.

Left: Corn leaves have parallel veins. Center: A leaf with netted veins. Right: Ginkgo tree leaves have palmate veins. L to R photo credits: Image by Hans Braxmeier, CC0; Thangaraj Kumaravel, a CC BY 2.0; Nate Hofer, CC BY-NC-SA 2.0.

Leaf segmentation

Simple leaves

Simple leaves have uninterrupted leaf margins. The leaf may have lobes like the oak leaf, but the blade has one continuous margin. The venation differs in the two examples below. The oak leaf is pinnate, with a major vein heading down the midrib of the leaf. The maple leaf is palmate, with major veins that radiate from the point of attachment to the petiole.

Oak leaves have pinnate venation (L), while maple leaves have palmate venation (R). Photo credits L-R: Dendroica cerulea, CC BY-NC-SA 2.0; Richard Skiba, CC BY-NC-SA 2.0.

Compound leaves

The sumac leaf is a good example of how a compound leaf has a blade that is completely interrupted and segmented into separate leaflets. What you see in the picture — the entire thing — is one leaf. The leaf is divided or segmented into leaflets.

Tree of heaven, (Ailanthus altissima) has pinnately compound leaves. Photo by Dr. Leslie J. Mehrhoff, CC BY 3.0.

The petiole extends from the point of attachment at the node to the first leaflet. The central axis from that point on — from the first leaflet to the tip of the leaf — is called the rachis.

Virginia creeper is an example of a palmately compound leaf. The stalk that connects the leaflet to the top of the petiole is called the petiolule. In this case there is no rachis; all leaflets are attached directly to the top of the petiole.

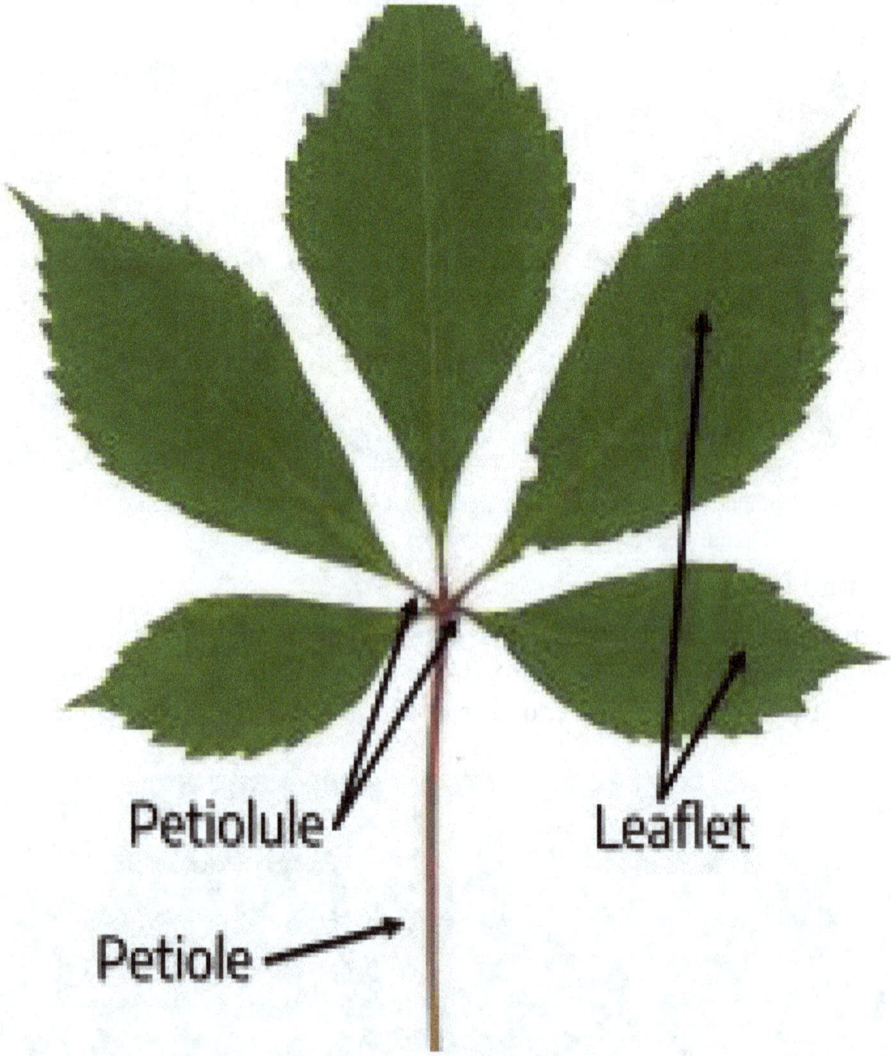

Virginia creeper (Parthenocissus quinquefolia) has palmately compound leaves. Photo by Gavatron, CC BY-NC-SA 2.0.

Below is a compound leaf with three leaflets, called a trifoliate leaf. Soybean, clover, and dry bean all have trifoliate leaves. In contrast to the palmately compound leaf above, there is a rachis to which the central leaflet is attached.

Poison ivy (Toxicodendron radicans) has trifoliate leaves. Photo by DaveSpier, CC BY-NC-SA 2.0.

To tell whether a leaf is simple or compound, ook for attachment to a node. If the point of attachment doesn't appear to be a node, it is likely a leaflet attached to the rachis of a compound leaf.

3.2 SHOOTS

Learning objectives

- Identify unique external features of shoots that distinguish them from roots.
- Locate the regions of origin for stems, branches, leaves, and inflorescences.
- Locate the regions of origin for stolons and rhizomes.
- Describe ways in which stolons and rhizomes are anomalies.

Small tomato plant in a pot. Photo by Daniel Morrison, CC BY 2.0

Shoots: Not only what's above the soil

While it's generally true that the shoot of a plant is the part above the soil, there are exceptions; these will be explored below.

A shoot is made up of a central axis called the stem, and components that grow from specific places on that stem. The stem can be tall and thick in diameter, like in a sunflower, or quite compact and compressed, like in an onion, where the stem may never get above the soil surface. The stem can be rigid like a corn stalk or floppy like a watermelon vine. In each case, the stem is the central axis to which the other shoot components attach. The stem doesn't need to be upright; it can grow horizontally.

Nodes and internodes

The distinguishing external feature of a stem, and of the branches that arise from the stem, is its repeated node - internode - node - internode construction. A node is the place of origin on the stem for branches, leaves, and inflorescences. Sometimes the node is slightly swollen and obvious, other times not; it depends on the type of plant. Located at the nodes are buds (a colloquial term; later we'll call them meristems) made up of cells that, given the correct biochemical signal, will rapidly divide and grow into branches, leaves, or inflorescences. More than one bud can grow from a node, so a node can support several structures.

This tomato plant has clearly distinguishable nodes and internodes. Photo by Greyerbaby, CC0.

In another chapter we will see how the plant stem has internal bundles of "pipes" that make up the vascular system through which water, nutrients, sugars, and other plant metabolites flow. At the nodes, some of these pipes diverge from the main bundle to provide nutrients, water, sugars, and other metabolites to the branches, leaves, and inflorescences that form at the node.

It's easy to see the nodes and internodes on motherword (Leonurus cardiaca).
Photo by Emilian Robert Vicol, CC0.

Between nodes are stretches of stem called internodes. One architectural function of the internode is to spatially orient the leaves, branches, and inflorescences. Long internodes, for example, will spread the leaves out along a stem so that they aren't shading each other as much as if the internodes were short. Nodes also help with leaf orientation.

The location of nodes determines whether the leaves are located, which is called leaf arrangement, and leaf arrangement is a characteristic of a particular type of plant.

- Alternate – the leaves are attached at nodes on alternate sides as they go up the stem.
- Opposite – the leaves grow directly opposite each other on the stem.
- Whorled – the leaves are oriented in a whorled formation in which their point of attachment appears to spiral up the stem.

Photos by (L to R) jlevinger, CC BY-NC-SA 2.0; John Tann, CC BY 2.0; Frank Vincentz, CC BY-SA 3.0.

Nodes are important because they are where leaves, branches, and inflorescences originate; the internodes are important because their length has a profound impact on plant architecture.

Goldenrod inflorescence. polyphemus_polly. CC BY-NC 2.0

With inflorescences, it can be difficult to tell where the stem stops and the inflorescence begins. One difference is that the stem has nodes and internodes, while the inflorescence does not.

What about vines?

The tip of the stem is called the apex, and the bud at the tip is called the apical bud, or apical meristem. In the image below, a bud has formed at the apex of the stem. This bud contains a meristem that will "break" or become active next spring and result in early season stem growth.

These cuttings of 'Northstar' cherry were collected from the tips of the twigs, and are called terminal cuttings. Photo by Emily Tepe.

The crotch formed between the leaf petiole (the stalk that attaches the leaf blade to the shoot stem) and the stem or branch is called the leaf axil, and the buds in that crotch are called axillary buds or axillary meristems. You can see two sets of axillary meristems in the image. If pruned above the axillary buds, the plant will grow from these nodes. Removing the apical meristem is one way to encourage branching. If you've ever grown basil, you know that snipping or pinching leaves off right above a node encourages more leaves to grow from that node. This function also gives plants a way to respond to feeding damage or injury by having dormant buds that can grow when prompted.

Review questions

- Looking a photo of a plant, can you identify the stem, nodes, and internodes?
- Can you recognize different types of leaf arrangements on different plants you see?
- Buds at the nodes can develop into one or more of three different structures. What are these three structures?
- What type of buds are found in the crotch between the leaf petiole and the stem?

Stolons and rhizomes

Stolons

Strawberry plants send out stolons which form new plants at the ends. Photo by Emily Tepe.

Some types of plants produce branches from nodes on the stem that are very close to, or right at, the soil surface. These branches, which have long internodes and lie on the surface of the soil, are called stolons. Not all plants produce stolons, but strawberry is a common example of a plant that does. The "runners" that extend out from the mature strawberry plant are in fact stolons. Stolons have the same typical repeating node/internode structure of a stem, but unlike other branches of the plant, at the nodes of the stolon, adventitious roots can form. Adventitious roots are roots that emerge from the stem rather than from roots. Even though they emerge from the stem above ground, they still act like roots. They anchor the stolon to the ground and absorb water and nutrients for the plant's use. Leaves, and even branches, can also form from the stolon nodes.

One of the main purposes of stolons is to propagate a plant. The mature strawberry plant sends out a stolon and a new plantlet forms at the stolon node. Once the plantlet is rootedm you can cut the stolon internode between the plantlet and the main stem and transplant the plantlet. For instance, you can start a new patch of strawberry plants by cutting off these plantlets and planting them in a bed.

Potato plants produce underground stolons which swell to form the tubers we eat. oliva732000. CC BY-SA 2.0

Potato is another plant that produces stolons; in this case the stolons are often below ground rather than above the soil. Potato tubers (the part that we eat) are formed from swellings of these stolons. When we grow potatoes we mechanically hill soil up around the potato stem during the growing season to cover the developing tubers, preventing them from turning green and bitter from exposure to sunlight.

Rhizomes

Rhizomes are another type of stem tissue originating from a node. In this case it is typically a node that is below the surface of the soil. A rhizome also grows horizontally and has nodes and internodes, but unlike stolons, it is underground. The ginger in the produce section of the grocery story is misnamed ginger "root," but is actually a rhizome, and you can tell that this is stem tissue because it has nodes and internodes. The nodes are the faint, slightly raised rings around the circumference of the rhizome. If you see nodes and internodes on a plant part, the tissue is stem tissue and not root tissue, even if it is underground.

Green shoots emerging from an iris rhizome. Science and Plants for Schools. CC BY-NC-SA 2.0

Rhizomes and stolons are the exceptions to the general notion that shoot tissue is above the soil surface. They are branches, complete with nodes and internodes, but are underground (rhizomes) or near the soil surface (stolons). They are stem tissue.

Turfgrass provides an example of rhizomes, stolons, and another structure typically associated with plants in the grass family called tillers. Here is an optional reading about those structures in turfgrass.

3.3 ROOTS

Learning objectives

- Locate and identify characteristics of the primary root, lateral or secondary roots, and root hairs.
- Distinguish two major types of root systems and how they develop.

Roots

The general perception is that roots are the parts of the plant that are found in the soil. While this is typically true, there are exceptions, just as there are with the notion that all shoots are above ground. To recognize roots, we'll look at more than whether or not they are in the soil.

Types of plant roots. Gordon Johnson from Pixabay. Pixabay license

Unlike stems and branches, roots don't have a node/internode pattern of construction. There are no nodes, no internodes, no leaves, and no branching from regular spots, as you find with stems and branches. Roots do have structures but, in contrast to stems and branches, these structures emerge irregularly from the root according to where they are needed, and to some extent according to inherited tendencies, rather than emerging at specific zones the way leaves, branches, and inflorescences emerge from nodes on the stem.

Purposes of roots

Anchorage

Roots keep the plant moored to the soil in a particular place. This "anchorage" not only facilitates other functions for the plant, but provides a benefit for the soil. An extensive root system helps hold the soil in place so that it is less likely to be eroded by wind or rain. Where there are roots the soil tends to be retained. No roots, and the soil easily gets washed or blown away. On the flip side, if conditions are poor the plant can't pull up roots and move somewhere else where prospects for growth are better.

Support

The roots, particularly the tap root that we examine below, provides the foundation for upright growth.

Absorption

Roots are the structure of the plant that absorb water and soluble nutrients..

Symbiotic interaction with other organisms

Roots of plants from the taxonomic family Fabaceae — which are commonly called legumes, and include plants like peas, beans, clover, and locust trees — can form a symbiotic relationship with Rhizobia bacteria. This results in nitrogen fixation, which allows for the conversion of nitrogen from the atmosphere into nitrogen compounds that the plant can use to produce proteins and other building-block molecules. For nitrogen fixation to occur, rhizobia require a plant host; they cannot independently fix nitrogen. These bacteria fix nitrogen after becoming established inside root nodules of plants in the Fabaceae family. Roots also develop associations with Mycorrhizal fungi. In contrast to the Rhizobia bacteria, which only symbiotically interact with a narrow range of plants, Mycorrhizae (mycorrhizae is the plural of mycorrhiza) are fungi that grow in association with roots of a wide range of plants — perhaps most plants. The mycorrhizae help acquire phosphorus from the soil and make it available to the plants' roots. Mycorrhiza may also enhance water uptake.

Nutrient storage

Roots of some plants can swell and store high-energy compounds like starch and sugar. Examples include carrots, beets, sweet potato — but not white potato. Roots also store some protein and other nutrients, but the focus is typically on high-energy carbohydrates.

Review questions

- What external features clearly differentiate roots from stems?
- What purposes do roots serve beyond absorption of water and nutrients?
- What are two examples of living organisms that symbiotically interact with roots? How do they differ in terms of the plants they infect and the benefits they provide?

Root systems

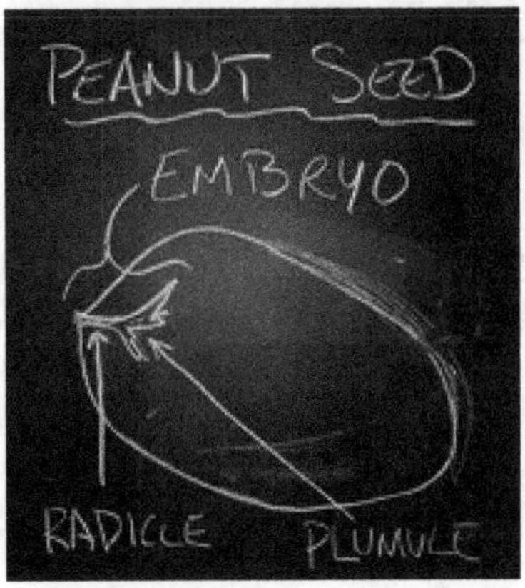

Peanut seed embryo. Image credit: Tom Michaels

A plant's root originates in the embryo formed within the seed. The section of the embryo that is root tissue is called the radicle (note the spelling). At the tip or apex of the radicle is a region of rapid cell division and growth called an apical meristem (you may recall that shoots have an apical meristem too). As a result of the apical meristem's rapid cell division, the radicle grows down into the soil. The root that forms from the embryonic radicle is called the primary root. This sketch of half of a peanut seed shows the radicle that is part of the embryo inside the seed. The plumule is the embryonic shoot.

Shortly after germination and establishment of the seedling, plants generally develop one of two types of root systems: tap root or fibrous root.

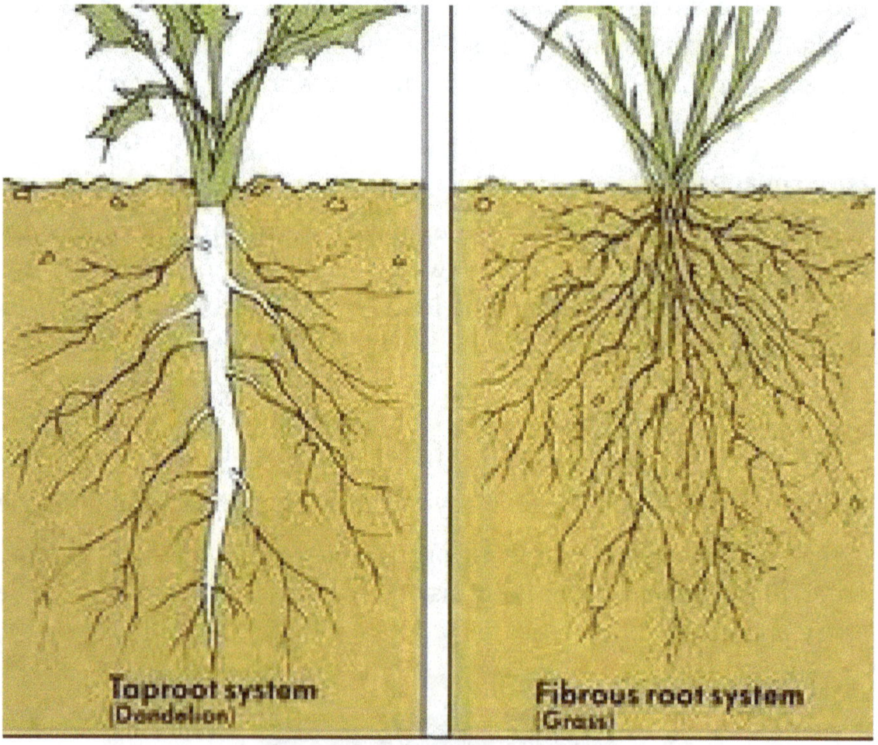

"File:Tap and fibrous root.jpg" by Cassandra gonzales is licensed under CC BY-SA 4.0

Tap root systems

The tap root is persistent, meaning that it is retained throughout the life of the plant; it is also defined as a strong primary root that grows downward into the soil. This tap or primary root is the central axis off of which lateral or secondary roots branch in irregular patterns in response to the availability of high-quality soil — soil with adequate moisture, nutrients, and favorable soil structure (proper particle aggregation and pore space that fosters gas exchange and moisture retention).

Dandelion is an example of a plant that has a tap root. See the taproot and many small secondary roots. Science and Plants for Schools, CC BY-NC-SA 2.0

Lateral or secondary roots typically grow relatively parallel to the soil surface, while the primary or tap root grows perpendicular to the soil surface. Tertiary roots branch off of secondary roots, again in response to nutrient and moisture availability.

A tap root system provides strong leverage and anchorage in the soil. If firmly connected to an upright stem, the tap root can resist uprooting by wind whipping at the shoot and herbivores yanking on the leaves and branches. Both the pigweed and the velvetleaf pictured here are tall, upright plants. The strong taproot helps provide the underground leverage to hold those plants upright.

Fibrous root systems

Fibrous root systems begin the same as tap root systems...with a radicle growing from the seed. However, after a period of early growth, the radicle or primary root stops growing (or slows its growth) and roots begin to form from the stem tissue that is underground, but just above the primary root. These roots emerging from stem tissue are adventitious roots – indicating that the roots emerge from the main stem.

Red and white onions. Alice Henneman. CC BY 2.0

In beans there are two types of adventitious roots. The roots that emerge from the region just above where the main stem stops and the root begins are called basal roots (basal because they are at the base of the main stem). The roots that emerge above these basal roots are called hypocotyl roots. As we'll see in a later chapter, the portion of the stem just above the root-shoot transition zone is called the hypocotyl. Adventitious roots that contribute to the fibrous root system stay close to the soil surface. Fibrous root systems are excellent at holding soil in place because they are thin, extensive, and weblike. This is why various types of grasses, which have fibrous root systems, are planted in areas that are subject to erosion from flowing water following rains. The fibrous root systems of the grasses, once established, hang on to the soil particles like a web of very thin, interlaced fingers.

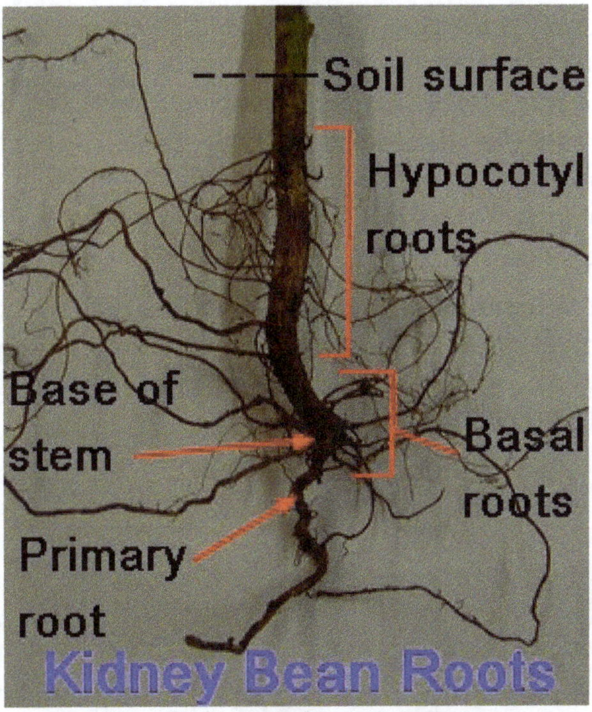

Photo credit: Tom Michals.

The kidney bean root system pictured above has been growing for a few weeks. You can identify the soil line on the stem by where the stem coloration transitions from green to cream or buff color. The root doesn't start immediately after the stem enters the soil. Instead, the stem continues underground for about an inch. Notice the point where the width of the stem drastically reduces; this is where the primary root tissue starts. Again, the primary root traces directly back to the radicle, which is part of the embryo in the seed. Just above where the root starts are the basal roots, and above these are the hypocotyl roots. As noted above, basal and hypocotyl roots are adventitious roots, as they emerge from the stem.

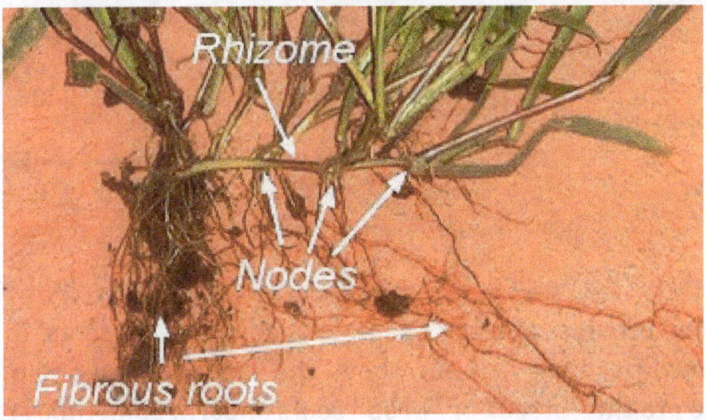

Rhizome of Echinopogon ovatus. The roots emerging from nodes indicate the rhizome is stem tissue.

In this picture of *Echinopogon ovatus*, known as hedgehog grass in its native Australia, you can see the rhizome or underground stem that is typical of spreading grasses. Note the nodes on the rhizome, which indicate that it is shoot tissue, not root tissue. Also note that the roots emerging from these nodes. The roots are adventitious because they emerge from shoot tissue rather than from the primary root (which disintegrates early in the growth of grasses). Shoot tissue often also emerges from these nodes.

Both tap root systems and fibrous root systems rely on root hairs to gather moisture and nutrients. Root hairs are extensions of the outer layer of cells (called the epidermis) of young roots. Root hairs live for only a few weeks, deteriorate, and are then replaced by fresh root hairs.

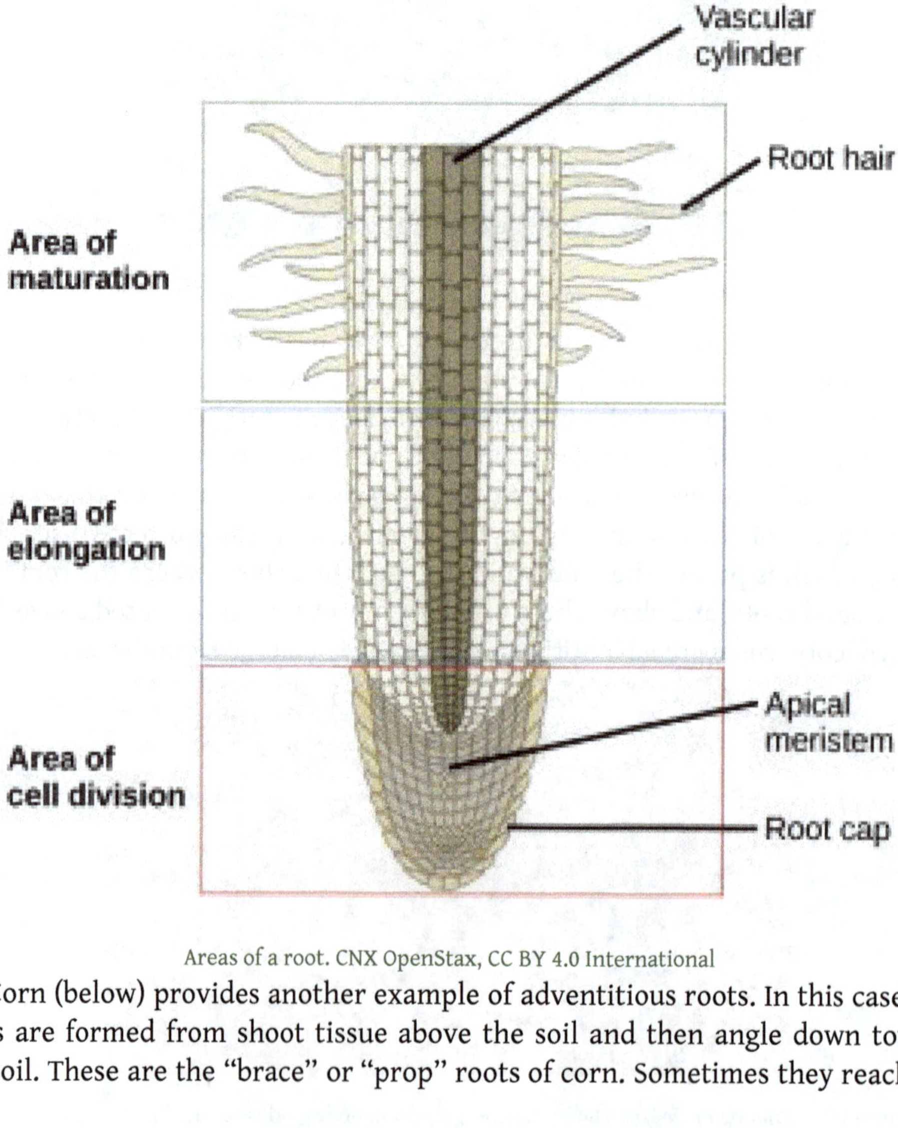

Areas of a root. CNX OpenStax, CC BY 4.0 International

Corn (below) provides another example of adventitious roots. In this case, the roots are formed from shoot tissue above the soil and then angle down toward the soil. These are the "brace" or "prop" roots of corn. Sometimes they reach the ground, and other times they hang in the air. The roots that reach the soil branch extensively and are important for moisture and nutrient absorption. Below is a photo of the adventitious brace roots of *Prunella vulgaris.*

Adventitious prop roots on corn plant. Krish Dulal, CC BY-SA 3.0

"Self-heal (Prunella vulgaris) showing stolons and adventitious roots" by Science and Plants for Schools, CC BY-NC-SA 2.0

TERMS

Here are the terms from this chapter that you will need to be familiar with.

Flashcards

Adventitious	Tissue arising from an organ other than expected.
Adventitious roots	Roots that emerge from the stem rather than roots.
Alternate leaves	Leaves are attached on alternate sides as they go up the stem.
Apex	Tip of the stem.
Apical bud	Bud located on the tip of the stem.
Apical meristem	Group of more or less continually dividing cells located at the tip of a shoot or root.
Axil	Upper angle between a lateral structure and the stem to which it is attached.
Axillary bud	Bud borne in the axil of a stem.
Axillary meristem	Group of more or less continually dividing cells located at the axils of a stem.
Basal root	Root that emerges from the region just above where the main stem stops and the root begins.
Bract	Leaf attached to the terminal node, which is part of the inflorescence rather than the stem. It may also be found at the base of each pedicel.
Branch	Vegetative growth coming from a node on the main stem.
Bud	Immature vegetative or floral shoot or both, often covered by scales; also called a meristem.
Chlorophyll	Green photosynthetic pigment found in plants, algae, and cyanobacteria that captures light for photosynthesis.
Compound leaf	Leaf with a blade margin that is completely interrupted and segmented into separate leaflets.
Fibrous root	Root system where the radicle grows and then rapidly slows or completely halts in growth. Once this happens, roots will emerge above the radicle and from the stem tissue located below the soil.
Hypocotyl roots	Roots that emerge above the basal roots.
Internode	Stem regions between nodes in plants.

Lamina	Another name for a leaf blade.
Lateral or secondary roots	Roots that extend horizontally from the primary root and serve to anchor the plant securely into the soil. This branching of roots also contributes to water uptake, and facilitates the extraction of nutrients required for the growth and development of the plant.
Leaf	A usually green, flattened, lateral structure attached to a stem and functioning as a principal organ of photosynthesis and transpiration in most plants.
Leaf axil	Upper angle between a leaf petiole and the stem to which it is attached.
Leaf blade	Broad portion of a leaf; does not include the petiole.
Leaf margin	Edge of the leaf blade.
Leaf primordia	Young leaves, recently formed by the shoot apical meristem, located at the tip of a shoot.
Leaf scar	Mark indicating the former place of attachment of petiole or leaf base.
Leaf sheath	Structure where the blade attaches to an envelope of leaf tissue that wraps around the shoot of the plant and attaches to a lower node on the stem.
Leaflet	Small leaf-like structure that is found on compound leaves. Multiple leaflets make up a single compound leaf.
Lenticel	Small opening in the cork of woody stems that allows for gas exchange.
Meristem	Group of continuously dividing cells; also called a bud.
Midrib	Main vein, generally in the center of the leaf, from which secondary veins emerge.
Node	Stem region of a plant where one or more leaves attach; location of lateral buds.
Opposite leaves	Leaves that grow directly opposite each other on the stem.
Palmate venation	Where several veins radiate from the point where the petiole attaches to the blade. The veins fork, travel a bit, fork again, travel, fork, and so on until they reach the margin of the leaf.
Palmately compound leaf	Compound leaf where the petiolules of the leaflets connect directly to the petiole (no rachis).
Parallel venation	Distribution or arrangement of a system of veins in a leaf blade in a non-intersecting network. The veins are parallel to each other and the long edge of the leaf.

Petiole	Stalk by which most leaves are attached to a stem; part of the leaf structure, not the stem.
Petiolule	Stalk that connects the leaflet to the top of the petiole.
Pinnate venation	Type of webbed venation where there is a strong midrib and the secondary veins fan out opposite one other.
Pinnately compound leaf	Compound leaf where the leaflets are arranged opposite one another on the rachis.
Primary meristem	Apical meristems on the shoot and root apices in plants that produce plant primary tissues.
Primary root	Root that forms from the embryonic radicle.
Prop root	Adventitious root that arises from the stem, penetrates the soil, and helps support the stem, as in corn.
Radicle	Embryonic root that breaks through the seed coat during germination and develops into the seedling's root system.
Rhizome	Horizontal stem growing just below the soil surface.
Root	Organ that anchors the plant into the soil, takes up water and nutrients, and stores food.
Root hair	Thin, hairlike outgrowth of an epidermal cell of a plant root that absorbs water and minerals from the soil. Root hairs live for only a few weeks, deteriorate, and are then replaced by fresh root hairs.
Sessile	When a leaf lacks a petiole; called a sessile leaf.
Shoot	Made up of a central axis (stem) with a repeating pattern of nodes and internodes.
Simple leaf	Leaf with an uninterrupted blade margin.
Stem	Supporting and conducting organ usually developed initially from the epicotyl and growing upward, and consisting of nodes and internodes.
Stipule	Usually a pair of appendages located at the base of a leaf but may be fused into a ring around the stem; variable in size, shape, and texture; serves for protection or to attract pollinators.
Stolon	Stem with long internodes that grows along the surface of the ground.
Storage root	Root that is modified for storage of nutrients, such as carrots and beets.
Tap root	Main root of a plant, usually stouter than the lateral roots and growing straight downward from the stem.

Terminal bud	Bud located at the apex of a stem.
Trifoliate leaf	Compound leaf with three leaflets that attach to a rachis.
Tuber	Swollen, underground, modified stems that store food.
Venation	Pattern of veins on a leaf.
Whorled leaves	Leaves oriented in a whorled formation in which their point of attachment appears to spiral up the stem.

CHAPTER-4

HOW PLANTS GROW, PART 2

In this chapter we explore the two main types of plant growth, determinant and indeterminate, and the vast diversity in form and number of inflorescences. As you learn the vocabulary, think about why each plant has these growth forms. A species with more flowers and seeds, for example, may be a better fit in its environment than a plant with a small number of flowers, but larger and more robust seeds.

Learning objectives

- Recognize the general patterns of plant growth and the diversity of flowers found in angiosperms.
- Distinguish the two types of growth patterns found in angiosperms – determinate and indeterminate.
- Compare the diversity and types of compound inflorescences.

4.1 GROWTH PATTERNS AND INFLORESCENCES

Learning objectives

- Recognize the general patterns of plant growth and the diversity of flowers found in angiosperms.
- Recognize the two types of growth patterns found in angiosperms – indeterminate and determinate.
- Identify the names of parts of simple and compound inflorescences.
- Recognize the diversity of arrangements of these parts.

Vegetative and reproductive growth patterns

Annual plants are produced from seed in the spring and die at the end of the growing season. These plants typically have a vegetative phase early in their lifespan, followed by a reproductive phase.

During the vegetative phase, stem growth takes place through the addition of new cells — including new nodes and internodes — by the apical meristem (the growing point or bud at the tip of the stem) and through elongation of those new cells. Leaves and branches then emerge at the nodes.

Eventually, and typically at very specific and predictable times, hormonal signals within the plant that are triggered by external agents such as night length or sustained temperatures stimulate the formation of reproductive buds or meristems at nodes. These reproductive meristems give rise to inflorescences. The appearance of these inflorescences signals the transition of the plant from vegetative to reproductive stages.

Determinate or indeterminate? Pumpkin vine by Krista76, CC BY-NC-SA 2.0

Determinate growth

In some plants, the apical meristem itself transforms into a reproductive meristem and produces an inflorescence at the end of the stem, called a terminal inflorescence. This determinate growth is beneficial in ornamental plants because it places flowers at the ends of the stems and branches, where they are clearly displayed and showy. Next time you walk past a bed of flowering plants, look at where the flowers are held on the plant. In most cases, they will be at the tip of the stems and branches and positioned on the outer periphery of the plant.

Grasses are determinate. Leonora (Ellie) Enking,

Determinate plants are also popular in agriculture because the plant produces all or most of its flowers at about the same time. The fruits ripen synchronously and can be harvested during a narrow window of time rather than ripening at various times during the season, which would stretch harvest out over many weeks. Corn is an obvious example, with the tassel at the top of the plant being a terminal male inflorescence and the corn ear a terminal female inflorescence. This type of growth pattern is called determinate because once the apical meristem of the plant transforms into a reproductive meristem, the number of nodes on that stem has been determined and will not increase. The stem may increase a bit in length through the elongation of the cells in the internodes, but no additional nodes will be formed.

Indeterminate growth

In other plants, the apical meristem remains a vegetative meristem that is capable of forming new nodes and internodes throughout the season. Once the hormonal signals are right, reproductive axillary meristems at the nodes below the apical meristem produce inflorescences. Even though the reproductive phase has begun, the plant can still grow new nodes and internodes from the apical meristems on stem and branches. As the season progresses and as new nodes are formed by the apical meristem, these nodes also mature to support the formation of more reproductive meristems and the growth of inflorescences.

Many tomato plants are indeterminate. square foot hydroponics, CC BY-NC-ND 2.0

This type of growth is called indeterminate since the number of nodes on the stem and branches is not determined, and more can be added throughout the year. For ornamental plants, indeterminate growth can mean highly desirable season-long flowering by an individual plant. For food plants, it means that fruits ripen at different times and so are available for a longer period of time through the year (many tomato cultivars are indeterminate). This may not be favored in large agricultural operations where the preference is for the efficiency of one simultaneous and large harvest, but it is great for gardeners who like small amounts of their produce to ripen throughout the season.

Inflorescences

The inflorescence is the flowering structure of the plant. Some inflorescences are very simple and support only one flower. These are called solitary or sole flower inflorescences, and the stalk attaching the flower to the node from which it grew is called the peduncle.

Salvia farinacea (Blue Sage). guzhengman. CC BY-NC-SA 2.0

Other inflorescences can support multiple flowers and display complex branching patterns.

The drawing to the right depicts a hypothetical inflorescence that supports multiple flowers. Toward the bottom are the last two nodes of the stem. The next-to-last node supports a leaf, and the terminal node supports an inflorescence and a leaf. Since that leaf is associated with the inflorescence, rather than emerging from a stem node, botanists call it a "bract," even though it may look like a smaller version of the plant's other leaves. Since the terminal node transforms to be reproductive, this plant must have determinate growth.

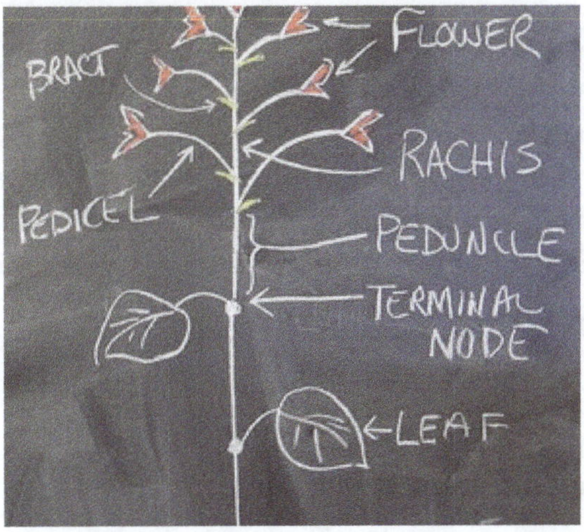

Inflorescence with labels. Image credit: Tom Michaels

The stem-like part of the inflorescence between the point where it emerges from the terminal node of the stem and the point where the branches of the inflorescence begin is called the peduncle. The central axis of the inflorescence, starting at the point where the first inflorescence branch is attached, is called the rachis. The rachis may subdivide one or more times, and these subdivisions are considered part of the rachis. Finally, there is a short stalk connecting the flower to the rachis; this is called the pedicel.

This type of inflorescence, with peduncle, pedicel, and single-flower structure, is called a simple inflorescence. An inflorescence with a group of flowers is called a compound inflorescence. A type of compound inflorescence with multiple flowers originating from a common point is an umbel. The illustration below shows a simple umbel compared to a compound umbel. On the left, in the simple umbel, the pedicels radiate out from a central point at the end of the peduncle and attach to a single flower. On the right, in the compound umbel, rachi radiate out from the end of the peduncle, and an umbel is attached to the end of the rachis.

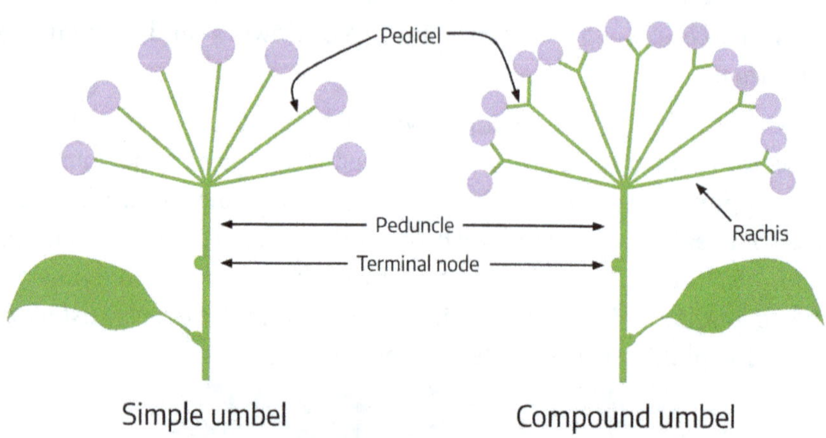

Simple and compound umbels. Illustration by Emily Tepe.

Inflorescences can be amazingly complex structures. This site about inflorescences **will help you recognize some of the possible arrangements and perhaps start looking for this complexity when you are out among plants.**

4.2 PLANT HORMONES

Learning objectives

- Understand the role of the five major hormone groups in plant growth and development.

- Recognize that cells, tissues, and organs have unique competency to respond to specific hormones.
- Connect specific hormones to plant responses and how they are used in plant propagation.

How plants respond to hormones

The five major groups of plant hormones — auxins, cytokinins, gibberellins, ethylene, and abscisic — acid are distinguished by their chemical structures and the response they evoke within the plant (see Table 4.1). For any cell to respond to a hormone it must be competent to perceive the chemical. Some cells simply lack the ability to "see" the hormone and do not respond to its presence. Competency to perceive a hormone depends on a cell's physiology when the hormone is present. If the hormone is perceived, its unique chemical structure causes a chain reaction or signal transduction that involves changes in gene expression and cell morphology. These cellular responses to hormones can lead to changes we see in the plant, such as movement towards light, a transition from vegetative growth to flowering, or the closing of leaf stomata due to drought stress.

The perception of the hormone occurs in cells and throughout a tissue or organ, depending on where the hormone is located, the concentration of the hormone, and the developmental state and physiological condition of the cell. Many cells within a tissue can respond in a coordinated manner, resulting in changes in the whole plant. Hormones are often made in one cell and translocated to other cells where they are perceived, and the response may occur far away from the site of hormone synthesis. Responses to hormones are studied through exogenous application of the chemical to a plant tissue — the hormone is applied to the outside (exo) of the plant and observations are made on how the plant responds. For hormones that are a gas, like ethylene, this means the hormone can be translocated from one plant to another plant. The plants are essentially talking to one another, using a wide variety of molecules.

Experiments in which hormones are exogenously applied to a plant reveal how plants respond to hormones; much of our knowledge about the role hormones play in plant growth is from this type of experiment. Whenever a hormone is exogenously applied, however, it is also interacting with all of the hormones present in the plant. These are endogenous hormones (endo means internal), and the cell responds according to the sum of all hormones in its presence. This can complicate the interpretation of responses to exogenous hormone applications.

The hormones used in plant propagation can be naturally occurring and found in many plants, or can be synthetic or synthesized to mimic the structure and response of a naturally occurring hormone. Synthetic hormones are often used instead of naturally occurring versions because they are less expensive to obtain, may cause greater or longer lasting responses, and can be less susceptible to degradation in the plant and during storage. Because exogenous application of hormones play a role in manipulating or disrupting plant growth, they are used extensively as herbicides (weed killers) and can be targeted to certain types of plants based on how certain species respond to the different structure.

Plant responses to hormones and their application in plant propagation

Auxin

Auxins are a group of related molecules that are involved in almost every aspect of the plant's life cycle. Auxins stimulate growth through cell elongation, which is integral to the plant's responses to environmental changes. Auxins are responsible for two types of growth responses: phototropism, the bending or growth of a shoot toward light, and gravitropism, a change in growth occurring after a change in gravitational force. The diagram below shows indoleacetic acid (IAA, illustrated with pink dots), a naturally occurring auxin, moving from the sunny to the shady side of a shoot tip. The differential accumulation of auxin on the shady side of the shoot causes those cells to increase growth and bends the shoot tip toward the light. Auxin's stimulation of cell growth is also important in healing wounds and forming calluses after pruning.

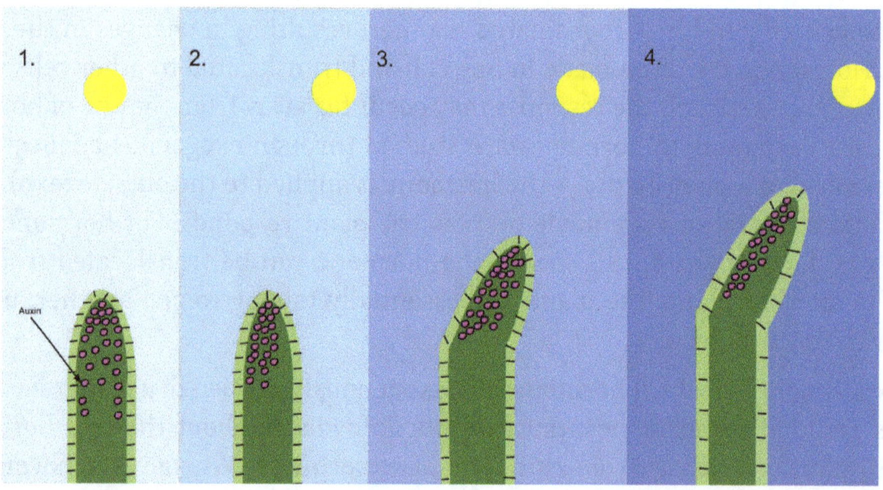

- This is a normal plant that has the sun positioned almost directly over the plant. During this time, the auxin (pink dots) that lies within the plant is evenly distributed.
- The sun is now positioned at an angle to the plant. The repositioning of the sun causes the auxin to move the other side of the plant, and becomes more concentrated. This overload of auxin next to these cells causes them to start to grow or elongate.

- This results in the the plant to look like it is growing toward the sun.
- If the sun moves to the other side of the plant, the auxin would again move to the other side of the plant and become concentrated on the side of the plant that is farthest away from the sun.
- The same growth or elongation of the cells on this side of plant would continue to grow towards the sun. Image by MacKhayman. CC BY-SA 3.0.

The ability of auxin to regulate growth can be turned against weeds (plants out of place). The synthetic auxin 2,4-dichlorophenoxyacetic acid, or 2,4-D, is a common herbicide that interrupts normal growth regulation when applied to the plant, causing leaf drop and death. Because dicotyledonous (dicot) plants have a higher competency to respond to 2,4-D, 2,4-D can be used as a selective herbicide to kill dicot weeds in lawns and corn fields, which are resistant, monocotyledonous (monocot) grasses. The grass is unharmed due to its lower competency to respond, while the dicot plants are killed.

In general, auxins are produced in the young leaves of a plant and translocated downward to older tissues. This downward translocation controls apical dominance, where growth of axillary buds is suppressed. Removal (pinching) of the shoot tip where auxin is being produced, as shown in the three photos of mint below, releases the axillary buds from apical dominance and they begin to grow. This is a common horticultural practice, increasing branching and flower production. Pinching is often used in seedling plants such as basil or zinnias to get globe forms in a pot instead of tall, single-stemmed plants.

One of the most important uses of auxin in plant propagation is to stimulate the growth of adventitious roots — roots that emerge from anywhere on the plant other than from the roots — on shoot cuttings. The photo below shows cuttings from two different Acer ginnala (Amur maple) plants that have different competencies to form adventitious roots. Both cuttings were treated with auxin, but only the competent plant forms adventitious roots (on the left). The cutting from a plant that lacks competency to respond to auxin did not form roots (right) and will eventually die.

Image by Matthew Clark

This form of asexual (clonal) propagation is used by both horticultural professionals and hobbyists. The competency for rooting cuttings can be species specific or seasonal. Collecting stems from a plant to use for cuttings can be more successful in the growing season, as with the Amur maples shown above. With plants such as grapes, however, cuttings are made and rooted during the winter when the vines are not actively growing. The video below demonstrates how shoot cuttings are taken from Amur maples, treated with auxin, and incubated in a high-humidity environment for several weeks to form adventitious roots. Exogenous application of auxin is not required for adventitious rooting of all plants. Some plants can form many adventitious roots without exogenous applications, because the endogenous auxin that occurs naturally in the shoot is sufficient for root formation.

Cytokinins

Like auxins, cytokinins are a group of related molecules that regulate growth and development. However, the plant's response to cytokinin is very different from the responses to auxin. Cytokinin comes from the word cytokinesis, which means cell division. You will learn about cytokinesis, specifically mitosis, in Chapter 13.

Cytokinins promote cell division, where one cell splits and two new daughter cells are formed. Cytokinins are important regulators of plant growth and development.

Cytokinins have an interesting interaction with auxin in plants. In the 1950s, Skoog and Miller were researching the growth of N. tabacum stems in tissue culture. They discovered that they could use specific ratios of an auxin (IAA) and a cytokinin (kinetin) to direct the growth of the stem tissue in culture. A high ratio of cytokinin relative to auxin led to shoot formation, a higher level of auxin led to root formation, and equal levels of each produced callus growth, which is undifferentiated plant cell growth. Skoog and Miller's transformational discovery formed the basis of the "MS" plant medium that remains popular for plant propagation using tissue culture.

Watch this video to learn more about the propagation of plants in synthetic media with exogenous hormones in tissue culture.

Gibberellins

Gibberellins, or gibberellic acid (GA), are a group of over 100 molecules that are primary regulators of stem elongation and seed germination. They were discovered during research on the cause of the "foolish seedling" disease of rice. The disease, characterized by tall plants with little grain, is caused by an infection with *Gibberella fujikora*, a parasitic fungus that produces GA in the rice shoots, causing increased stem elongation.

In Chapter 9.2, on seed physiology, you will learn that some seeds are dormant and do not germinate even when the proper environment is provided. Seed dormancy, which has several causes and evolutionary advantages, always has the common feature of preventing seed germination until the time, season, or seed physiology is correct. Planting a dormant seed or a dead seed gives the same result: no germination. For plant propagators, dormancy can be confusing, raising the question "are my seeds dead or are they dormant?" Either condition prevents germination and plant propagation. Treating seeds with GA is a common method to break dormancy and facilitate germination. GA treatment of Gentiana lutea (bitter root) seeds, for example, increases germination from 0% (no germination) to over 80% when treated with 100 parts per million (ppm) GA (see the graph of germination on the left). If a propagator of G. lutea had not known about seed dormancy, they may have assumed their bitter root seeds were dead. For most plants, GA is the endogenous hormone that triggers seed germination.

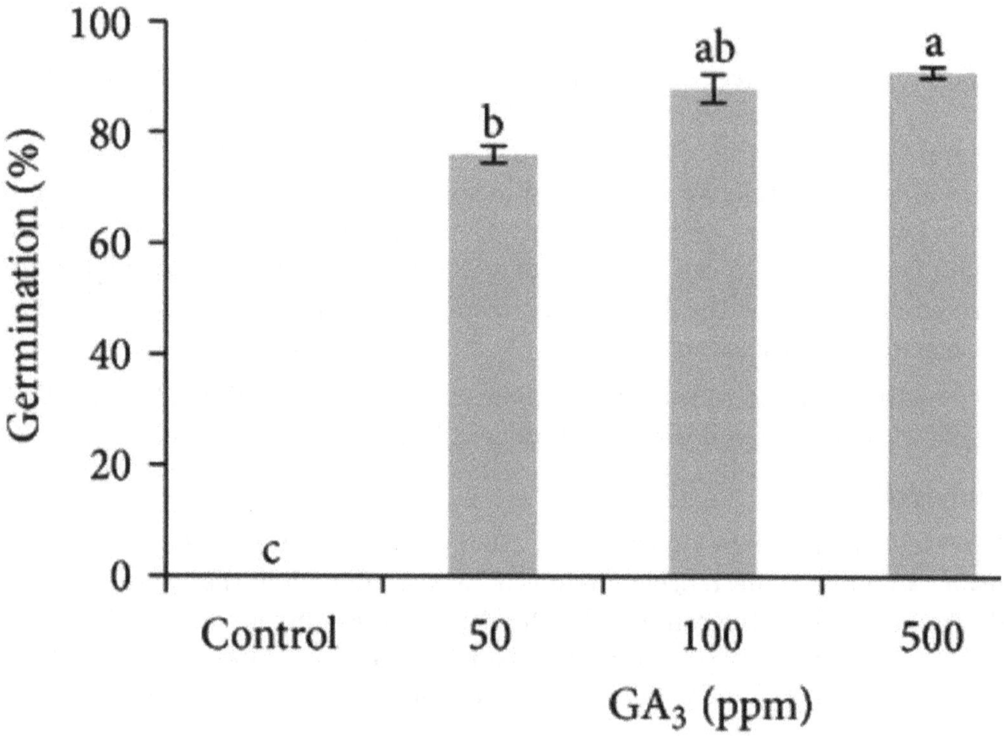

Abscisic acid

While GA facilitates seed germination, abscisic acid (ABA) inhibits it. Abscisic acid is a single molecule that regulates germination and the response of a plant to reduced water availability during drought stress. ABA levels increase as water becomes less available to the plant, evoking several responses, including the closing of stomates. Closing stomata slows transpiration (also called evapotranspiration), the movement of water in the plant from the root to stem to leaf and out through the stomata into the atmosphere. You'll read more about stomata and the movement of water in Chapter 11, Plants and water.

Reducing water content is one of the final steps in seed maturation and is important for seed longevity by reducing metabolism to a minimum, which is the quiescent nature of mature seeds. Increasing endogenous ABA levels in seeds prepares them to survive lower water content, is important to seed maturation, and prevents precocious germination (vivipary). Seeds with low levels of ABA during seed development may prematurely germinate. Low ABA levels may result from a genetic mutation or environmental causes. Vivipary in some fruits is not uncommon and may occur during storage of fruit in the grocery store.

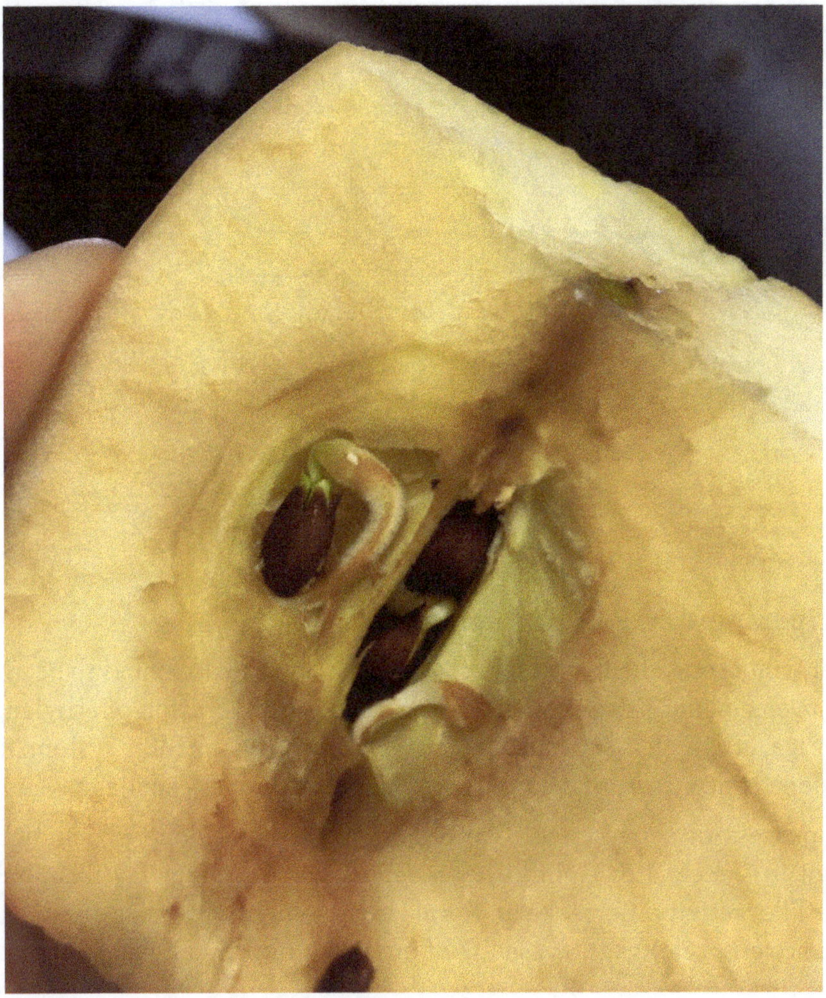

Apple in vivipary. Image by Matthew Clark.

Ethylene

Ethylene is well known as the gaseous, ripening hormone. It also regulates seedling growth and the formation of root hairs, and can lead to epinasty – the bending of branches downwards.

Many plants are sensitive to the effect ethylene has on fruit ripening. The iconic examples are tomato and banana. These fruits are climacteric – they continue to ripen after harvest. The perception of ethylene by the cells that make up the fruit triggers the ripening process and the production of more ethylene. As the concentration of ethylene increases, so does the speed of the ripening. Picking immature or green fruit enables shipment over long distances, because the fruit is firmer and less likely to be damaged in transit. The green fruit can then be treated with ethylene from an ethylene generator (right) to accelerate ripening.

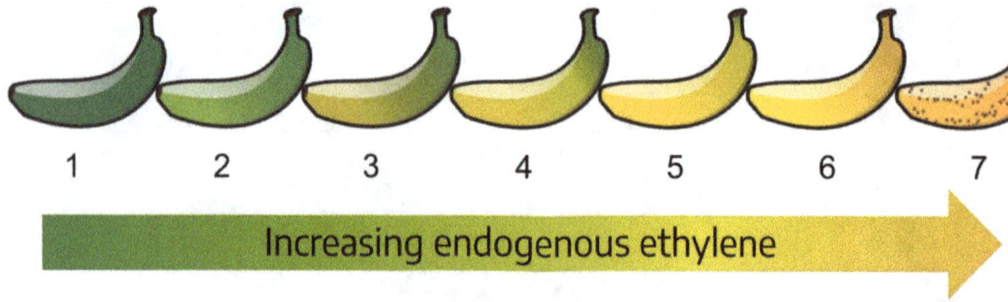

Seven ripening stages of banana.

For other fruit crops, the introduction or production of ethylene is to be avoided to prevent over-ripening and spoilage. To prevent the generation of ethylene during fruit storage, ethylene is scrubbed from the air using an air filter system. Reducing the ethylene concentration means slower ripening and less spoilage.

The process of senescence is also triggered by ethylene production and is important in the cut flower industry. Keeping cut flowers away from gases with ethylene-like activity helps keep floral arrangements looking fresh. Reducing ethylene action prolongs the vase life of many cut flowers as well as the storage of fruits.

Review

The five major groups of plant hormones control many aspects of plant growth and development and have important applications in plant propagation. However, many other molecules are also key to the plant's response to its environment. These highly diverse signal molecules modulate the plant's physiology through complex interactions. A cell's response to the many different hormones is a sum of its genetic makeup, its physiology, and the environment.

TERMS

Here are the terms from this week's lessons that you will need to be familiar with for your assignments and for the quiz.

Flashcards

Abscisic acid	A hormone that regulates seed maturation and responses to changes in water availability.
Adventitious root	A root that emerges from anywhere on the plant other than from the roots.
Angiosperms	A group of flowering plants whose seeds develop inside an ovary.
Annual plant	A plant that is produced from seed in the spring and dies at the end of the growing season.
Apical dominance	Where a shoot suppresses growth of floral or vegetative axillary buds below the growing point.
Auxins	A group of related hormones that regulate many aspects of plant growth and development and are key to stimulating adventitious rooting.
Competency	The ability to respond to a signal, such as a plant hormone.
Compound inflorescence	An inflorescence with a group of flowers and includes a rachis.
Cytokinins	A group of related molecules that regulate cell division and are key to stimulating adventitious shoot formation.
Determinate	When the stem of a plant terminates in a flowering stalk and new stem growth continues from subterminal lateral buds.
Endogenous hormone	A hormone that occurs within the plant.
Ethylene	A gas that regulates fruit ripening and plant senescence.
Exogenous hormone	The application of a hormone to a plant.
Floret	A single flower in a compound inflorescence.
Flower	A reproductive structure in a flowering plant.
Gibberellins	A group of related molecules that regulate seed dormancy.
Gymnosperms	A group of plants whose seeds are produced without the protection of an ovary.
Indeterminate	When the apical meristem remains a vegetative meristem capable of forming new nodes and internodes throughout the season. Once the hormonal signals are right, reproductive axillary meristems at the nodes below the apical meristem produce inflorescences.
Inflorescence	The complete flower structure of a plant; includes the flower, pedicle, rachis, and peduncle.
Natural hormone	A hormone made by a plant.
Pedicel	The short stalk that holds up the flower.
Peduncle	The large, central stalk that attaches the rachi to the stem of the plant.
Perception	The ability of a plant cell or tissue to detect a hormone that depends on a cell's physiology at the time the hormone is present.
Perennial	A plant that lives for more than two growing seasons (more than two years); perennials may be woody or herbaceous (the latter with underground perenniating structures).

Plant hormone	A signal molecule that regulates growth, development, and responses to environmental and other signals, also known as a plant growth regulator or phytohormone.
Rachis	The stalk of a flower that is situated between the peduncle and the pedicel on a compound leaf. Also the name for the central axis on a compound leaf where the leaflets are attached.
Reproductive meristem	The apical meristem that transforms into the reproductive tissues (the inflorescence) of the plant.
Response	The action taken by the plant after perception of a signal.
Senescence	A regulated process that results in cell death and is associated with leaf fall and death of the plant.
Signal transduction	The process in which the perception of a signal, such as a hormone, is moved within a cell, cell to cell, or throughout a tissue.
Simple inflorescence	A type of inflorescence with a peduncle, rachis, pedicel, and single flower structure.
Synthetic hormone	A hormone made by people; can mimic the response of a naturally occurring hormone.
Tropism	A growth or turning response to an environmental or other signal such as phototropism (response to light) or gravitropism (response to gravity); can be controlled by auxin and other hormones.
Umbel	An inflorescence with multiple flowers originating from a common point.
Woody perennial	A plant that lives for more than a year, has hard rather than fleshy stems, and bears buds that survive above ground in winter. Trees, shrubs, many vines, and bamboo are examples of woody perennials.

CHAPTER-5

INSIDE PLANTS

The panda eats shoots and leaves. Maybe a comma is out of place here or there, but in the plant each part has an orderly place and function. The fundamental vegetative (non-reproductive) organs of the plant are stems, roots, and leaves. These organs have unique tissues with unique arrangements that can be used in propagation.

Learning objectives

- Recognize the internal cellular structure of leaves, shoots, and roots, and describe their specializations.
- Connect structure to processes involving gas, water, and photosynthate movement.
- Understand how and where photosynthesis occurs.

5.1 INSIDE LEAVES

Learning objectives

- Recognize the internal cellular structure of leaves.
- Describe how the structure facilitates photosynthesis.
- Describe how the structure facilitates gas exchange.
- Explain why leaves in autumn turn from green to yellow, orange, and red.

Leaf anatomy

Learning leaf anatomy is a bit like taking a sandwich apart and seeing what's inside. We'll start with the upper surface and progress down through the leaf. Here are the layers you will find:

Upper epidermis with cuticle. The cuticle is a protective waxy coating of cutin on epidermis cells, restricting water loss and preventing disease.

- **Palisade mesophyll** — densely packed, columnar-shaped, elongated cells full of chloroplasts. Chloroplasts are structures inside plant cells that contain chlorophyll and are the site of light capture during photosynthesis. These are analogous to cortex parenchyma cells in the stem, but in the leaf they are specialized for light energy capture.
- **Spongy mesophyll** — loosely packed cells with large air spaces between them. The air spaces allow movement and exchange of gases, specifically oxygen, carbon dioxide, and water vapor. Spongy mesophyll cells also contain chloroplasts.
- **Vascular bundle** — xylem, phloem, and bundle sheath cells along with nearby parenchyma and either sclerenchyma or collenchyma fibers for support.
- Lower epidermis with guard cells that regulate the size of the stomata, the gaps in the epidermis that allow gas exchange between the atmosphere and internal parts of the leaf.

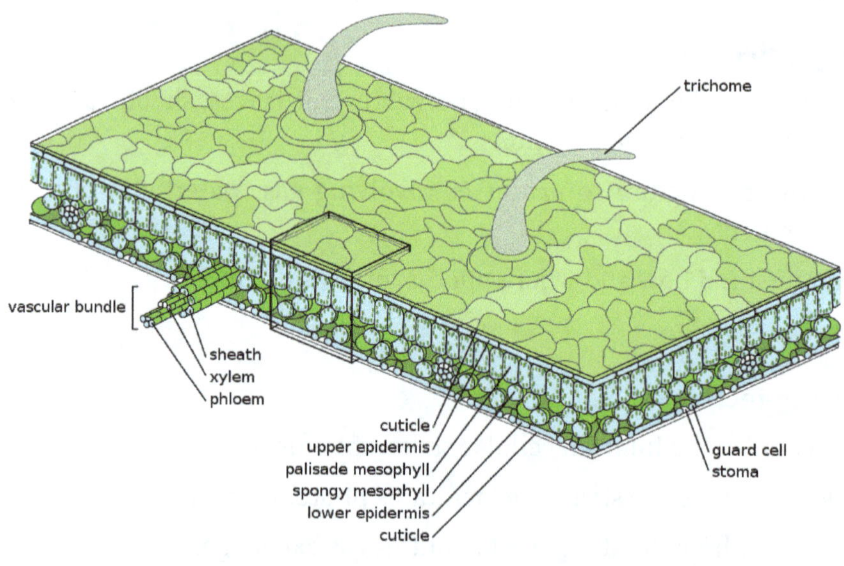

Leaf structure. Zephyris.

In the graphic above, you can see the spongy and palisade mesophyll in detail. During the growing season, these cells are packed with chloroplasts containing the pigment chlorophyll. Chlorophyll absorbs red and blue wavelengths of light to power photosynthesis, and reflects green light back to our eyes, making leaves appear green to use in spring and summer.

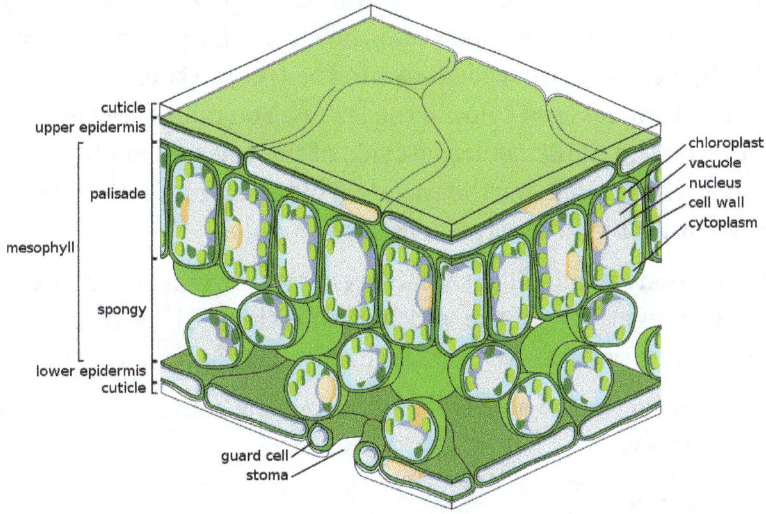

Leaf tissue structure. Zephyris. CC BY-SA 3.0.

Chromoplasts (not to be confused with chloroplasts) are cellular organelles that contain types and colors of pigments other than the chlorophyll found in chloroplasts. In late summer and fall, the mesophyll cell chloroplasts produce chlorophyll at a slower rate than earlier in the year. As the existing chlorophyll reaches the end of its functional lifespan and eventually degenerates, the amount of chlorophyll declines over time. Because chlorophyll replacement doesn't make up for chlorophyll loss, the green color fades from the leaves. Meanwhile, the non-green pigments in the chromoplasts hold their own or increase in quantity, including the yellow and orange from the carotenoid pigments and the red from the anthocyanin pigments.

Carotenoid pigments are present in the leaf all growing season, but during the warm part of the season they're hidden by the high concentration of green-colored chlorophyll. When the cool autumn weather comes, the days shorten and the chlorophyll concentration in the chloroplasts declines. The carotenoids in the chromoplasts don't degenerate as quickly as the chlorophyll does, so we get a beautiful display of oranges and yellows in the leaves once the green is gone. These orange and yellow colors can be counted on every year because the carotenoids are produced and present all year long.

In contrast to the carotenoid pigments, the amount of red color in leaves from the anthocyanins differs from autumn to autumn. Anthocyanins are produced primarily in the autumn in response to bright light and excess plant sugars in leaf cells. Warm, sunny days encourage sugar production in leaf mesophyll, and cool nights hinder the transport of these sugars out of the cells, so the sugars stay in leaves. The high sugar then stimulates anthocyanin production. Warm, bright days

+ cool nights = red leaves. In years with overcast days and warmer nights, there is less sugar production in the leaves and the sugars translocate (move out) from the leaf mesophyll cells to other locations in the tree, so there is less anthocyanin pigment produced and less red color in the leaves. You might see this happening on individual trees: on the sunlit, southwest side of the tree you might find leaves with more red than on the shaded, northeast side. This effect of weather on anthocyanin is particularly noticeable in trees such as sugar maples.

Some other woody plants, like sumac, seem to be bright red in later September and early October regardless of the weather conditions. Other trees, like birch and oak, will be yellow or tan even with warm bright days and cool nights.

Here is a cross section of a pear leaf showing the same cells illustrated in the graphic shown earlier.

Micrograph cross section of leaf. Berkshire Community College Bioscience Image Library. Public domain

The palisade mesophyll is highly adapted for capturing light energy. As noted earlier, the cells are packed tightly together, filled with chloroplasts, and make up the cell layer just under the protective epidermis on the top surface of the leaf oriented toward the sun. The spongy mesophyll cells below the palisade layer are less densely packed together, so the region is laced with air channels. This makes the spongy mesophyll highly adapted for gas movement and air exchange around the cells.

Guard cells surround a gap in the epidermis. This gap is called a stoma or stomate (plural = stomata), and is the portal through which gas exchanges between the interior and exterior of the leaf. Gases in the intercellular spaces between spongy mesophyll cells are free to move out of the stomata into the atmosphere, and atmospheric gasses are free to move in. As we will see later when we address photosynthesis, healthy plants must have the ability to move oxygen (a waste product of photosynthesis) out of the leaf, and carbon dioxide (a raw material for building sugars) into the leaf. Water vapor must also be able to move out of the leaf as part of the transpiration process.

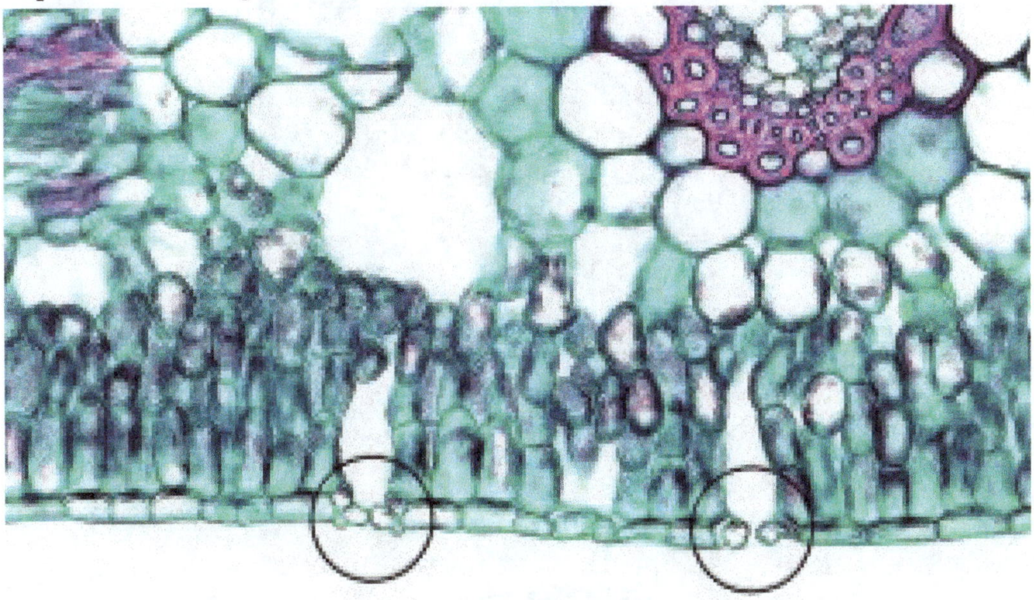

The circles show the guard cells around the stomata on the lower surface of the leaf. bccoer. CC0.

The vascular bundle (the xylem and phloem in the graphic above) is highly adapted for transport of fluids. Water and dissolved minerals move up the xylem from the roots and into the leaf. Sugars produced in the leaf during photosynthesis move out of the leaf through the phloem and are translocated to other parts of the plant. The thick-walled bundle sheath cells that surround the xylem and phloem provide mechanical support, and contribute to some types of photosynthesis. Wikipedia has a good illustration showing the tissues associated with the vascular bundles. Take a look.

5.2 INSIDE STEMS

Learning objectives

- Discuss how stems grow by elongation and by increasing their girth or diameter.
- Identify the tissues that are initially produced by apical meristems and the cells into which they eventually mature.
- Distinguish monocot and dicot stems based on the arrangement of the vascular bundles.
- Describe the functions of some of the specialized cells in the stem.

Introduction to stems

Notice this image of a palm tree because it will come up again later. Think about where and how it is growing. Travel Aficionado – CC BY-NC 2.0

Herbaceous plants are plants whose above-ground plant parts die back to the soil surface at the end of the growing season. Herbaceous plant shoots grow in length and diameter in the same way that roots do: the apical meristem at the tip of the stem produces new cells and then those new cells enlarge in length and volume in the region of elongation just behind the meristem. Stems of herbaceous plants typically do not thicken very much and rely instead on branching to grow laterally (although there are exceptions **like** sunflower and tomato whose stems do thicken noticeably). Branching of the shoot is initiated at nodes where there are axillary buds containing axillary meristems that grow into branches.

Continued growth of the stem in diameter, like you find in a tree where the trunk and branches increase in diameter each year, requires an active lateral meristem called the cambium. Cambium is a meristem in the vascular tissue positioned between the xylem and phloem that continues to produce new xylem cells toward the interior of the stem and new phloem cells toward the exterior. Production of xylem and phloem from cambium cells is called secondary growth and is typical of woody perennial plants. We will explore plants with secondary growth in a different lesson.

Some herbaceous plants are annuals, like beans and pansies, where the whole plant dies over the winter. These plants complete their life cycle from seed to flower to seed in one year. Other herbaceous plants, like Kentucky bluegrass and chrysanthemums, are herbaceous perennials where only the above-ground growth dies over winter (unless it is a really nasty winter). The root, or in some cases underground stem tissue, stores nutrients for the next season's growth. In spring when soil and air temperatures are sufficiently warm, new shoots emerge from nodes on compact stem tissues often called crowns. The cells in the crown tissue survive the harsh winter temperatures because they accumulate sugars and other compounds that act as antifreeze compounds. They are located very close to the soil surface, or even protected a bit by the soil. This position on or in the soil, plus snow cover, results in much warmer temperatures around the crown than the air temperature above the snow. In fact, the air temperature can be -20°F **above** the snow and only a few degrees below freezing (32°F) around the crown.

Dicotyledonous stems

This micrograph of a herbaceous dicot stem shows four basic parts (in order from outside to inside): epidermis, cortex, vascular bundle, and pith. Notice how the vascular bundles of dicots are arranged in a ring around the circumference of the plant stem with the cortex to the outside and pith to the inside.

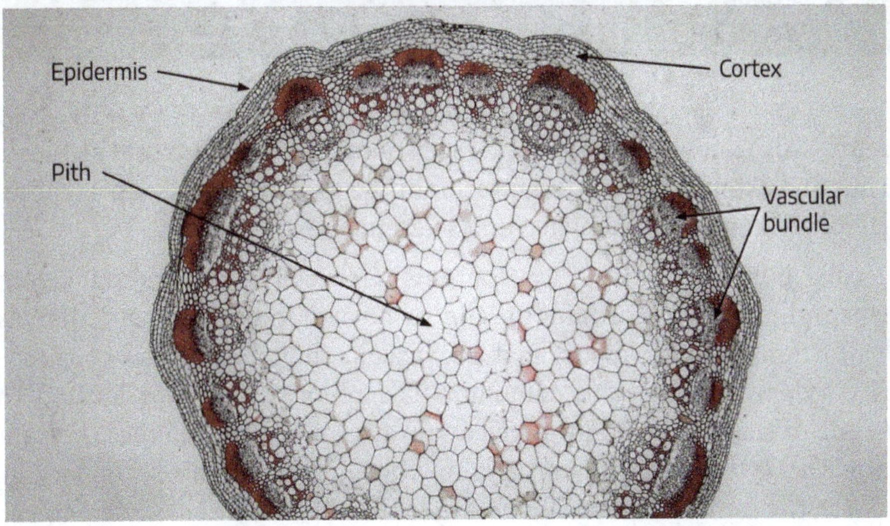

Dicot stem. Berkshire Community College Bioscience Image Library. Public domain

Epidermis – This tough covering is a single layer of living cells. These cells are closely packed and function to protect the internal parts of the plant. The walls are thickened and covered with a thin waxy waterproof layer called the cuticle that reduces water loss from the plant. Stomata with guard cells are found in the epidermis, and these function to allow gasses into and out of the stem, similar to their function in leaves. In some stems either unicellular or multicellular hair-like outgrowths, called trichomes, appear from the epidermis. Below is an image of trichomes on a tomato stem, inflorescence, and leaf.

Tomato stems, leaves, and inflorescences have abundant trichomes. Quinn Dombrowski –

Cortex – Also known as the ground meristem, the cortex is found just inside the epidermis and extends toward the interior of the stem. It is made up of three types of cells: parenchyma, collenchyma, and sclerenchyma.

Vascular bundles – This region contains sclerenchyma fibers that strengthen the stem and provide protection for the vascular bundle. In dicots, the vascular bundles form a distinct ring. A mature vascular bundle consists of three main tissues – xylem, phloem, and cambium. The phloem is always located toward the outside of the bundle and the xylem always toward the center. The cambium separates the xylem and phloem, and, in those plants where secondary growth takes place, the cambium produces new xylem and phloem cells: xylem toward the center, phloem toward the outside.

The pith occupies the central part of the stem and is composed of thin-walled parenchyma cells often with larger intercellular spaces than you would find in the cortex.

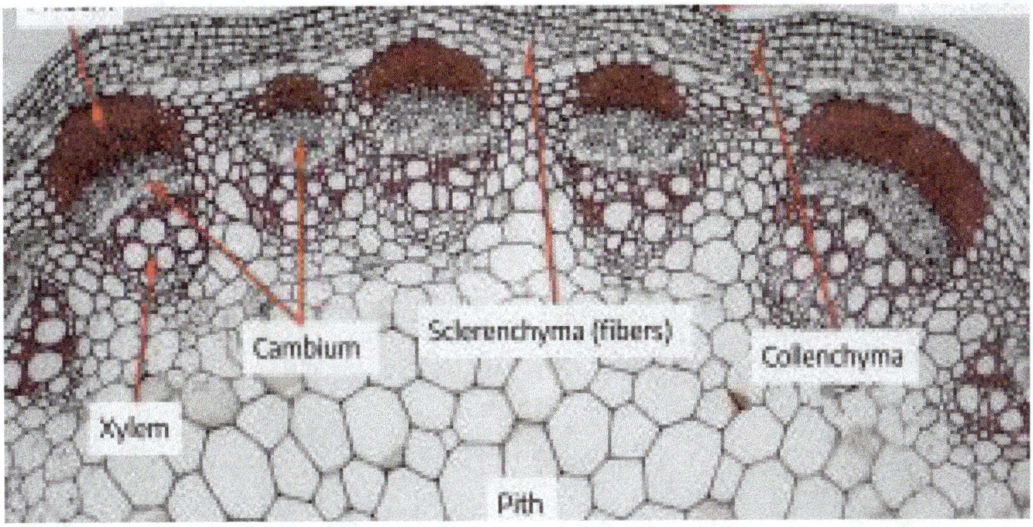

Cross section of a young sunflower (Helianthus) stem, magnified 100x. Berkshire Community College Bioscience Image Library. Public domain

In this cross section of a dicot stem note the collenchyma cells in the cortex just under the epidermis. Sunflower stems are quite tough, and this toughness is in part due to the layer of collenchyma cells positioned to give the stem mechanical stability. Toward the outside of each vascular bundle are fibers of sclerenchyma that also contribute to the sunflower stem's toughness. Then, moving from the outside to the inside, you will find phloem, a layer of cambium, and then xylem. The pith is in the center.

Now you might recall that just above we said that herbaceous annuals like sunflower don't have cambium to increase the girth of the stem, so what's with the cambium in this picture of a sunflower stem? Well, you will see a layer of cambium between the xylem and phloem of dicot stems, and it is active for a while and produces a modest amount of xylem and phloem which contributes to the sunflower's larger diameter stem. However, this cambium tissue is not continuously active as in woody stems, and the plant does not survive through the winter.

Monocotyledonous stems

The tissues making up monocot stems are essentially the same as what we saw in dicots with the main difference apparent in the placement of the vascular bundles. In monocots, vascular bundles are scattered throughout the stem instead of being oriented in a ring. Since there is no ring of vascular bundles, there is no "inside" pith and "outside" cortex. All the ground tissue is considered to be cortex.

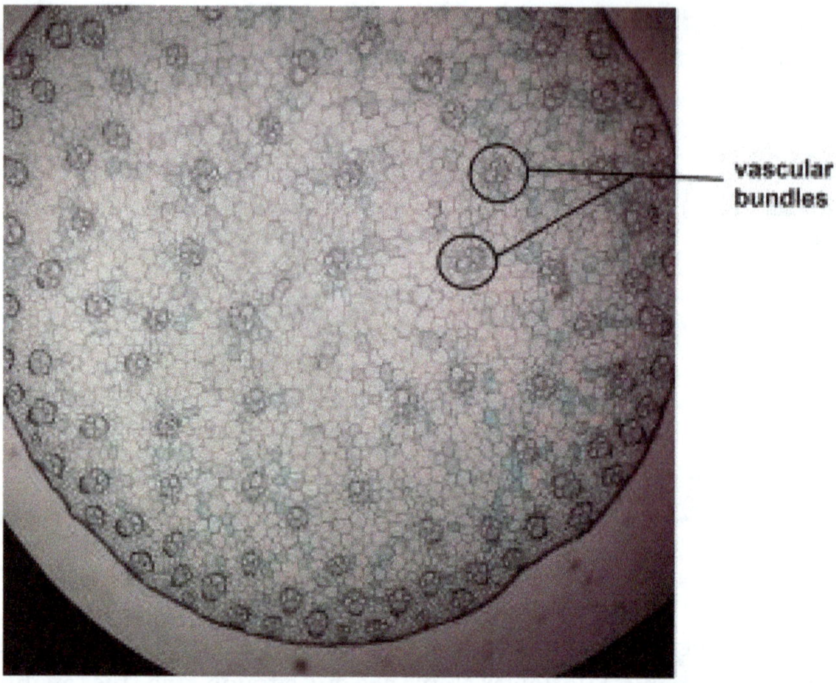

Monocot corn stem cross section showing vascular bundles. Melissa Ha. CC BY-NC 2.0

In monocot vascular bundles the phloem is always oriented toward the outside of the plant and the xylem toward the inside. There is no cambium and no secondary growth. Around the outside of the vascular bundle is a layer of parenchyma cells called the bundle sheath. This layer of cells is very important in photosynthesis. For now we will consider it a protective covering and supportive sheath around the vascular bundle.

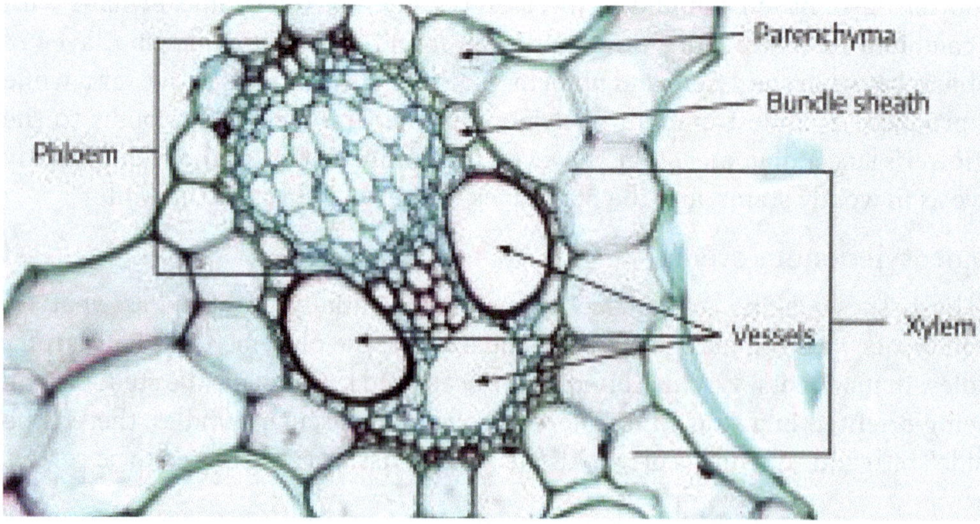

Monocot stem showing vascular bundles. Berkshire Community College Bioscience Image Library. Public domain.

One final note

So why the palm tree? Because it is an exception.

A palm tree is a monocot, but unlike other monocots whose stems are quite thin (like grasses, for instance), their primary stems do increase in girth from year to year even though they do not have secondary cambium. Palms have a special layer of meristematic cells, called the primary thickening meristem, oriented toward the outside of the stem that each year can initiate new vascular bundles and new parenchyma cells. Each year the stem expands in girth as a result of the palm's production of new parenchyma and vascular bundles. You don't need to memorize this – it is just an example of an interesting exception to the rule.

5.3 INSIDE ROOTS

Learning objectives

- Explain how roots elongate and increase in diameter via primary and secondary meristems.
- Identify the tissues in the root that originate from the root meristems and the cells into which they eventually mature.
- Distinguish between monocot and dicot roots.
- List the functions of the cells in plant roots.

- Please read the OpenStax pages summarizing root biology. This resource reinforces several of the topics addressed below, so you may read it either before or after you complete this chapter. You don't need to memorize the information there; use it to reinforce the content below.

Roots. Leonard J Matthews – CC BY-ND 2.0

Root review

The functions of roots include:

- Anchoring the plant in the soil (and stabilizing the soil).
- Supporting the upright growth of the plant.
- Providing a site for the symbiotic relationship of the plant with particular beneficial fungi and bacteria.
- Absorbing water and dissolved minerals from the soil.
- Storing nutrients like starch for subsequent use by the plant (tap roots and tuberous roots are examples of geophytes that store nutrients in roots).

Angiosperm (flowering) plants are often classified by whether they rely on a primary root system an adventitious root system. A plant doesn't necessarily have only primary or just adventitious roots. One of these systems, however, will be dominant. Whether a root is considered primary or adventitious depends on whether the root traces back to the radicle (embryonic root) or arises from (normally underground) stem tissue.

Tap root (or primary root) system

Tap or primary roots arise as a continuation of the embryonic radicle tissue and persist into maturity. Secondary roots (also called lateral roots) arise from the primary root, and tertiary roots arise from the secondary. This primary -> secondary -> tertiary root formation is the usual rooting system for dicots. Next time you see a dandelion or other weed that doesn't look like a grass, yank it up and look at the root system. The main root you see (assuming that it didn't break off when you pulled it up) is probably a tap root. The photo on the right, above, demonstrates this point. It shows a young taproot system with many secondary roots branching off the primary root.

This Desmanthus spp. plant has a strong central taproot with many lateral or secondary roots. Btcpg - CC BY-SA 4.0

Adventitious or fibrous root system

In plants of this type, the primary root, which originates from the radicle, weakens prior to maturity and new, vigorous, adventitious roots arise from stem tissue. Adventitious roots may grow from nodes or might arise from the internodes. They originate from parenchyma cells in the cortex of the stem. Dig up a clump of turfgrass and look at the roots. Turfgrass has a fibrous root system, as does corn.

Adventitious corn roots. NY State IPM Program at Cornell University. CC BY 2.0.

As noted above, although adventitious roots originate from the stem they don't have to emerge from a node. While it's not apparent in the photo, the adventitious corn roots you see above do trace back to a node. In contrast, in the photo of a tomato stem, below, we see emergence of adventitious roots from both node and internodal regions of the stem.

Tomato stem with adventitious roots. Mark. CC BY 2.0

Internal root structure

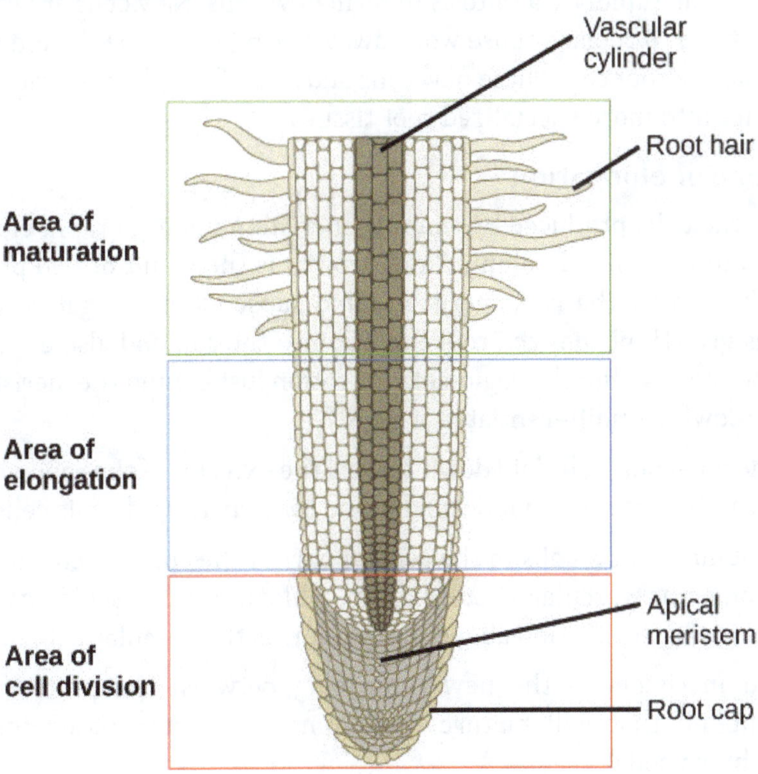

Root with areas labeled. CNX OpenStax. CC BY 4.0 International

Root cap

Shaped like a thimble, this structure covers the tip of the root and provides protection as the root grows into the soil. These parenchyma cells are produced by the root's meristem which is just behind the root cap. The outer cells of the root cap are continuously worn away through contact with the soil, and new cells are added to the inner portion.

In addition to protecting the interior of the root, the cap secretes a mucilage which stabilizes the water content of the surrounding soil, ensuring longer-lasting nutrition to the root system and making for easier root probing. Finally, the root cap contains statocytes, specialized cells that help the plant sense gravity and grow accordingly. These cells are full of starchy organelles which settle at the lowest part of the cell and encourage growth in that direction. If the root cap, with these statocytes, is removed, a plant may grow in random directions because it has lost the ability for gravitropism (growth in response to gravity).

Root meristem

The cells here divide rapidly via mitosis to form new cells. New cells are laid down toward the root cap to replace those worn away during root growth, and also laid down away from the root cap. These new cells laid away from the root cap elongate and then mature into more specialized root tissues.

Region or zone of elongation

In this region, the cells produced by the root meristem undergo rapid elongation – they expand in length and volume. Root growth is the result of two processes: new cell production by the root meristem, and subsequent elongation of those new cells. This growth pushes the root further into the soil and also expands the diameter of the root. Within the region of elongation just behind the meristem you will find the following undifferentiated tissues:

- Protoderm – new cells laid down toward the exterior of the root which will mature to become the root dermal tissue (primarily epidermis cells).
- Procambium – new cells in the central part of the root which will mature to become the vascular tissue (xylem, phloem, and vascular cambium), labeled in the illustration above this section as the vascular cylinder.
- Ground meristem – the new cells lying between the protoderm and procambium that will mature to become the cortex tissue (primarily parenchyma cells).

Region of differentiation (also called the region of maturation)

Here the root becomes thicker, and secondary or lateral roots are initiated. In this region the protoderm, procambium, and ground meristem cells undergo differentiation into the specialized cells associated with the dermal, vascular, and cortex tissues, as noted above.

Root-hairs begin to form in the region of differentiation; these are fine outgrowths of epidermis cell walls and membranes, and increase the area of absorption of the root.

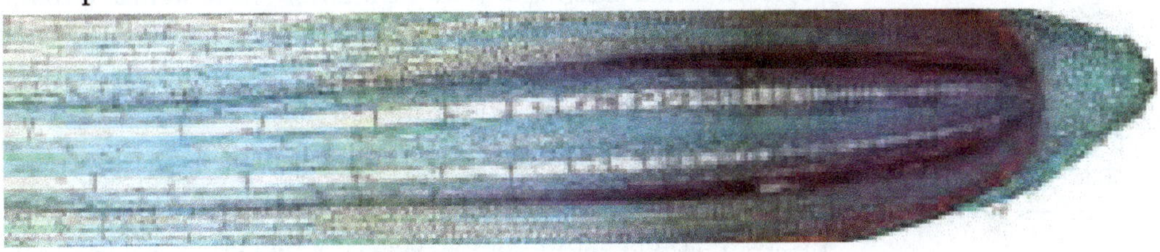

Here is a magnified image of a root, and a drawing to point out the various components. Jen Dixon –

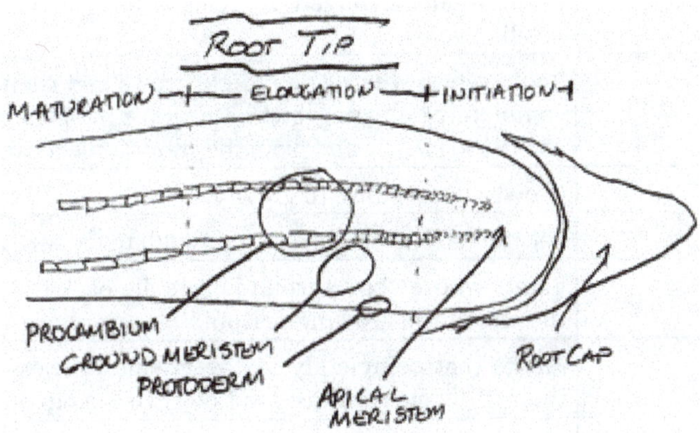

Root tip. Image credit: Tom Michaels

TERMS

Flashcards

Adventitious roots	Roots that emerge from the stem rather than roots.
Anthocyanin pigments	Red pigments that are produced primarily in the autumn in response to bright light and excess plant sugars in leaf cells.
Apical meristem	Group of more or less continually dividing cells located at the tip of a shoot or root.
Axillary bud	Bud borne in the axil of a stem.
Axillary meristem	Group of more or less continually dividing cells located at the axils of a stem.
Cambium	Lateral meristem in vascular plants, including the vascular cambium and cork cambium, that forms parallel rows of cells resulting in secondary tissues.

Carotenoid pigments	Yellow and orange pigments that are present in the leaf all growing season, but during the warm part of the season these colors are hidden by the high concentration of green-colored chlorophyll. They take longer to break down than chlorophyll.
Cortex	Also known as the ground meristem, is found just inside the epidermis and extends toward the interior of the stem and root, and is made up of three types of cells: parenchyma, collenchyma, and sclerenchyma.
Crown	Compact stem tissue at or near the soil surface.
Cuticle	Protective waxy coating of cutin on epidermis cells that restricts water loss.
Cutin	Water-resistant substance that coats the wall of the cell exposed to the environment and helps limit the loss of water that is inside of the plant to the atmosphere.
Dicotyledon (dicot)	Seed plant that produces an embryo with paired cotyledons, floral organs arranged in cycles of four or five, and leaves with net-like veins.
Fibrous root	Root system where the radicle grows and then rapidly slows or completely halts in growth. Once this happens roots will emerge above the radicle and from the stem tissue located below the soil.
Gravitropism	Growth in response to gravity.
Guard cells	Located on the epidermis and regulate the size of the stomata.
Herbaceous	Plants whose above-ground parts die back to the soil surface at the end of the growing season.
Herbaceous annual	Plants that completely die over winter. These plants complete their life cycle from seed to flower to seed in one year.
Herbaceous perennial	Plants where only the above-ground growth dies over winter. The underground portion lives for more than two growing seasons (two years).
Mitosis	Cell division where a cell divides into two identical daughter cells.
Monocotyledon (monocot)	Seed plant that produces an embryo with a single cotyledon and parallel-veined leaves; includes grasses, lilies, palms, and orchids.
Phloem	Tissue consisting of sieve tube and companion cells in the vascular system of plants that moves dissolved sugars and other products of photosynthesis from the leaves to other regions of the plant.
Pith	Occupies the central part of the stem and is composed of thin-walled parenchyma cells often with larger intercellular spaces than you would find in the cortex.
Primary root	Root that forms from the embryonic radicle.
Procambium	New cells in the central part of the root that will mature to become the vascular tissue (xylem, phloem, and vascular cambium).
Protoderm	New, primarily epidermis, cells laid down toward the exterior of the root which will mature to become the root dermal tissue.

Root cap	Thimble-shaped mass of cells that covers and protects the root apical meristem from rocks, dirt, and pathogens.
Secondary growth	Production of xylem and phloem from cambium cells.
Secondary root	Root that forms off of the primary root.
Statocytes	Specialized cells that help the plant to sense gravity and grow accordingly.
Stomate/Stoma/Stomata	Gap in the epidermis that allows gas exchange between the atmosphere and internal parts of the leaf.
Tap root	Main root of a plant, usually stouter than the lateral roots and growing straight downward from the stem.
Translocation	Movement of a substance from one place to another.
Trichome	Either unicellular or multicellular hair-like outgrowths arising from the epidermis; found on stems.
Vascular bundle	System containing vessels that carry or circulate fluids and dissolved minerals in the plant; composed of xylem, phloem, and bundle sheath cells.
Xylem	Supporting and water-conducting tissue of vascular plants.
Zone of Differentiation	Area in roots where tissues are formed (expand in width).
Zone of Elongation	Area in roots where recently produced cells grow and elongate prior to differentiation.

CHAPTER-6

CELLS, TISSUES, AND WOODY GROWTH

Cells are the building blocks of the plant. Even the mightiest oak begins as a single, microscopic cell produced by fertilization. That single cell replicates, differentiates, and develops into the complex tissues, organs, and systems of the plant. In this chapter we'll will learn about the internal workings of plant cells and how those cells, some living and some dead, orchestrate the plant's life cycle.

Learning objectives

- Compare the functions of the diverse cell types and their organelles.
- Understand how cells build the plant and their origins in meristematic tissue.

6.1 PLANT CELLS AND TISSUES

Learning objectives

- Label the parts of a plant cell.
- List the types of tissues in a plant and describe where they are located and the specialized cells that make up each of these tissues.
- Summarize the key functions of those tissues.

Plant cell

The graphic below illustrates the key parts of the plant cell.

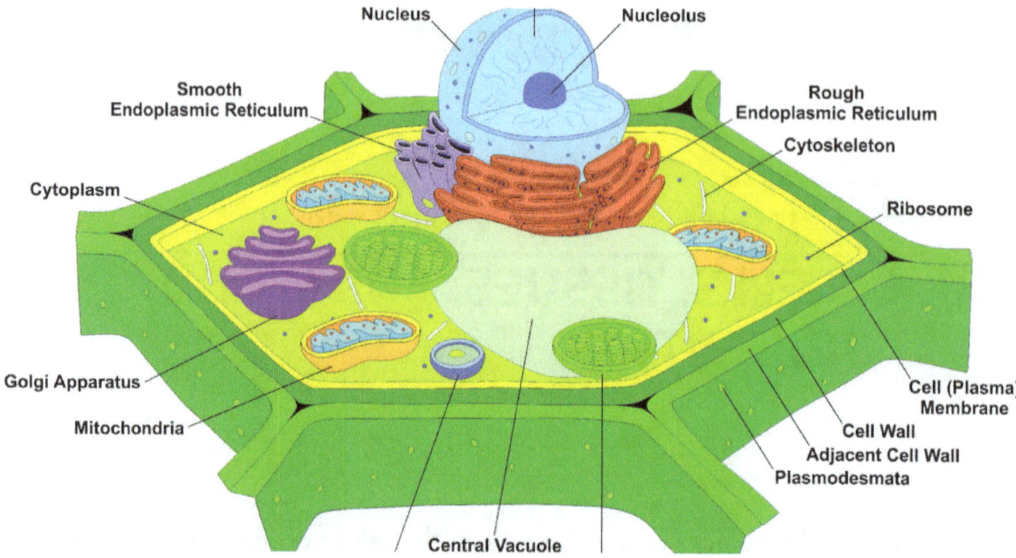

Diagram of a plant cell. Licensed from Shutterstock.

Cell wall

The outer covering of the cell, the cell wall is a rigid membrane that contains cellulose (a carbohydrate that is indigestible for humans). The cell wall protects the parts inside, and the cellulose molecules in the wall provide the support and rigidity needed to maintain the cell's three-dimensional structure.

Cell membrane

The cell membrane is made up of layers of protein and lipid (fats and oils are examples of lipids). The cell membrane is semi-permeable — it allows select compounds in and out, but blocks other types of compounds. If the cell were like a bicycle tire, the cell wall would be the thick, protective outer tire tread and the cell membrane would be the inner tube.

Chloroplast

An organelle ("organelle" is the generic name for a plant organ) that contains chlorophyll. In the chloroplast, light energy is captured and the first steps are taken in the chemical pathway that converts the energy in light into forms of energy that the plant can transport and store, like sugar and starch. Chloroplasts are not evenly distributed throughout the plant but, as you might expect, are concentrated in parts of the plant that are exposed to and oriented toward the sun. A plant cell in the leaf blade will have many chloroplasts, while cells in the middle of the stem will have few or none.

Mitochondria

(Singular = mitochondrion)

The mitochondria is where stored sugars from photosynthesis are metabolized to produce forms of energy that the plant can use for growth. This metabolism is known as respiration and uses oxygen to convert sugars (and other carbohydrates) to energy and carbon dioxide. This is the cell's power plant. All cells have numerous mitochondria.

Nucleus

An organelle that contains the chromosomes. Chromosomes contain the genetic material (deoxyribonucleic acid; DNA) that is carried within each cell and that directs which chemical reactions are turned on and off in the cell. Chromosomes are the hereditary material passed on to new cells and to subsequent generations. Each cell has one nucleus.

Vacuole

An organelle containing various fluids, ions, chemical energy, and waste products from the cell. The vacuole takes up much of the cell volume and gives shape to the cell.

Cytoplasm

The fluid inside the cell membrane in which the organelles and other plant cell parts are suspended.

Middle lamella

A material containing pectin that forms between cells and cements the cell wall of one cell to the cell wall of an adjacent cell. If bricks in a wall are like cells in a plant, the middle lamella in the plant is like the mortar between bricks in the wall.

Plant cells have other parts as well, but these are the key ones to know and understand now.

Review questions

- What is the difference in function between the cell wall and the cell membrane?
- What is the mortar that holds cells together? If lettuce is grown in a soil with low calcium content, the outer edges of leaves can degenerate and die, causing tip burn. Could this involve the mortar that holds cells together?
- Where is light energy captured?
- What happens in the mitochondria, and what is the connection between that function in mitochondria and the function of chloroplasts?

Tissues

A tissue is a group of cells that share a function. The cells within a tissue may differ from one another, but they all contribute to a particular function. We're going to look at three types of tissues: dermal, cortex, and vascular.

Dermal tissue

Dermal tissues (derma is Greek for "skin") are on the outside of the plant and provide protection for the plant cells they surround. The cells making up dermal tissues are tough so that they can protect against mechanical challenges to the plant, like abrasion. They have thick cell walls. In the shoot, the epidermis cells, which are the main cell type in dermal tissue, secrete a water-resistant substance called cutin (a waxy polymer), which coats the wall of the cell exposed to the environment. This coating helps limit the loss to the atmosphere of water that is inside the plant. Cutin is absent or greatly reduced in root tissue because roots need to reach out into the soil to absorb water and nutrients.

A cluster of mite eggs just beneath the epidermis of a leaf. Scot Nelson. Public domain

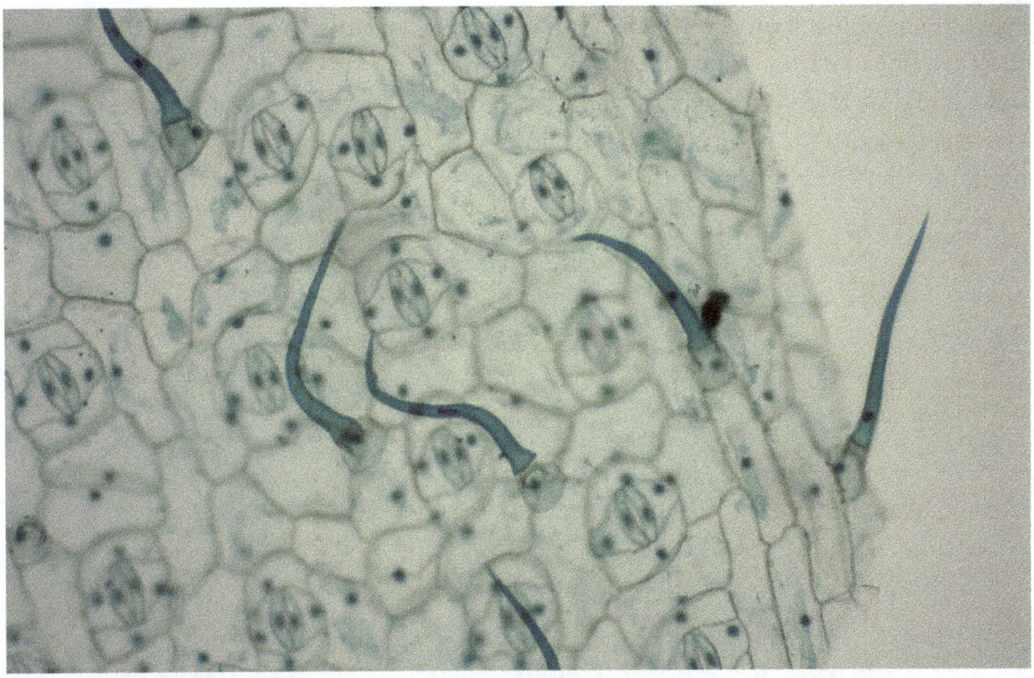

Trichomes extending from the surface of a leaf. Marc Perkins, CC BY-NC 2.0

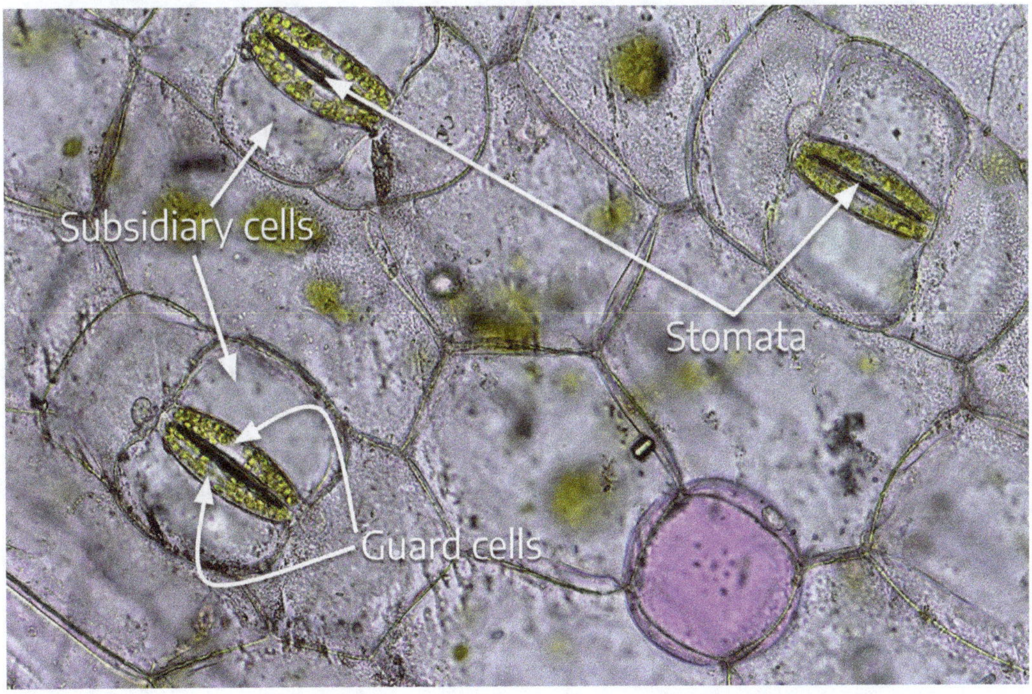

Stomata guard cells. BlueRidgeKitties, CC BY-NC-SA 2.0

The epidermis is the outermost layer of cells in the plant. It is normally only one cell thick, but in some cases the epidermis can be a few cells thick. Epidermis cells typically have few if any chloroplasts. They are often called pavement cells because they are flat like tiles or puzzle pieces. Depending on the plant, the epidermis may have hairs, or trichomes, that extend out from the plant. Some of these trichomes are associated with glands that contain oils or other substances secreted by the plant.

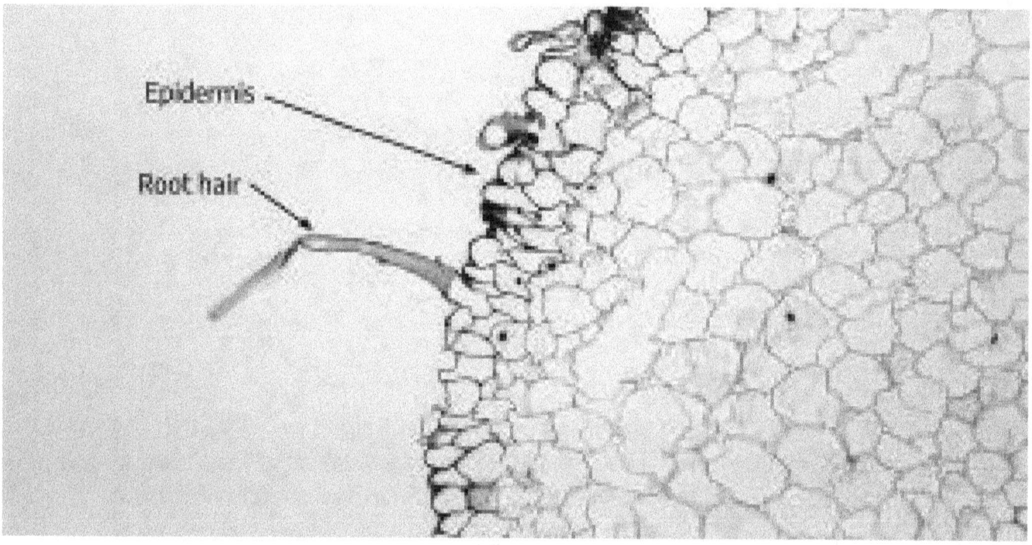

Notice that the root hairs are an extension of the epidermal root cell. Berkshire Community College Bioscience Image Library. Public domain

The epidermis contains pairs of guard cells that will open to form stomata (Greek stoma = mouth; an opening in the leaf surface) through which gasses can move into and out of the deeper cell layers in the leaf. These guard cells are found most abundantly on the underside of leaves, but may also be on the upper leaf surface and on the stems.

The root has dermal tissue as well. The predominant cell type, like in the shoot, is also epidermis, but as noted above there is no cutin covering the root epidermis because the root is underground and less prone to dehydration. There are no guard cells or trichomes, but there are root hairs . The root hair is a very small-diameter extension of the epidermis cell wall and cell membrane that extends out into the growth medium. Water and nutrients enter the plant through absorption into the root hairs.

Review questions

- What unique feature of the epidermis is found in roots and not shoots?
- What is the function of the waxy cutin layer? Why don't you find it on the root epidermis?

- Are stomata found in roots, shoots, or both? Why does this make sense?
- What's the difference between a cell and a tissue?

Cortex or ground meristem tissue

The cortex (sometimes called "ground meristem") tissue is found just inside the epidermis and extends toward the interior of the stem and root. Some types of plants also contain cortex tissue at the very center of the stem called the pith, but you won't find pith in roots or in all plant stems. Cortex cells provide structural support for the stems. In leaves, this tissue just inside the epidermis is called the mesophyll ("middle of the leaf"). Mesophyll tissue is the site of most photosynthesis reactions in the leaf.

Three types of cells make up the cortex:

Parenchyma

- The most common type of cortex cell.
- Has thin cell walls (called a primary wall in the graphic below).
- The mature cell is alive.
- Has the ability to begin dividing to help heal wounds (by covering the wound with plant tissue called callus).
- Will also divide to initiate adventitious roots on stem cuttings.
- Site of many other functions, such as photosynthesis and storage of starch and other chemical compounds.
- Leaf mesophyll tissue is a type of parenchyma that is packed with chloroplasts.

Collenchyma

- A living cell at maturity.
- Cell walls are thicker than the thin parenchyma cell walls, which give collenchyma strength. However, these cells remain somewhat flexible compared to sclerenchyma, which you will read about next.
- The cells can connect together to form resilient strands, like the strands of a celery stalk. These strands provide support for young tissues.
- Because the cells are alive, they can respond to external stimuli. If the plant is regularly shaken by wind, for example, the collenchyma cells will respond by producing thicker cell walls for greater support of the plant stem so that it can remain upright.

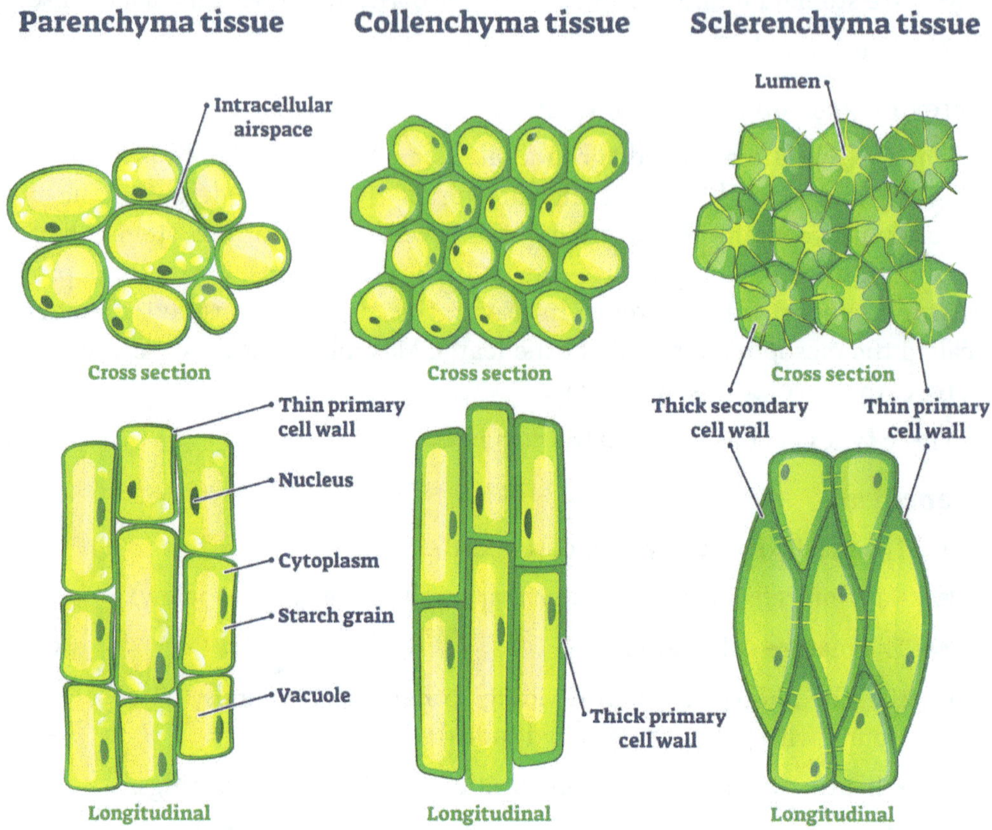

Types of plant tissue. Licensed from Shutterstock.

Sclerenchyma

- This type of cell has a primary and secondary cell wall. The primary cell wall, on the outside of the cell, is rich in cellulose, just like other plant cell walls. Once the cell has reached its final size, a secondary cell wall is deposited just inside the primary wall. The secondary wall has a high concentration of lignin that gives the cell rigidity. This rigid, lignified secondary cell wall is responsible for sclerenchyma's hardness and strengthening properties. Sclerenchyma comes in two types:
    - Fibers (see below) formed from long strands of sclerenchyma. These tough fibers give the plant rigidity. We extract these fibers from plants and use them in fabrics, carpets, and rope. Examples of plant fibers made up of sclerenchyma cells include jute, hemp, and flax (the fabric made of flax fibers is called linen). Cotton is not in this list; it is an epidermal fiber produced by the plant's seed coats.

  - Sclereids are cells with hard, tough cell walls. Sclereid cells can coalesce and cover other plant parts. For instance, they form the hard covering around the seeds (the endocarp) of stone fruits like cherries, the hard shell around walnuts, and the hard covering of coconut. Sclereids also make up the grit that crunches between your teeth when you eat a pear.
- Sclerenchyma cells are dead at maturity. They don't thicken in response to external stimuli the way collenchyma can.

Review questions

- Which cortex cells are alive and which are dead when mature?
- Which cells make up the tough fibers from which rope and fabrics can be made?
- Which cells divide to initiate adventitious roots?
- What tissue in the leaf corresponds to the cortex in the stem?

Vascular tissue

Vascular tissues form the plumbing system in the plant through which water, nutrients, sugars, and other compounds flow. These plumbing pipes and associated cells are bundled together in the plant in a structure called the vascular bundle. There are three main types of vascular tissue: xylem, phloem, and vascular cambium. Xylem and phloem are composed of different types of cells, listed below.

Xylem

- Moves water in the plant.
- The water flow is unidirectional. Water in xylem heads from root to stem to leaf and then out of the plant stomates through a process called transpiration.
- The part of the tree that we call "wood" is made up of xylem.
- These cells are dead at maturity, and they are hollow.
- Xylem tissue is composed of four different types of cells:

Vessels

Elongated cells that connect end to end to form tubes. The cells are dead at maturity. The end walls of the vessels are perforated, so water can move freely through the holes and flow from cell to cell. Vessels have a relatively large diameter compared to other xylem cells and allow greater movement of water.

Tracheids

These cells are elongated and narrower than vessels, and connect by overlapping at their ends. These cells are also dead at maturity and contain pits through which water can move. Tracheids appear earlier in the paleontological record of plant evolutionary development than vessels and are thus considered "primitive" (not inferior, but appearing earlier in evolutionary time). Vessels are a subsequent evolutionary adaptation that allow for greater water flow because of their larger diameter.

Xylem fibers

Sclerenchyma cells lying near the vessels and tracheids, and thus part of the vascular bundle. They are strung together end to end like the vessels and tracheids, but unlike those water carriers they have no pits or perforations and instead have thick primary and secondary cell walls. They provide flexible support for the plant from within the vascular bundles.

Xylem parenchyma

Back of a ukulele that shows parenchyma rays. Del-Uks, CC BY-NC-ND 2.0

In woody plants there are parenchyma cells around the vascular bundles that extend horizontally through the xylem (the woody part of the plant) and develop into rays moving laterally from the center to the exterior of the plant. Most of the vascular cell types are arranged in a linear fashion parallel to the long axis of the stem, but parenchyma rays are arranged laterally from the middle of the stem out toward the epidermis. They function to conduct water through the wood (xylem). Oak furniture for example, it will have a "grain" which is caused by the annual rings of xylem, and will have rays that, on edge, look like small pits in the wood. We will see this in later lectures when we deal more extensively with wood and secondary growth. As you can see in the photo to the right, some of the natural markings you see in an instrument's wood are from parenchyma rays.

Phloem

- Moves some nutrients taken up by the roots to other parts of the plant.
- Moves sugars manufactured in leaves by photosynthesis, and other plant compounds such as plant hormones like auxin, to other parts of the plant.
- The flow in the phloem is multi-directional among leaf, stem, and root.

Phloem tissue also has four types of cells:

Sieve tube members

Elongated cells that join end to end to form tubes for passage of liquids. The end walls have pores. Unlike xylem cells, these cells are still alive. They have a thin cell membrane containing a layer of living protoplasm that hugs the wall of the cell.

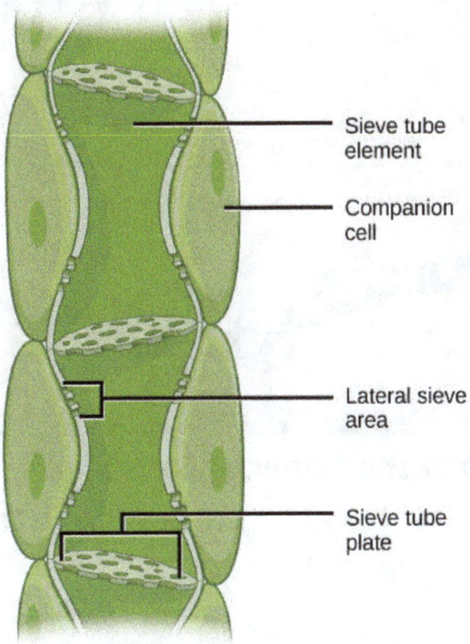

Companion cells

Associated with sieve tube members. Contain a nucleus, may direct the metabolism of the sieve tube member, and are alive.

Phloem fibers (sclerenchyma cells)

Provide support, same as for xylem.

Phloem parenchyma cells

Adjoin the sieve tube cells, same as for xylem.

Vascular cambium

This third type of vascular tissue is a meristematic region (meaning that the cells can actively divide to form new growth) where new vascular tissues originate in plants with secondary growth, like trees. We will study secondary growth in Chapter 6.2.

6.2 WOODY GROWTH

Learning objectives

- Understand primary and secondary growth of trees.
- Explore the factors that affect the rate of tree growth.

Introduction

Alan at the Science Museum of Minnesota with a cross section from a 600 year old tree

Image credit: Alan Smith

Long-lived trees like bristlecone pines can live more than 5,000 years! Understanding how trees grow can unlock a record of the environment a tree has experienced through its lifetime, and provide a record of the climate conditions during that period. In this exercise you will compare how trees grow in height (primary growth) and diameter (secondary growth).

A longer and more detailed video on secondary tree growth can be found here: How Trees Grow You can also review the previous lessons on apical meristem growth. As you review the lessons and videos, think about the environmental and genetic factors that affect the rate of secondary growth of trees.

The study of tree rings is called "dendrochronology," – the science of determining environmental change using annual growth rings in trees. Here's a short video on Dendrochronology (Tree Ring Dating). Here's another optional video on the nitty gritty of collecting a tree ring "Dendrochronology: How to Core a Tree." The photograph below shows a grafted kiwi vine. Notice the bright green vascular cambium on the outside edge of the cut branch, just below the brown bark. These are the actively growing cells, where cell division and production of xylem and phloem in each growing season are produced. This fast growth often causes the bark to "slip" as it is expanding and making room for the new growth under it. During the summer, you may take a young branch and easily peel the bark away from wood below. You will notice that it is quite wet. This time of year is generally good for propagation techniques like grafting, especially T-budding (you will learn this method later) because the plant tissues used are at the right stage of growth. Plant propagators take advantage of these natural processes for the best results.

You can see the slightly green cambium just below the bark where a branch was cut off. Adapted from and image by Te Tumu Paeroa, CC BY-SA 4.0.

Below the cambium, working to the center of the tree, is the sap wood. Sap wood is still functional for moving water from the roots. You can identify it because of its color, and it may be noticeably wet. The next layer inside is the heart wood. This is what is typically used in lumber. The wood is functioning to support the tree, but it no longer has the capacity to move water.

Cross section of a Betula pendula truck showing heart wood. Beentree.

TERMS

Flashcards

Callus	Growing mass of unorganized parenchyma cells produced in response to wounding.
Casparian strip	A band-like deposit of waterproof suberin that wraps around each cell in the endodermis and forces water to move through the cells rather than the intercellular spaces.

Cell membrane	Made up of layers of protein and lipid (fats and oils are examples of lipids) and semi-permeable, meaning that it allows select compounds in and out, but blocks other types of compounds.
Cell wall	Rigid membrane that contains cellulose (a carbohydrate that is indigestible for humans) and is the outer covering of the cell.
Chloroplast	Organelle that contains chlorophyll where light energy is captured and where the first steps are taken in the chemical pathway that converts the energy in light into forms of energy that the plant can transport and store, like sugar and starch.
Chromoplasts	Cellular organelles that contain types and colors of pigments other than the chlorophyll found in chloroplasts.
Collenchyma	Elongated cell type with thicker walls; usually arranged in strands; provides support.
Companion cells	Associated with sieve tube members (direct the metabolism) and containing a nucleus (alive).
Cytoplasm	Fluid inside the cell membrane in which the organelles and other plant cell parts are suspended.
Dermal tissue	Tissue on the outside of the plant; provides protection for the plant cells it surrounds.
Endodermis	Innermost cells of the cortex.
Epidermis	Outermost layer of cells in the plant.
Ground meristem	New, primarily parenchyma, cells lying between the protoderm and procambium that will mature to become the cortex tissue.
Mesophyll	Site of most photosynthesis reactions in the leaf; located in the middle layer of the leaf.
Middle lamella	Material containing pectin that forms between cells and that cements the cell wall of one cell to the cell wall of an adjacent cell.
Mitochondria	Organelle where stored sugars are metabolized to produce forms of energy that the plant can use for growth; the cell's power plant.
Nucleus	Organelle that contains the chromosomes. Chromosomes contain the genetic code that is carried within each cell and that directs which chemical reactions are turned on and off in the cell.
Organelle	Generic term for a plant organ.
Palisade mesophyll	Densely packed, columnar-shaped, elongated cells full of chloroplasts. Analogous to cortex parenchyma cells in the stem, but in the leaf are specialized for light energy capture.
Parenchyma	Cell type with thin cell walls; unspecialized, but carries on photosynthesis and cellular respiration and can store food; forms the bulk of the plant body.
Pericycle	Single layer of tightly packed cells located in the vascular cylinder that retain the ability to divide and produce new cells; source of lateral roots.
Sclerenchyma	Cell type with thickened, rigid, secondary walls that are hardened with lignin; provides support for the plant.

Sieve tube members	Elongated cells that join end to end to form tubes for passage of liquids. The end walls have pores. Unlike xylem cells, these cells are still alive. They have a thin cell membrane containing a layer of living protoplasm that hugs the wall of the cell.
Spongy mesophyll	Loosely packed cells with large air spaces between the cells, allowing movement and exchange of gases, specifically oxygen, carbon dioxide, and water vapor. Also contain chloroplasts.
Tissue	Group of cells that share a function.
Tracheids	Elongated and narrower than vessels, connected by overlapping at their ends, dead at maturity, and containing pits through which water can move.
Vacuole	Organelle containing various fluids including stored chemical energy like starch and waste products from the cell. Takes up much of the cell volume and gives shape to the cell.
Vessels	Elongated xylem cells that connect end to end to form tubes, are dead at maturity, and have perforated end walls so water can move freely through the holes and flow from cell to cell. Vessels have a relatively large diameter compared to other xylem cells and allow greater movement of water.

CHAPTER-7

MERISTEMS AND FLOWERS

Plants have meristematic cells that are not differentiated, but that are destined to divide and produce other cells that may also divide, elongate, and differentiate into specialized cells and tissues. This process is called growth. A highly specialized structure produced after the transition from a vegetative meristem to a reproductive meristem is called a flower. Floral organs are arranged in concentric circles or whorls and are the site of sexual reproduction.

Learning objectives

- Understand the meristems of primary and secondary growth and the specialized tissues they produce.
- Understand flower morphology and how sexual reproduction via pollination occurs.

7.1 MERISTEM MORPHOLOGY

Learning objectives

- Differentiate between primary growth from apical meristems and secondary growth from lateral meristems.
- Describe two types of lateral meristems and the types of tissues that are derived from these meristems.

Primary growth from meristems

You'll recall that the apical meristem is the site of cell division and new cell production at the tips of the plant stems and roots. The cells that make up the meristem are undergoing mitotic cell division to produce more cells. These new cells result in growth and development of plant tissues. (If you haven't previously studied mitosis, you'll have the opportunity to do so during this class.) For now it is sufficient to know that mitosis is the process of cell division where one plant cell divides into two identical cells.

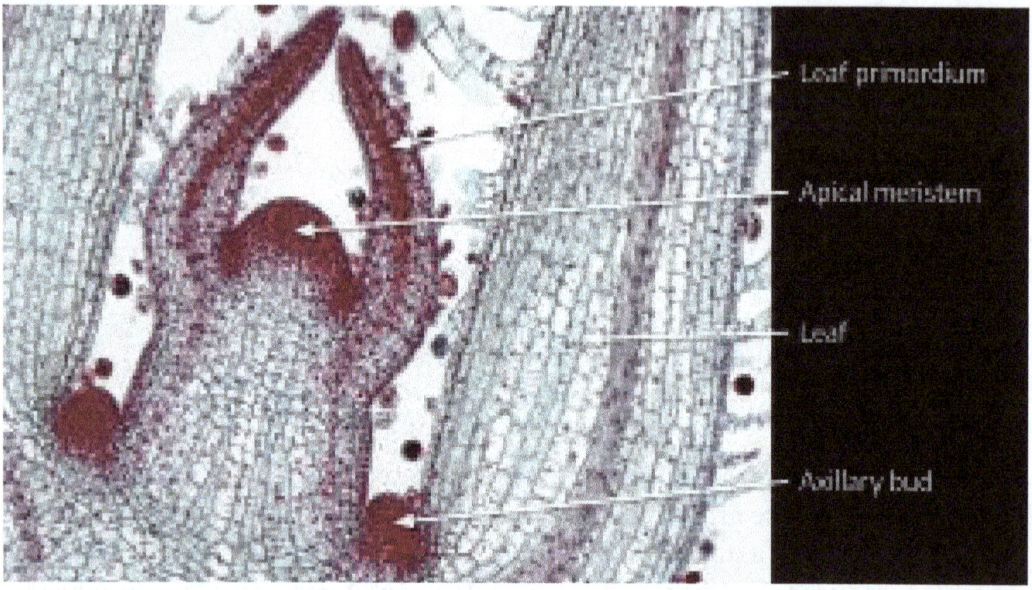

Microscope view of coleus shoot tip with labels. BlueRidgeKitties, CC BY-NC-SA 2.0

Above is a micrograph of a coleus shoot tip. You can see the dome of the apical meristem at the very tip of the shoot surrounded by leaf primordia (rudimentary leaves). On the far left and far right are the cells of two growing leaves. You can see a trace of vascular tissue on the left leaf near the "L" of the leaf label. There is another red stained area called the axillary bud, which we've studied previously. The axillary bud is another very small shoot tip with a meristematic area. Axillary buds are found at a node and typically occur where a leaf petiole attaches to a stem. The axillary buds in this stage of growth are inactive, but in time may begin active cell division and develop into new branches off of the main stem.

The coleus micrograph is clearly stem tissue because you can see leaves and leaf primordia, so where are the nodes and internodes? The region where the leaves are attached, and where you find the axillary buds, is a node. Above this is the internode, and at the top where you find the leaf primordia is another node.

The root meristem looks very different from the shoot apical meristem. Recall that, unlike branches that develop at nodes, lateral roots are formed adventitiously, as the result of meristematic activity in the pericycle cells of the root's vascular system in the zone of maturation. We don't see a node-internode structure like we saw with the coleus shoot tip.

When meristem cells divide, whether in the shoot or the root, one of the two resulting sister cells typically continues to be a meristem cell. The other sister cell divides a few more times and then differentiates into dermal, cortex, or vascular tissue in the stem or root. Meristem cells that remain meristematic are called initials because they continue to divide, producing new cells. The other sister cells that divide once or twice more and then differentiate are called derivative cells. The xylem and phloem tissues that result from differentiation of derivative cells are called primary xylem and primary phloem, where the word "primary" signals that the cells originated from cell divisions of the apical meristem.

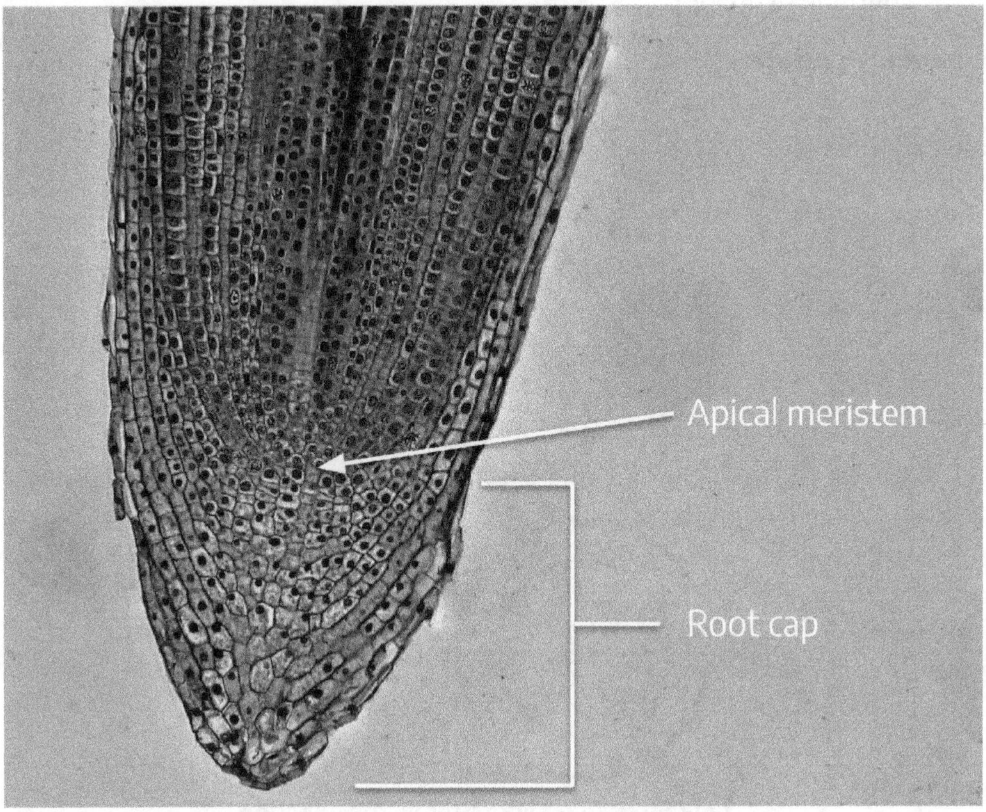

Apical meristem with root cap. Berkshire Community College, Public Domain.

To reiterate, young stems and roots have primary xylem and primary phloem that formed as a result of differentiation of derivative cells. Primary xylem and primary phloem cells trace back to an apical meristem.

Earlier you learned the arrangement of the vascular tissues in monocot and dicot stems and roots. Remember that mitotic cell divisions in the apical meristem result in lengthening of the root or shoot through production of new cells plus the elongation of those cells. With a few exceptions, this is the only type of growth — growth that is initiated by cell division in the apical meristem — you'll find in monocots. Dicots, however, have another type of growth — from a different type of meristem — that results in thickening of the stem.

Review questions

- If shown a micrograph of an apical meristem, how would you determine whether it is from a root or a shoot?
- What happens to the initial cell mentioned in the question above? Does it continue to divide?

Secondary growth (thickening): Introducing lateral meristems

Vascular cambium

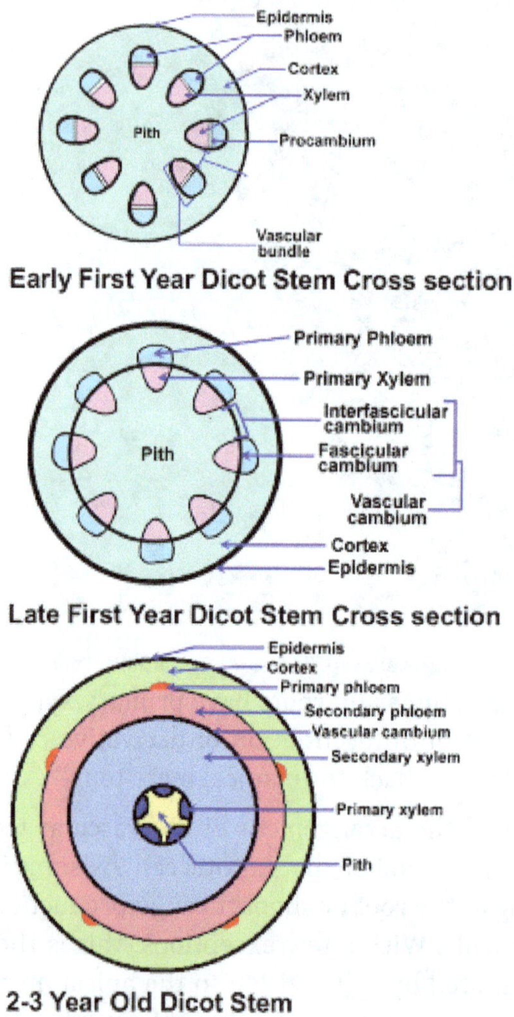

Images by Dr. Thomas L. Rost, emeritus professor at UC-Davis.

Unlike annual herbaceous plants that only survive for one growing season, and whose stem and root cells trace back to cell divisions of the apical meristems, woody plants and shrubs are perennial dicots that have the capacity for secondary growth and can survive from year to year.

Some annual herbaceous dicot plants, like tomatoes, can have secondary growth, but for now let's consider those the exceptions and focus on perennial dicot woody plants. Secondary growth is the result of activity by a special type of meristem called a lateral meristem. As with apical meristems, lateral meristems are made up of cells that undergo mitotic cell division. Mitosis in lateral meristems results in lateral growth (thickening of the stem or root) and adds to the girth of a plant rather than its length. Remember that length is the outcome of cell division in the apical meristem plus elongation of those cells. Girth or thickening is the result of lateral meristems. We'll learn about two types of lateral meristems: vascular cambium, and cork cambium.

Let's start with the vascular cambium.

The three drawings on the right show a cross section of a stem for an imaginary woody dicot. The top drawing illustrates the stem early in the first year of growth, and shows the vascular cylinders arranged in a ring around the stem. The phloem is oriented to the outside, the xylem to the inside. A thin layer of parenchyma cells between the xylem and phloem has differentiated into the fascicular cambium (fascicular refers to bundles, in this case, cambium in the vascular bundles). The fascicular cambium is meristematic and can divide to produce new phloem toward the outside and new xylem to the inside. The new xylem and phloem produced by the cambium are called 2o (secondary) xylem and 2o phloem. Recall that the original xylem and phloem that differentiated from the apical meristem's derivative cells are called the 1o (primary) xylem and 1o phloem.

The middle drawing is of the same stem later in the year. The cortex (cortical) parenchyma cells that lay between the vascular cylinders directly in line with the fascicular cambium begin to differentiate into a type of cambium called interfascicular cambium (cambium between the bundles). This is symbolized by the line connecting the vascular cylinders. This cambium is also meristematic, and produces 2o xylem and 2o phloem.

The cross section on the bottom illustrates the stem in its second or third year of growth, when there is a noticeable buildup of 2o xylem and 2o phloem with remnants of 1o xylem and 1o phloem.

In summary, the vascular cambium is a lateral meristem formed by differentiation of parenchyma cells located between the primary xylem and phloem into fascicular cambium, followed by differentiation of cortical parenchyma between the vascular cylinders into interfascicular cambium. After a few years of secondary growth, fascicular and interfascicular cambium can no longer be distinguished, and it is all simply known as vascular cambium. This layer of cambium runs vertically (assuming that the stem is oriented vertically) and parallel to the surface of the woody stem.

The illustration below shows how the cambium divides to produce 2o xylem and 2o phloem, with the outside of the stem toward the top of the page. Frame #1 shows a single cambium cell (C). This cell divides mitotically (M) to form two cambium daughter cells (Frame #2). Frame #3 shows that the cambium cell on the top differentiates (D) into a phloem cell (P-toward the outside) and the other cambium cell divides mitotically (M).

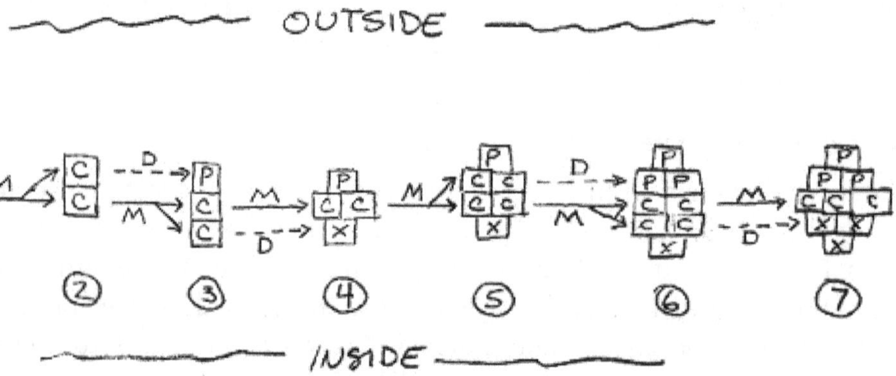

Cell division and differentiation from cambium to xylem/phloem. Tom Michaels

This type of cell division, in which new cells are formed either to the outside or inside, and the cell wall that separates the two new cells is parallel to the outside of the stem, is called periclinal division. Periclinal division by the cambium makes new cells that add girth to the plant. The cells that are added subsequently differentiate into xylem and phloem depending on their location to the outside or inside of the cambium. The meristem needs to divide periclinally to add girth to the plant stem.

In Frame #4, pay particular attention to a different type of cell division, where the cambium cell has divided so that the wall between the two cells is perpendicular to the outside of the stem. This is called anticlinal division. The meristem occasionally needs to divide anticlinally because as the stem is growing in girth, the diameter of the ring of vascular cambium must expand to keep up, or it will split into pieces and no longer form a continuous ring around the stem. Frame #4 also shows that the cambium cell to the inside has differentiated into xylem (X). In Frame #5, the two cambial cells that formed from anticlinal division now each divide periclinally.

Review questions

- How might you recognize a plant that has secondary growth?
- In a perennial woody dicot, how do the discrete vascular bundles found in the new seedling stem become continuous rings of xylem and phloem in the three-year-old woody stem?

- Explain what is happening in Frames #6 and #7 in the drawing above. Note that there are two important changes: differentiation and anticlinal division.

Cork cambium

Let's look at the second type of lateral meristem, cork cambium. Cork (called "phellem" in this image) provides a protective covering around the expanding trunk of the woody plant.

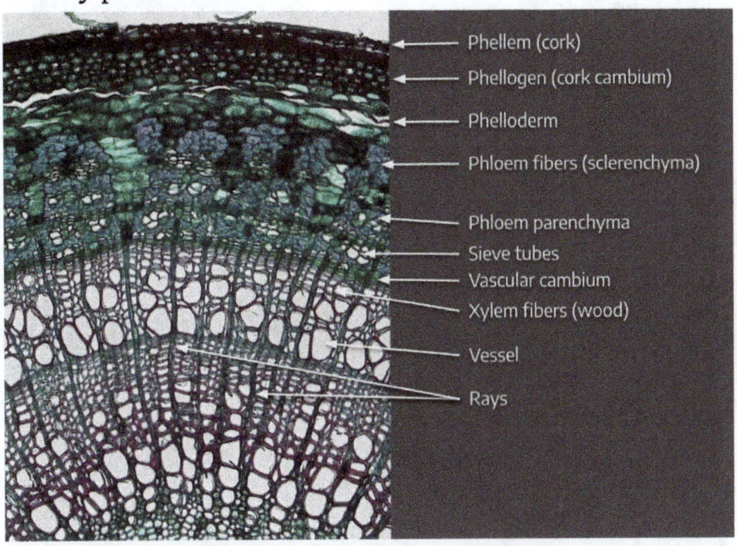

Public domain image labeled by Emily Tepe.

Cork develops in plants with secondary growth after the initiation of secondary xylem and phloem and the expansion of the stem and root's girth. Cortex parenchyma cells next to the epidermis of the young stem differentiate into the cork cambium (also called phellogen), which is meristematic. The cork cambium lays down some new cells toward the inside called phelloderm, but lays down most of its new cells to the outside, and these derivatives of the cork cambium differentiate into the cork cells. The cork cells are lined with a waxy substance called suberin (we first saw this substance in conjunction with the Casparian strip around endodermis cells) that make the cells impermeable to water and gases. Breaks in the cork cells, called lenticels, allow gas and water exchange. You can see these lenticels in corks from wine bottles (wine corks are made from the thick cork of the Cork Oak, *Quercus suber*). Cork cells die when they mature. These cells replace the protective function provided by epidermis in young roots and stems. Cork cambium and its derivatives (phelloderm and cork) are called the periderm.

Botanically speaking, the word refers to all of the tissues exterior of the vascular cambium. So bark includes:

- Primary and secondary phloem
- Phelloderm if present
- Cork cambium
- Cork

Look again at the list above and note that the phloem is part of the bark. This is often overlooked!

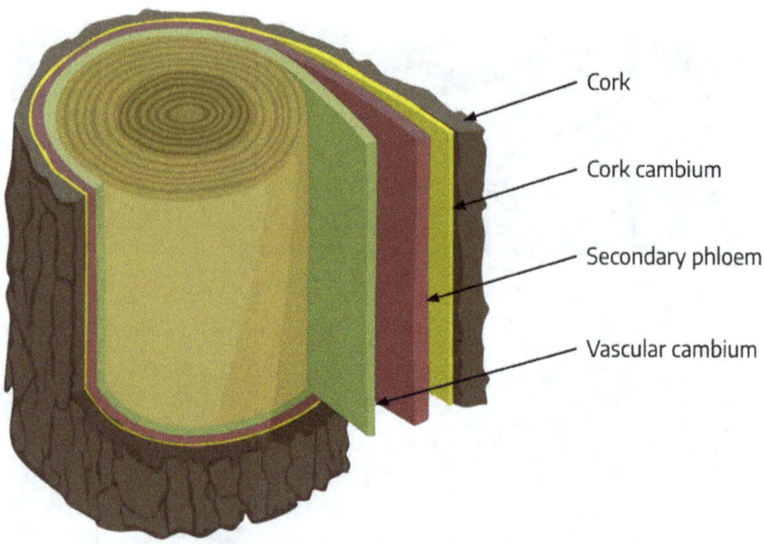

Cross section of trunk. Dpaczesniak. CC BY-SA 4.0.

This illustration summarizes the layers introduced above, but omits the phelloderm. If you were to add a label for the phelloderm, where would it be? In mature trunks the phelloderm is quite a thin layer in the bark though, particularly relative to the cork layer, so that is probably why it is absent in this illustration.

If you peel bark from a living tree, you are exposing the surface of the vascular cambium of the tree, and on the inside of the bark will be the phloem. Peeling off bark will kill the plant because you are removing phloem, which disrupts the ability of the plant to move sugars through the trunk of the tree, exposes the tree to moisture loss and predator invasion, and kills the vascular cambium through desiccation. Killing the vascular cambium halts production of new xylem which will subsequently interfere with water transpiration up the stem. This is what happens when rodents girdle the trunks of young fruit trees (left) just below the snow line as they feed on the tender tissues over winter.

Photo of a girdled trunk. Emily Hoover

A tree's annual rings are found in the secondary xylem. In the spring, the newly produced xylem cells have thin walls and are large in diameter so they can accommodate the abundance of soil moisture that is typical of April and May.

Trunk cross section showing sapwood and heartwood. MPF, CC BY-SA

As the growing season progresses and soil moisture levels are depleted, the cell walls of newly produced xylem are thicker and the diameter of the xylem cells diminishes. In the next spring, the newly produced xylem cells are once again thin-walled and large. The distinct contrast between the small-celled late summer wood from the previous year and the large-celled spring wood from the following year is noticeable, and this line of demarcation between late summer and spring xylem is called the annual ring. In the image to the right you can see the darker xylem, called heartwood, and the lighter xylem, called sapwood. Heartwood is older xylem that is clogged with resins that darken the cells and limit their ability to transport water. Sapwood is younger, resin-free, and still functioning to conduct water up the trunk.

There isn't nearly as much phloem as xylem in a woody plant. Phloem isn't produced as rapidly as xylem, and is crushed between the vascular and cork cambium layers, so we don't see annual rings in phloem.

Review questions

- Describe the origin of annual rings. If a woody dicot is growing in a tropical climate where the weather is the same day in and day out, will you find annual rings in the wood?
- What is the most exterior cell layer in an herbaceous stem called? What is the most exterior layer of cells in a five-year-old woody perennial plant stem called.
- If some kids in your neighborhood get hold of a little hatchet and chop off thin slices of bark all the way around the base of one of your trees, what will happen to the tree? Why?

What about roots?

Roots of woody perennial dicots also have vascular and cork cambium. The vascular cambium arises from parenchyma cells lying between the xylem and phloem in roots, just as in stems. The illustration below shows the initiation of the vascular cambium in a young root. It looks like a trace of white between the primary xylem in the middle and the primary phloem in each of the four lobes of the vascular cylinder. With age, the cambium encircles the root and produces (secondary) xylem to the inside, and (secondary) phloem to outside, again just as in stems. Older roots with secondary growth also have a periderm similar to that found in the stem that replaces the function of the epidermis. If you've ever tried to dig in the soil surrounding a mature tree, you know that roots are as tough and woody as the branches above ground.

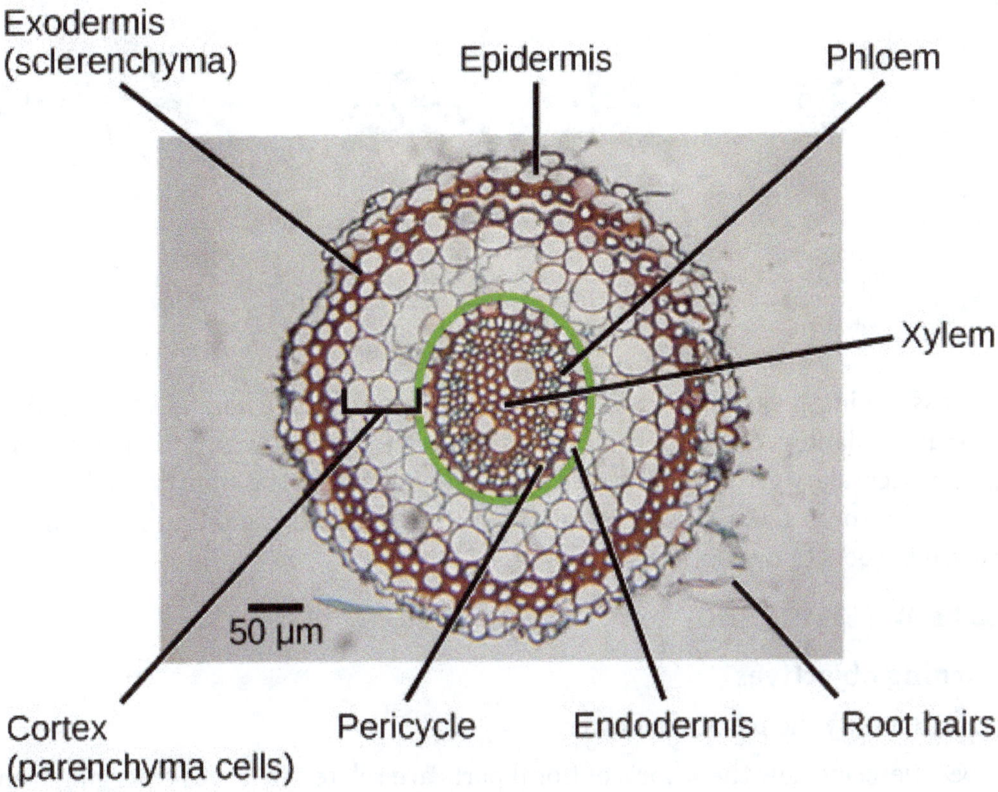

Root cross section with cell types. CNX OpenStax, CC BY 4.0

Here is a diagram to help you visualize lateral meristems including cell division.

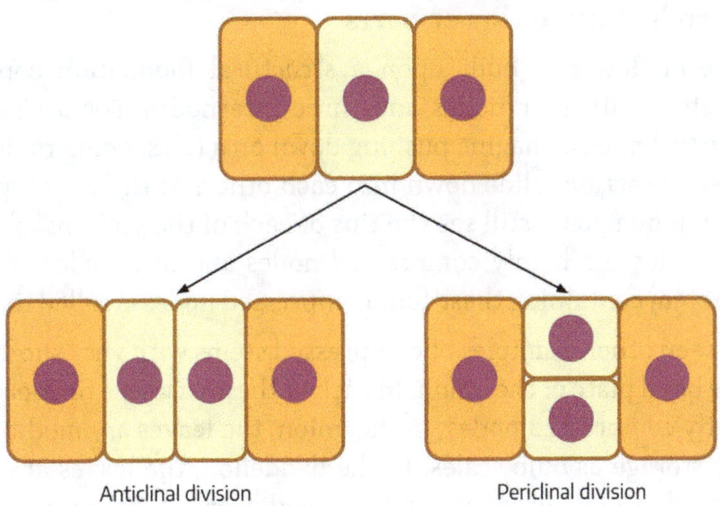

Anticlinal and periclinal division.

The next diagram shows a vascular bundle as it grows by dividing both periclinally and anticlinally.

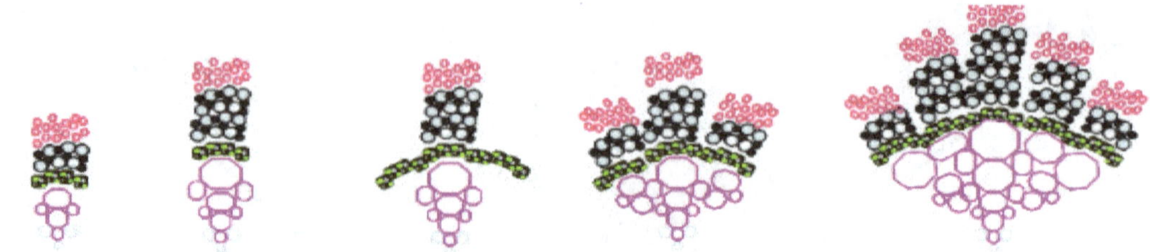

Primary vascular bundle

Cambium makes new ring of xylem and phloem

Cambium grows circumferentially

Cambium generates more secondary xylem and phloem

Cambium generates a second ring of xylem and phloem

Ross Koning at Eastern Connecticut State University has a helpful site regarding secondary plant growth (optional reading) with descriptions and images that explain secondary growth in the stem vascular column, in the stem dermal tissues, and in the roots. L. Rost from U.C.-Davis has this nice explanation of secondary growth in roots (more optional reading).

7.2 FLOWER MORPHOLOGY

Learning objectives

- Identify the parts of a flower.
- Describe how the whorls of floral parts are related to leaves emerging from nodes on a stem.
- Describe ways in which flower structure affects pollination.

General introduction to flower parts

The angiosperm flower is built upon a structural foundation consisting of a compressed stem with four nodes and three internodes. For a visual image of these compressed nodes, imagine pushing down on a telescoping radio antenna so that the antenna sections slide down into each other. At the very top of the fully compressed antenna you'll still see the tips of each of the sections of the antenna, and this resembles the highly compressed nodes and internodes of a stem. The region of the stem containing these four compressed nodes is called the receptacle.

We have seen other examples of compressed stems with very short internodes, including the basal plate of the onion. In each of these cases we find leaves attached to these tightly compressed nodes. In the onion, the leaves are modified for water and nutrient storage as bulb scales. In the dandelion, the leaves are unmodified, but are arranged in a low-growing rosette. The flower is another example of a very compact stem with four nodes and three internodes. The leaves on this stem are highly modified to serve a reproductive function. They are so highly modified that, except for the structures at the fourth node, the parts don't resemble leaves at all.

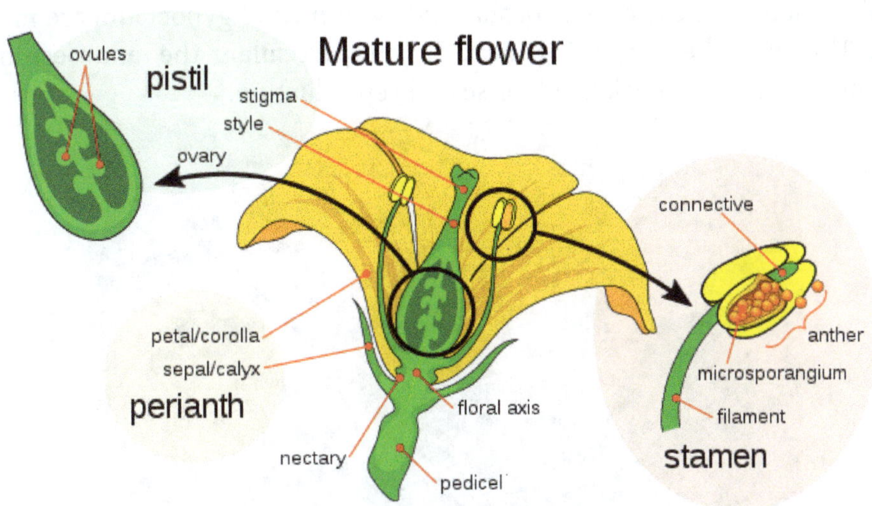

Main parts of a mature flower. Lady of Hats, Public Domain

Complete the flower parts tutorial (strongly recommended) from the science department of John Burroughs School in St. Louis; it will introduce you to, or help you review, the parts of a flower using a graphic like the one shown here.

In #8 in the tutorial the author calls the structure the carpel and notes that pistil is an archaic term. You can use either carpel or pistil in this course, but we'll usually say carpel. Pistil is sometimes still for the structure if it is composed of two or more fused carpels. After completing the tutorial you know that the flower is actually a shortened branch containing a stem with four very compact nodes. This short stem is called the receptacle. From the nodes on the receptacle emerge four different kinds of modified leaves that collectively have these names:

- Calyx: the fourth node, at the base of the receptacle, individually called sepals; these are the parts that still bear some resemblance to leaves
- Corolla: the third node, individually called petals
- Androecium (Greek derivation meaning "men's house"): the second node
- Gynoecium (Greek derivation meaning "women's house"): the first node at the tip of the receptacle

For clarity, instead of using a number to refer to a particular receptacle node, we'll use the name of the modified leaves attached to that node. For instance, we'll call the fourth node where the calyx is attached the "calyx node," the third node where the corolla is attached the "corolla node," and so on.

We've noted that the calyx, corolla, androecium, and gynoecium are modified leaves. This may be surprising because, in particular, the androecium and gynoecium are highly specialized for sexual reproduction.

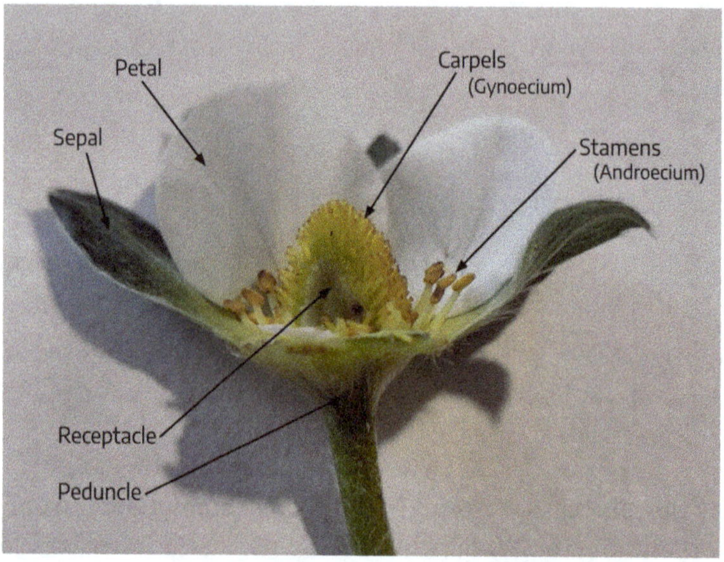

Strawberry flower. Image by Andrew Petran and Emily Tepe.

The nodes where these modified leaves are attached are often called whorls since usually two or more of these modified leaves are attached radially around the node. Monocots generally have three modified leaves attached at each node, while dicots usually have four or five attached at each node. One interesting feature of the strawberry flower shown above is the enlarged gynoecium node of the receptacle that is covered by carpels. The receptacle will become the red culinary "fruit," while the actual botanical fruits will be the achenes (you've considered them seeds, but they are a type of fruit) embedded into the outer surface of the receptacle.

Watch the next two videos to learn more about flower parts and about the specialized florets of an Asteraceae flower.

Review questions

- What is the receptacle?
- Draw a flower and label the parts, including the nodes.
- What is the relationship between a whorl and a node? In what way is this relationship connected to the concept that a flower's corolla is made up of modified leaves?

Calyx

The fourth whorl at the base of the receptacle is the calyx whorl. The calyx is made up of modified leaves called sepals. In some species the sepals look like miniature leaves; they are green and photosynthetic. In other species, like the lily, they are showy and almost indistinguishable from the petals. When sepals and petals are showy and indistinguishable, they are called tepals.

The saffron flower, below, is the source of the rare spice gathered from the long, red-orange stamens that you can barely see inside the opening flower. Notice that there are apparently no sepals. That is because the three sepals and three petals of this monocot plant look the same; they are therefore tepals.

Photo of a saffron flower. Faruk Nafiz Ertürk, CC BY-SA 2.0

It may interest you to know that saffron is a geophyte that grows from a corm. We introduced geophytes in a previous lecture as plants that develop underground organs that allow the plant to survive during periods of hostile environmental conditions.

Watch this video **on tepals.**

Corolla

The next whorl toward the tip the receptacle bears the corolla. The corolla is composed of highly modified leaves called petals. Petals attract pollinators through their bright colors and showy patterns. Petals may also exude nectar near their site of attachment to the receptacle to reward insects who visit the flowers and, when doing so, spread pollen from flower to flower. Color patterns on the petals may simulate a "bulls eye" or landing strip that provides the insect with a visual guide pointing to the location of the nectar, as in the violet, below.

Photo of a violet. Inga Munsinger Cotton, CC BY 2.0

The calyx and corolla are collectively named the perianth. Some flowers lack a perianth. Corn is an example; it is wind pollinated and has no need for a showy perianth to attract insect pollinators.

Review question

- Sepals and petals are modifications of what other plant structure?

Androecium

Rose of Sharon flower Tim Green CC BY 2.0

Tomato androecium. pxfuel.com Public domain.

The third whorl as we move towards the tip of the receptacle is the androecium whorl. The androecium is composed of modified leaves called stamens. Stamens are found in many different arrangements. The picture of the Rose of Sharon flower, in the top photo above, shows an androecium composed of an abundance of stamens in an open arrangement. In contrast, the 10 stamens in the tomato androecium (bottom photo above) are fused into a cone that surrounds the gynoecium. The tomato's cone of stamens is a structure through which the tomato encourages self-pollination. It is difficult for pollen from another plant to get to the stigma of the carpel before pollen from the flower's own stamen.

Stamens have two distinct components:

Filament (a long stalk)

- The filament has an architectural function, in that it lifts the anther to a position where it can effectively release pollen grains into/onto the pollinator.
- It also has a physiological (or some might say plumbing) function, in that it connects the anther to the plant's vascular system so that it can receive water and nutrients.

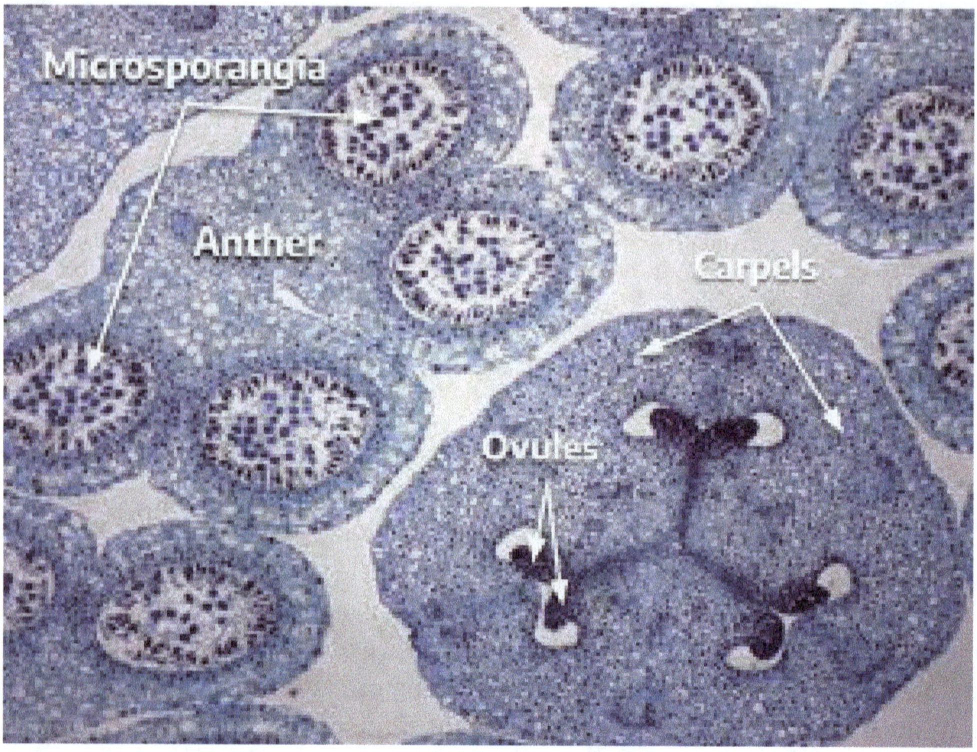

Cross section of a developing flower. Marc Perkins. CC BY-NC 2.0

Anther (usually four sacs containing pollen grains).

- Inside the pollen sacs are microsporangia and microspore mother cells where a special type of cell division called meiosis takes place (to be covered later). Meiosis in the microsporangia leads to formation of the male gametes (sperm) that will be packaged in a pollen grain.

The image above shows a cross section through a developing flower, showing components of the androecium and gynoecium.

Gynoecium

The whorl at the tip of the receptacle supports the gynoecium. The gynoecium is composed of carpels. Several carpels may be fused into a compound carpel (which may also be called a pistil). The Berberis (Oregon Grape) flower on the right has a fused carpel; the photo clearly shows the locule (inner chamber) with the ovules.

Photo of Berberis (Oregon Grape) Simon Garbutt CC BY-SA 3.0

As you saw in the tutorial, the carpel consists of three parts:

- Stigma
- A tip on the end of the structure called the stigma.
- Style
- The stalk that elevates the stigma.
- Ovary

The swollen base, which includes:

- A chamber called a locule
- Inside the locule, one or more ovules
- Inside the ovules, an embryo sac — the megasporangia and megaspore mother cells, which when fertilized by pollen the ovule become a seed
- Meiosis of megaspore mother cells in the embryo sac leads to the formation of the female gametes (eggs)

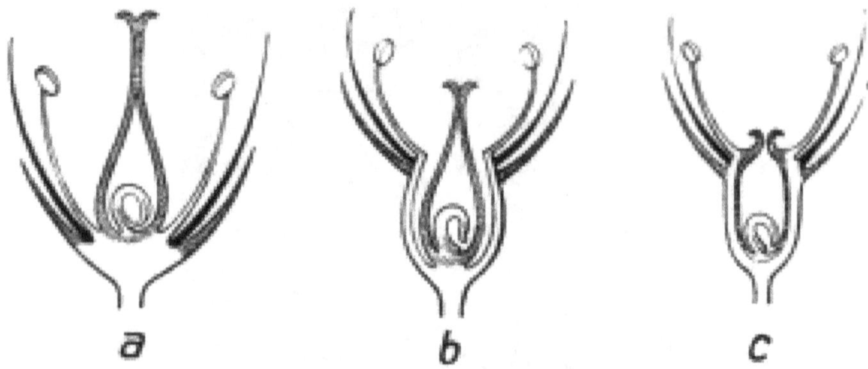

Drawing of a hypogynous (a), a perigynous (b) and an epigynous flower (c). Gustav Hegi. Public domain

When you begin to compare structural differences among flowers, you will find that the position of the calyx and corolla relative to the ovary differs between some species. A more complete description of that relative alignment can be seen in the picture above, and is described here:

Epigynous flowers

Other flower parts are attached ABOVE the ovary.

This is called an inferior ovary because the gynoecium node isn't positioned right at the tip of the receptacle, but rather is sunken down into the receptacle below where the androecium, calyx, and corolla are attached. This means that the ovary is surrounded by other tissues, primarily receptacle tissue. This will have an impact on whether any accessory plant tissues make up the fruit. More on that later.

Perigynous flowers

The ovary is surrounded by the fused bases of flower parts (calyx, corolla, androecium) that surround the ovary.

Hypogynous flowers

Other flower parts are attached BELOW the ovary. This is called a superior ovary because the ovary sits above the point of attachment of the top whorl.

A flower that has all four of the parts described above — Calyx, Corolla, Androecium, and Gynoecium — is called a "complete" flower. Flowers missing one or more parts are described as "incomplete." We noted earlier that corn produces a flower without perianth. Such a flower could be described as "incomplete."

Review questions

- Sperm are produced from microspores contained in what plant part? To which whorl is this part attached?

- Eggs are produced from megaspores contained in what plant part? To which whorl is this part attached?
- Which type of ovary is shown in the tutorial graphic at the beginning of this lecture from which you learned the parts of the flower?
- In what way does the filament of the stamen have an architectural function? How might this make a difference to the plant's reproduction?

Pollination patterns

A flower with both androecium and gynoecium — that is both male and female parts — is called perfect or bisexual or hermaphroditic. Perfect flowers may be capable of self-pollination. Pollen produced within the flower may fall on a stigma in the same flower, and the sperm that it carries may fertilize the egg in the ovule. Sometimes, the timing of events during the stages of flower maturation encourage self-pollination. For instance, in some species the anther matures and pollen is shed, and the stigma is receptive, before the flower even opens. This is called cleistogamy, and is common in self-pollinating agricultural crops. Self pollination is encouraged because it is difficult for pollen from another flower to get to the stigma within a closed flower before pollen from the flower's own anther gets there first. Flowers that are fully open when mature are chasmogamous and facilitate pollination from other flowers.

On the other hand, the development timing of male and female organs in a different species might encourage outcrossing by releasing pollen when the stigma in the same flower is not receptive, or vice versa. This relative timing is called Protandry when the pollen is shed before the stigma is receptive, and Protogyny when the stigma is receptive prior to pollen shed.

Male and female watermelon flowers. Pollinator. CC BY-SA 3.0

There are also genetic mechanisms called self-incompatibility, through which the stigma and style recognize pollen produced by the same plant and stop fertilization, thereby avoiding self pollination and promoting mixing genetic material (DNA) from diverse individuals.

Some plants produce imperfect male and imperfect female flowers on the same plant. The flowers containing only androecium are called staminate (male) flowers while the flowers with only gynoecium are called pistillate (female) flowers. Squash and melons, such as the watermelon shown above, are examples of plants with imperfect flowers. Corn and cucumber are others. Notice the enlarged receptacle and inferior ovary at the base of the pistillate flower of the watermelon. These flowers, because they are missing one of the four parts, could also be described as incomplete.

For the ultimate avoidance of self pollination, some plants have only staminate or only pistillate flowers. A single-sex plant like this is called dioecious (plants with both sexes, whether perfect or imperfect flowers, are called monoecious). Hops, asparagus, and hemp are examples of dioecious crop plants.

TERMS

Flashcards

Androecium	One of the whorls of a flower and is all of the male reproductive parts; stamens.
Annual rings	Demarcation between small-celled later summer and large-celled spring secondary xylem.
Anther	Pollen-bearing component of the stamen.
Anticlinal division	Type of cell division in which the new cells have divided so that the wall of the cells is perpendicular to the outside of the stem.
Axillary bud	Bud borne in the axil of a stem.
Bark	All tissues exterior of the vascular cambium, including the primary and secondary phloem, phelloderm (if present), cork cambium, and cork.
Bisexual	A flower that has both the androecium and gynoecium; also called hermaphroditic or perfect flower.
Calyx	One of the whorls of a flower; located at the base of the receptacle, and contains all the sepals.
Carpel	Composed of three parts: stigma, style, and ovary.
Cell division	Process in mitosis where one plant cell divides into two identical cells.
Chasmogamy	When the anther matures after the flower opens and pollen is shed before the stigma becomes receptive; common in non self-pollinating crops.
Cleistogamy	When the anther matures, pollen is shed, and the stigma is receptive before the flower opens; common in self-pollinating crops.

Complete flower	Where all four whorls are present: calyx, corolla, androecium, and gynoecium.
Cork	Outer protective tissue of bark; also called phellem.
Cork cambium	Lateral meristem responsible for secondary growth that replaces the epidermis in roots and stems; also called phellogen.
Cork cells	Cells located in the cork; lined with suberin and dead at maturity.
Corolla	One of the whorls of a flower consisting of all the petals.
Cortex	Also known as the ground meristem; found just inside the epidermis, extends toward the interior of the stem and root, and is made up of three types of cells: parenchyma, collenchyma, and sclerenchyma.
Derivative (cells)	Other sister cells that, after the initial meristematic initial cells are created, divide once or twice more and then differentiate.
Dermal	Outside of the plant; provides protection for the plant cells they surround.
Dioecious	When an entire plant has only male or only female flowers; means "two houses."
Epigynous	When the perianth and androecium are positioned above the ovary; also called an inferior ovary.
Fascicular cambium	Cambium within the vascular bundle.
Filament	Stalk that holds up the anther so that pollen grains can be effectively released.
Gynoecium	One of the whorls of the flower and is all of the female reproductive parts; carpels.
Heartwood	Older, darker xylem in the stem that is clogged with resins that limit the transport of water.
Hermaphroditic	A flower that has both the andreocium and gynoecium; also called a perfect flower or bisexual.
Hypogynous	When the perianth and androecium are attached below the ovary; also called a superior ovary.
Imperfect flower	Flower that has only the androecium OR only the gynoecium present.
Incomplete flower	Flower missing one or more of the four whorls.
Inferior ovary	When the perianth and androecium is positioned above the ovary; also called an epigynous flower.
Initials (cells)	Meristem cells that remain meristematic because they continue to initiate new cells.
Interfascicular cambium	Cambium between the vascular bundles.
Lateral meristem	Specialized meristems made up of cells that undergo mitotic cell division.
Lenticels	Breaks in the cork cells that allow gas and water exchange.
Locule	Chamber in the ovary.

Monoecious	When an entire plant has both male and female parts (can be perfect or imperfect); means "one house."
Ovary	Part of the carpel; contains ovules which develop into seeds.
Ovary wall	Provides protection to the ovules; also called the pericarp.
Ovule	Part of the ovary that contains an embryo sac; surrounded by the nucellus, which develops into a seed after fertilization.
Peduncle	Large, central stalk that attaches the rachi to the stem of the plant.
Perfect flower	A flower that has both the andreocium and gynoecium; also called hermaphroditic or bisexual.
Perianth	Both the calyx and corolla.
Periclinal division	Type of cell division where the new cells are formed either to the outside or inside and the cell wall that separates the two new cells is parallel to the outside of the stem.
Periderm	Consists of the cork cambium, phelloderm, and cork.
Perigynous	When the ovary is surrounded by the fused bases of the perianth and androecium.
Petals	Modified leaves that make-up the corolla; showy, and attract pollinators.
Phellem	Another name for cork.
Phelloderm	New cells that are laid down toward the inside of the stem or root by the cork cambium.
Phellogen	Another name for cork cambium.
Pistil	Term used when several carpels are fused together.
Pistillate flower	An imperfect flower that contains only the gynoecium.
Primary (cells)	Cells that originate from cell divisions of the apical meristem.
Primary growth	Growth that results from activity by an apical meristem; causes the elongation of the cells in the apical meristem region, which leads to increasing plant length.
Primary phloem	Phloem tissue that results from differentiation of derivative cells (procambium).
Primary xylem	Xylem tissue that results from differentiation of derivative cells (procambium).
Protandry	When the pollen is shed before the stigma is receptive.
Protogyny	When the stigma is receptive prior to the pollen shedding.
Receptacle	Base of the flower where the floral parts are attached.
Sapwood	Younger, lighter xylem in the stem that is resin-free and transports water up the trunk.
Secondary growth	Growth that results from activity by a lateral meristem; causes thickening of the stem or root rather than elongation.
Secondary phloem	New phloem formed on the outside and produced by the fascicular cambium.

Secondary xylem	New xylem formed on the inside and produced by the fascicular cambium.
Self-incompatibility	When there are genetic mechanisms that inhibit self-pollination of a flower.
Self-pollination	When the pollen from the plant pollinates the stigma of the same plant.
Sepals	Outermost whorl of the flower that protects the flower and photosynthesizes.
Stamen	Modified leaf; collectively makes up the androecium. A stamen is made-up of the anther and filament.
Staminate flower	An imperfect flower that contains only the androecium.
Stigma	Receptive apex of the carpel of a flower, on which pollen is deposited at pollination.
Style	Part of the carpel that elevates the stigma to a position for reception of pollen; conduit for pollen tube growth.
Suberin	Waxy substance present in the cell walls of corky tissues; impermeable to water and gases.
Superior ovary	When the perianth and androecium are attached below the ovary; also called a hypogynous flower.
Tepal	When the sepals and petals are showy and indistinguishable.
Vascular cambium	Lateral meristem producing vascular tissues.
Vascular tissue	System containing vessels that carry or circulate fluids and dissolved minerals in the plant; composed of xylem, phloem, and bundle sheath cells.
Whorl	Node on the receptacle where the four types of modified leaves are attached (four whorls of a flower).

CHAPTER-8

FRUIT

Fruit are important for their culinary importance and provide calories, nutrition, and pleasure. They also are the location for the development of seeds — the most important means to propagate plants and the source of genetic variation.

Learning objectives

- Define simple, aggregate, and multiple fruit.
- Explain the general characteristics of fleshy and dry fruits.
- Identify the difference between a true fruit and an accessory fruit based on structure and tissues.

8.1 FRUIT MORPHOLOGY

Learning objectives

- Define "fruit" from a botanical point of view.
- Describe the differences among simple, aggregate, and multiple fruits.
- Explain the general characteristics of fleshy and dry fruits.
- Identify the difference between a true fruit and an accessory fruit based on structure and tissues.

The graphic to the right shows a cross section of the carpel. Recall that the carpel is the female reproductive structure that is attached to the top whorl of the flower — the gynoecium node. The carpel has three basic parts:

- The stigma, at the tip, and to which pollen grains adhere.
- The style, the channel of tissue through which the pollen tube grows.
- The ovary, at the base, and housing the ovules that contain the plant's eggs.

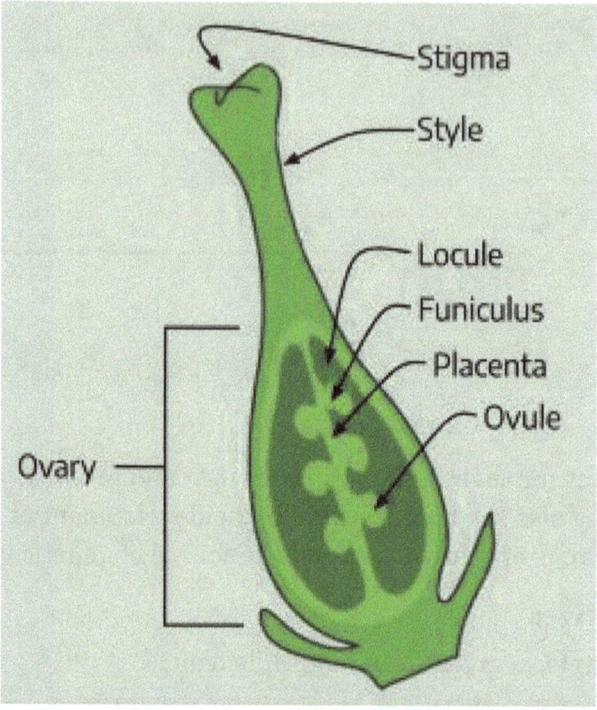

Carpel structure. OpenStax Biology. CC BY 4.0

The photo below shows a bit more detail about the carpel — in particular, the attachment of the ovules within the ovary via a stalk called the funiculus, emerging from the placenta. It also names the chamber in which the ovules hang: the loculus (or locule). Remember the locule, but you won't need to remember funicule, funiculus, or placenta. If you do come across those names in the future, then recall that they relate to the attachment of the ovule within the ovary.

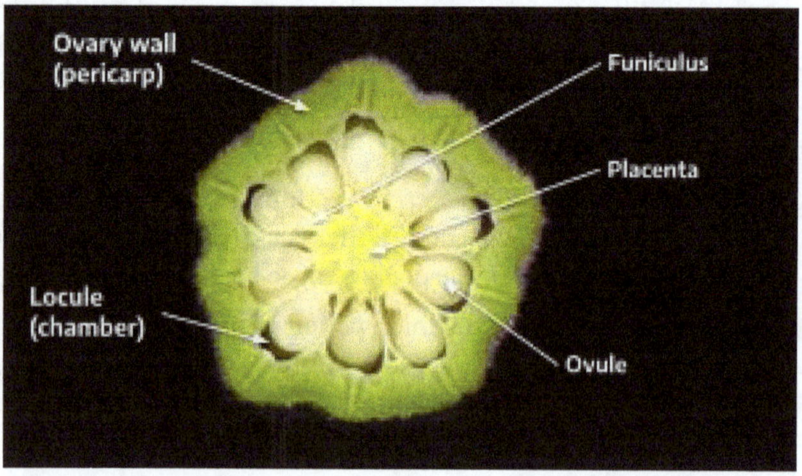

Carpel structure, cross section. Jmprouty. CC BY-SA 3.0

A fruit, in the botanical sense, is the ripened ovary together with the seeds within the ovary. People often think of a fruit from the culinary point of view, considering it to be the part of a plant that has seeds and when ripe is ready to eat, and think of vegetables as a savory food that is any edible part of a plant not associated with seeds (these include roots, stems and leaves). But some plant parts that, when we wear our chef's hat, we think of as vegetables, are really botanical fruits. That green pepper chopped up on your pizza is a botanical fruit, as is the tomato that's pureed to make the pizza sauce. A squash is another botanical fruit that we treat like a vegetable in the kitchen. Adding to the confusion is the legal decision that tomato fruits are vegetables (Nix v. Hedden 1893; The Washington Post). In Plant Propagation, we'll define fruits from the botanical standpoint: the ripened ovary of a flower, together with the seeds within that ovary. Review the diagram below of the tomato flower and fruit to be sure you know exactly which parts of the flower develop into the fruit. In the flower, the ovary wall provides protection to the ovules that contain the egg — the female gametes. The egg cells within the ovules are fertilized by the sperm from the pollen, the ovules develop into the seeds, and the ovary matures into the fruit. For some species, fruits are brown, green, starchy, bitter, proteinaceous, dry, and durable. They may be inedible, unpalatable, or so small as to be terribly inconvenient as a food. The ovary wall doesn't always become a sweet and fleshy fruit like a peach. In kidney beans, for instance, it becomes a dry and brittle protective pod. Nevertheless, both the peach and the bean pod are botanical fruits.

The illustration shows a cross section of a tomato, with the fleshy ovary wall around the outside, locules (chambers) within the tomato, and seeds forming within the locules. The seeds are attached to the central placenta tissue by the funiculus.

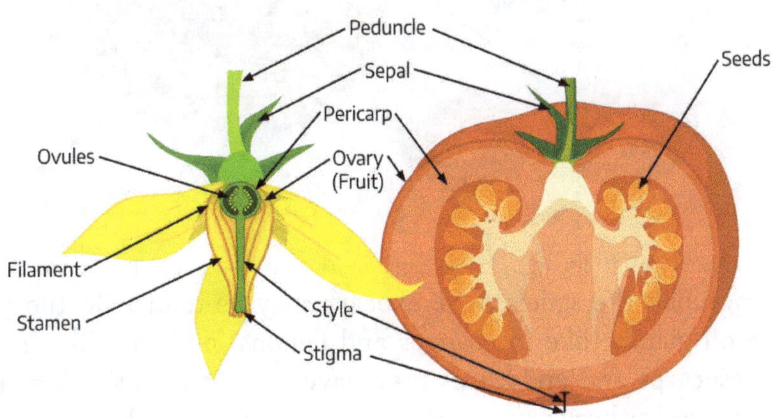

Tomato flower and fruit. © Shutterstock, used with permission. Adapted and labeled by Emily Tepe.

Review activity

- Draw a diagram of a flower showing which parts become the fruit.
- Based solely on the botanical definition of a fruit, would the receptacle, petals, sepals or stamens be part of a true botanical fruit?
- If you cut open a green pepper where is the locule?

Parts of a fruit

The ripened ovary wall is called the pericarp — peri meaning around and carp referring to the carpel. Pericarp = around the carpel. The pericarp can be dry, as with bean pods, or fleshy like a peach, or sometimes both, as in an avocado, where the outer layer is leathery and the inner layer is fleshy.

The pericarp of a fleshy fruit typically has three layers, and each may have distinct characteristics. The photo of a peach below shows the layers making up the pericarp:

- Exocarp: the outer layer of the pericarp)also called the epicarp)
- Mesocarp: the middle layer (fleshy in this example)
- Endocarp: the inner layer (hardened with sclerenchyma cells in the coconut example below)

What type of cell makes up this hard shell of a walnut or peach?Sclerenchyma!

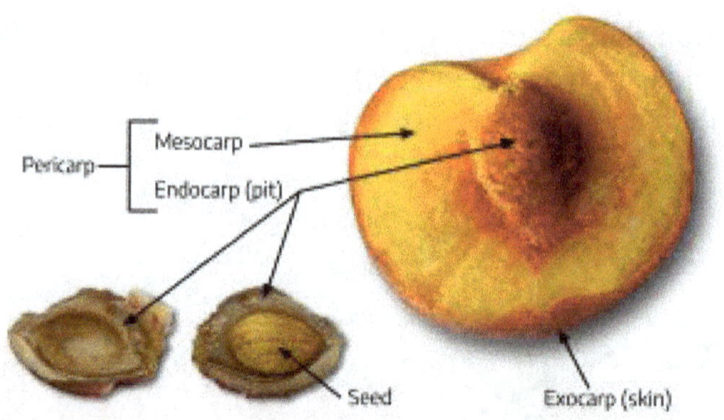

Peach with fruit parts. Liz West and An.ha. CC BY 2.0 and CC BY-SA 3.0

Depending upon the species, the exocarp may be tender, leathery, or hard. It may have oil glands (like in oranges and lemons) or hairs (like in kiwifruit). Similarly, mesocarps and endocarps may have various modifications that make them hard, soft, or leathery depending on the species. Remember that the pericarp is ripened ovary wall tissue and the exo-, meso- and endocarp are layers of that ripened ovary wall.

The coconut is a fruit. The photo below shows that the hard shell around the coconut that you see at the grocery store is actually the endocarp (the inside layer of the ovary wall). The fibrous mesocarp and the leathery exocarp have been removed from the coconut fruit by the time it reaches us. The fibrous mesocarp is used for other purposes like floor coverings in high-foot-traffic areas where a tough, durable fiber is needed.

Cross section of a coconut. James St. John. CC BY 2.0

A "nut" is a botanical name for a particular type of fruit. From a culinary perspective, there are many foods that we call nuts, but to a botanist a nut is a fruit with a particular structure.

You likely already knew that a peanut isn't a nut; it's a type of fruit called a legume, and is related to peas, beans, and locust trees. A cashew isn't a nut either; it's a type of fruit called a drupe. If you are interested in seeing other unusual fruits, visit the Fruits called Nuts **page posted by W. P. Armstrong at Palomar College.**

Review activity

- Next time you're grocery shopping, look around the produce section and try to figure out what parts of the fruits are the exo-, meso-, and endocarp. For instance, what are the parts that make up the banana? Where are those seeds?
- Why are fruits important to plant propagation?

Types of fruits

Simple fruit

A simple fruit is formed from a flower with one carpel, or multiple carpels fused together so that it looks like just one carpel. The ovary wall surrounding the carpel or carpels ripens into an independent fruit (independent in the sense that it isn't fused together with other ovaries). The photo below shows a grape, which is a simple fruit.

Cross section of a grape. Yelkrokoyade. GNU Free License 1.2

Aggregate fruit

In an aggregate fruit, the fruit is formed from the ripened ovaries present in one flower with numerous simple carpels. The ripened ovaries from that one flower coalesce into one larger unit, but you can still see evidence of the individual carpels. The raspberry, for instance, comes from one flower with many carpels. As the pericarps mature they mature together to form the thimble of the raspberry that we eat. You can still see the mosaic of individual ruby red carpels that fuse together to form the thimble. If you've ever harvested raspberries you know that the thimble pulls off a firm white structure, and that structure is the receptacle of the flower. Below is an illustration of the fruit structure.

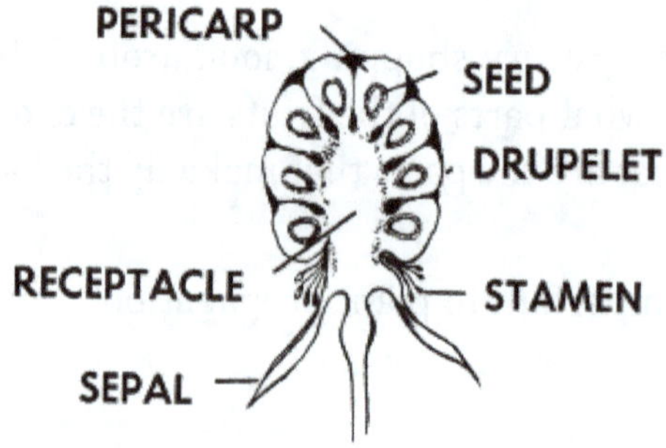

Cross section of an aggregate fruit. Pearson Scott Foresman. Public Domain

Multiple fruit

A multiple fruit is formed from the ripened ovaries present in one flower with numerous simple carpels. The ripened ovaries from that one flower coalesce into one larger unit, but you can still see evidence of the individual carpels. In the photo of the pineapple, below, you can see individual flowers, some of which are still open and showing purple-pink petals. The pericarps of these individual flowers coalesce into one large multiple fruit.

Pineapple flowering. Peta Hopkins. CC BY-NC-ND 2.0

The distinction between aggregate fruit and multiple fruit has to do with the number of flowers involved in the fruit. An aggregate fruit is from one flower with many ovaries, and the multiple fruit is made up of multiple flowers.

Fruit types

Fruits are also categorized according to whether the pericarp at maturity is:

- Fleshy: accessory parts of the ovary develop into succulent tissues with a high moisture content.
- Dry: at maturity the fruit has a low moisture content. Dry fruit that opens and releases the seeds from the pericarp is called dehiscent and dry fruit that remains closed retaining the seed within the pericarp is called indehiscent.

A dry, dehiscent pericarp may split open along sutures in various ways, and these ways of splitting open are also characteristic of particular types of fruits.

Simple fruits

Simple fruits with fleshy pericarp (exocarp, mesocarp, endocarp):

Drupe

A stone fruit, derived from a single carpel and containing usually one or two seeds. The exocarp is a thin skin, the mesocarp may be fleshy, and the endocarp is hard (i.e., "stony") as shown in the photo of the peach, below. Examples of drupes include peach, plum, cherry, apricot, and almond.

Apricot cross section. Fir0002/Flagstaffotos. CC BY-NC

Berry

A simple fruit formed from one flower with a superior ovary. The fruit has a fleshy pericarp, one carpel or multiple fused carpels, and many seeds. A tomato (below) is a berry, a grape is a berry, blueberries and cranberries are berries...but a raspberry is not. (Remember that a raspberry is an aggregate fruit where the carpels do not fuse the way they do in multiple-carpel berries).

Cross section of a tomato. FoeNyx. CC BY-SA 3.0

Pepo

A simple fruit formed from one flower with an inferior ovary. The fruit has a fleshy mesocarp, a rigid or leathery exocarp, one carpel or multiple fused carpels, and many seeds. The photo of squash below shows the fusion of three carpels to form the fruit, each carpel having many seeds. The fleshy interior that we eat is the mesocarp. Other examples include zucchini, cucumber, summer squash, and winter squash such as acorn and butternut squash.

Cross section of squash. RoRo. CC0 1.0

Hesperidium

Like a berry, but with a leathery exocarp instead of a fleshy exocarp. Each section of the hesperidium represents one carpel in the flower, but in the mature fruit the exocarp and mesocarp form an uninterrupted cover. The interiors of the carpels are packed with fluid-filled vesicles that are actually specialized trichomes. The exocarp contains volatile oil glands in pits. The orange, below, is an example of a hesperidium. All citrus fruits are this type of fruit.

Citrus cross section. Forest and Kim Starr. CC BY 3.0 US

Simple fruits with dry pericarp, dehiscent

Legume

Dry fruit made up of a single, folded carpel, multi-seeded, dehiscent along two sutures. It is easy to see the funiculus in peas, When you open the pod to shell out the peas there is a small stalk attaching the pea seed to the pod; that's the funiculus. Beans are also legumes.

Pea pod. Inorikof. CC BY 2.0

Capsule

A dry, dehiscent fruit made up of several fused carpels. The photo below shows the exterior of the poppy capsules and a cross section showing the locules with seeds inside. The capsule may split open in several ways depending on the species. In a poppy, the cap pops off to eject the mature seeds.

Poppy capsule. Katharina N. Pixabay license

Simple fruits, dry pericarp, indehiscent

Caryopsis

A fruit from one carpel containing a single seed. The pericarp is fused to the seed. A corn kernel is a caryopsis. The outside of the corn kernel is the pericarp.

Maize kernel. Xochiquetzal Fonseca/CIMMYT. CC BY-NC-SA 2.0

Achene

Like the caryopsis (one seed per ovary), but the seed can be threshed so that it is free of the pericarp. You can buy sunflower seeds "in the shell." The "shell" is the pericarp. We also discussed achenes earlier when looking at the actual fruits on a strawberry.

Sunflower seed. Benjamint444. CC BY-SA 3.0

Nut

A dry, indehiscent, one-seeded fruit with a hard exocarp. The ovaries that produce nuts have more than one carpel, but through abortion, only one seed matures. In the photo below, the pericarp of the acorn is partially encased in a tough covering called the involucre. True nuts will always have a hard exocarp and just one seed, while this is not always the case with culinary nuts. Peanuts, for instance, have a hard exocarp but multiple seeds. Horse chestnuts have a leathery exocarp and hard endocarp. True nuts have a hard exocarp.

Acorn. Piqsels. Public Domain

Review questions

- What features distinguish simple, aggregate, and multiple fruits?
- What is meant by dehiscent and indehiscent? How would you classify the corn caryopsis and the sunflower achene in this regard?

Accessory fruits

Earlier, we sawhow one or more of the other flower parts (androecium, corolla and calyx) can be attached below the ovary (hypogynous parts, superior ovary), above the ovary (epigynous parts, inferior ovary), or around the middle of the ovary (perigynous parts). In the case of epigynous and perigynous parts there are tissues surrounding the ovary to which the other flower parts attach and that can also adhere to the outside of the ovary and become part of the fruit. One example of a fruit from this type of flower where there is accessory tissue (hypanthium) adhering to the outside of the ovary wall is the pome, examples of which are the apple, pear, and quince. Another example is the strawberry, where together the receptacle tissue and ovary wall tissue form the strawberry fruit. Since it isn't part of the ovary wall, the receptacle tissue is considered an accessory tissue, so the strawberry, instead of being a true fruit, is a type of fruit called an accessory fruit.

Apple. Piqsels. Public Domain

Apple is an example of a species where the culinary fruit part we eat is actually hypanthium tissue rather than ovary wall tissue. In an apple, the ovary is the papery core that encloses the seeds. Since the part we eat isn't the ripened ovary wall, the apple is called an accessory fruit, signifying that we are really enjoying hypanthium tissue (an accessory tissue), not ovary wall. The hypanthium is also the tissue that makes up the fleshy part of rosehips (fruit of roses) — logical, because apple and rose are in the same taxonomic family: Rosaceae. Plants in the same family have common flower (and therefore fruit) morphologies.

Strawberry. Robert Owen-Wahl. Public Domain

The situation can be very complex with plants like strawberry where the juicy part we eat is the swollen receptacle of the former flower and the actual botanical fruits are the brown specks sticking to the outside of the receptacle. The strawberry is an accessory fruit because the red fleshy part we eat is made up not of the ovary wall, but primarily of receptacle tissue. The strawberry is also an aggregate fruit formed from multiple ovaries present in one flower. The fruit is an achene that contains the single seed from a single ovary attached to the outside of the receptacle.

Review activity

- Cut open an apple and identify where the hypanthium ends and the ovary wall begins.
- Next time you are at the grocery store, identify which "fruits" are true fruits, accessory fruits, aggregate fruit, or multiple fruits.

You should know the types of categories, such as simple fruits, multiple, aggregate, dehiscent, indehiscent, and some examples, but you do not need to keep an exhaustive list in your mind. You should also know the definition of an accessory fruit and the relationship of these fruits to epigynous and perigynous flowers.

There are many more types of fruits. If you are interested in some optional reading you can discover more about fruit types. Or do a Google search on "types of fruits" or on any of the specific fruit types mentioned above.

TERMS

Flashcards

Accessory tissues	Tissue of the fruit that is from non-carpel origin, usually in epigynous and perigynous flowers – e.g., the flesh of an apple is hypanthium tissue and the ovary is the papery core that encloses the seed.
Aggregate fruit	Fruit formed from the ripened ovaries present in one flower with numerous simple carpels.
Dehiscent	Used to categorize fruits with seeds that separate from a dried pericarp.
Endocarp	Inner layer of the pericarp.
Exocarp	Outer layer of the pericarp.
Fruit (botanical sense)	Ripened ovary together with the seeds within the ovary.
Funiculus	Stalk that connects either an ovule or a seed to the placenta.
Indehiscent	Used to categorize fruits with seeds that are retained within the dried pericarp.
Mesocarp	Middle layer of the pericarp.
Multiple fruit	Fruit formed from the ripened ovaries from a cluster of flowers that are in close proximity in an inflorescence and that coalesce into one unit.
Pericarp	Ripened ovary wall; made up of three parts: exocarp, mesocarp, and endocarp.
Placenta	Part of an ovary where the funiculus attaches.
Seed	Ripened ovule containing a seed covering, food storage, and an embryo.
Simple fruit	Fruit formed from a flower with one carpel or multiple carpels fused together so that it looks like just one carpel.

CHAPTER-9
SEEDS

Seeds are key to feeding the world's population as a nutrition and calorie source and for the propagation of most crops. In addition, being a fusion of paternal (pollen) and maternal (egg) genetic material, seeds create diversity; evolution acts on this diversity, and plant breeders take advantage of it to improve crops.

Learning objectives

- List three functions of a seed and identify its structures.
- Describe the process of seed germination.
- List the external factors required for seed germination.
- Understand dormancy and how the skilled plant propagator overcomes it.

9.1 SEED MORPHOLOGY

Why a coconut? Because inside the coconut's fibrous husk is a seed (image: Coconut. Filo gèn. CC BY-SA 4.0).

Learning objectives

- List three functions of a seed and name the seed part that has that function.
- Identify the parts of the embryo and the structures they become.
- List the types of nutrients that are stored in seeds.
- Describe the differences between the tissues that provide protection for the seed.

Seeds and diversity

To review, the two fundamental ways of propagating plants and how they differ in their outcomes are sexual reproduction through seeds or spores, and asexual or vegetative reproduction through manipulation of various plant parts, including cuttings from leaves, roots, and stems, or grafting.

Asexual reproduction, also called vegetative propagation, normally results in progeny (offspring) that are an exact genetic copy of the parent plant that donates the vegetative parts used in propagation. All of the progeny propagated from the same plant contain the same genes as the parent plant, and the progeny are also identical to each other. Differences among progeny or between parent and progeny may arise as a result of mutation, but this is rare. More likely, any differences among plants that have been asexually propagated from the same parent occur because the growing environment differs from plant to plant in some important way (fertility, water, light). The way the plant looks and performs in a particular environment is called its phenotype. The collection of genes the plant contains is called its genotype. Vegetative propagation results in identical genotypes, but two plants with identical genotypes may have different phenotypes if they are grown in contrasting environments.

Sexual reproduction in weedy, native, or undomesticated plants typically results in seeds that are genetically different from either parent, and in progeny that are all genetically different from one another. Sexual reproduction in domesticated plants can also result in genetically diverse seed, but for those domesticated plants that are highly inbred, like peas, beans, and cereals other than corn (like barley, wheat, oats, and rice), sexual reproduction can also result in seeds that are genetically identical to the parent. For now, remember that sexual reproduction means that a seed is formed as a result of fusion of a sperm and egg cell, and that there is the potential for progeny to differ from the parent(s).

Sexual reproduction, and the genetically variable progeny that result, give a plant species a great deal of flexibility to adapt to new habitats and environmental conditions. Some progeny might be extremely well adapted to new niches and can thrive there, while some will not fit and will not survive and reproduce. To assist the species in spreading, the seeds of the plants that successfully reproduce often have strategies for dispersal away from the parent plant. This spread the species to diverse locations and new ecological niches, and also reduces the likelihood of too many plants of the same species competing with each other in a limited space for the same scarce resources.

Seed formation is normally assumed to be the result of sexual reproduction rather than vegetative propagation. In rare cases, however, an embryo can develop solely from maternal tissue with no fusion of egg and sperm. This is called apomixis. Apomixis is really a form of asexual reproduction disguised as sexual reproduction. The process results in a seed, but the embryo in the seed isn't the result of sex — the fusion of male and female gametes. All of the progeny are identical to the parent because the embryo is actually a clone (genetically identical offspring) of the plant on which the seed is produced. Kentucky bluegrass reproduces this way, as does dandelion. It is ironic that our lawns are often inhabited by two apomictic species ... the Kentucky bluegrass we struggle to grow, and the dandelions many homeowners go to great extremes to eradicate.

Basic seed morphology

Seeds have three main functions:

Propagation of the plant

This is accomplished by the embryo, which is the nascent (new, young) plant resulting from the combination of genes from the male sperm, transmitted by the pollen, to the female egg, held in an ovule in the ovary. The embryo has an axis with one end differentiating into the shoot and the other into the root.

Nutrient storage

Two types of structures can store nutrients in the seed — the cotyledon and the endosperm. The nutrients fuel growth of the embryo.

Protection

The embryo and nutrient source need a tough covering for protection from the environment and predators, and this is typically, but not always, provided by a structure called the seed coat (sometimes called the testa).

The drawing above shows a simple cross section of a kidney bean seed (a dicot, but only showing one cotyledon), illustrating these three functions. The kidney-shaped part is the outline of the cotyledon, the main mass of the bean seed and the site of stored nutrients. Sandwiched between the cotyledon halves is the embryo, which is the nascent plant. On the outside of the cotyledon is a thin layer that is the protective seed coat. The kidney bean seeds we see in the grocery store have a seed coat that typically has a light or dark maroon color. In peanuts, the different structures are easy to see when you pull the two cotyledons apart.

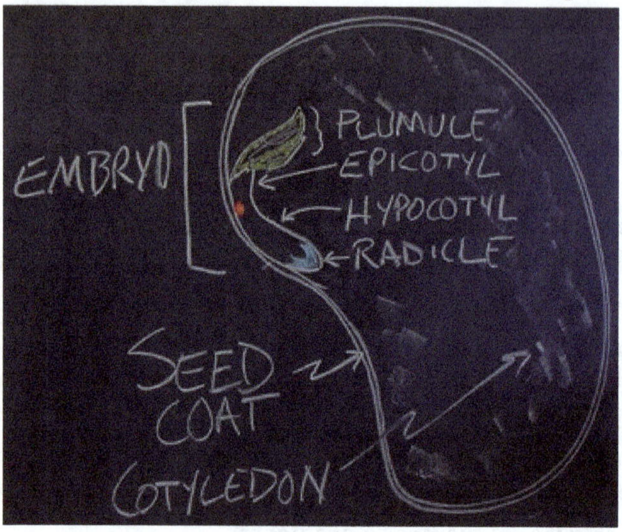

Cross section of kidney bean. Tom Michaels

First function: Propagation of the plant by the embryo

In flowering plants the embryo is normally the result of fusion of egg and sperm. The egg is held within an ovule, which in turn is held within the ovary, which can hold several ovules, depending on the species. The egg is typically fertilized by sperm from pollen. The maturing ovule develops within the ovary of the maternal plant.

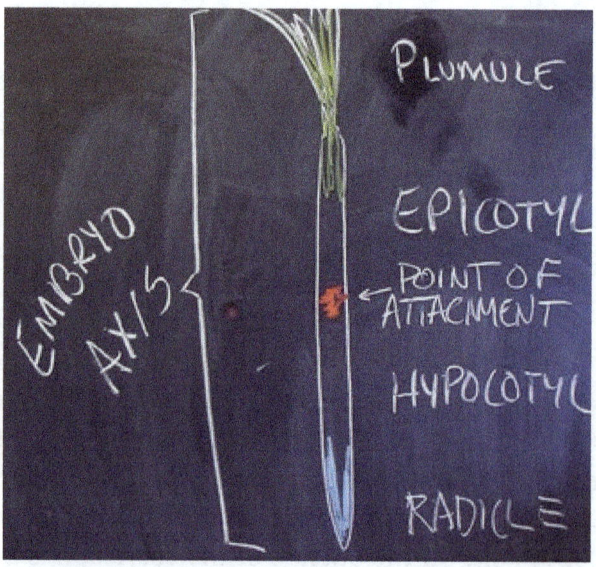

Embryo. Tom Michaels

A mature seed has an embryo with a linear arrangement of parts. This arrangement is called the embryo axis. The drawing above shows the embryo axis from the kidney bean, straightened out to show the individual structures.

- Embryo axis — the embryonic root and shoot
- Parts making up the shoot tissue of the embryo axis:
  - Plumule — the first true leaves of the plant that you can sometimes see already attached to the embryo. These leaves will emerge from the seed, rise above the soil surface, and start to collect energy from the sun.
  - Point of attachment — the spot shown in red in the diagram (it's not red on a real embryo) on the embryo axis where the cotyledon attaches. The cotyledon is attached to the embryo, and is actually part of the embryo. In the case of the bean, the cotyledon is a nutrient storage organ and the nutrients flow to the embryo through the point of attachment of the cotyledon to the embryo axis.
  - Epicotyl — the part of the embryo axis that is above (epi-) the point of attachment of the cotyledons
  - Hypocotyl — the part of the embryo axis that is shoot tissue below (hypo-) the point of attachment of the cotyledons, but above the radical. The hypocotyl is the part of the shoot between the attachment of the cotyledon and the start of the root (radicle).
- Part making up the root tissue of the embryo axis:
  - Radicle — the embryonic root tissue

At the tip of the epicotyl is the shoot apical meristem that will produce new nodes and internodes. If you're counting nodes on the embryonic axis, the first node on the stem starting from the point of transition of root to shoot is the point where the cotyledons attach. The cotyledons are actually embryonic leaves. The second node is where the plumule is attached.

At the tip of the radicle is the root apical meristem that will produce the primary root.

Review questions

- Differentiate among embryo axis, epicotyl, hypocotyl and cotyledon. Which are derived from the cell resulting from the fusion of egg and sperm?
- Describe why apomixis results in a seed that is actually a form of asexual propagation.
- Of the tissues making up the embryo axis (plumule, epicotyl, hypocotyl, radicle), which are shoot tissue and which are root tissue?

Double fertilization and the endosperm

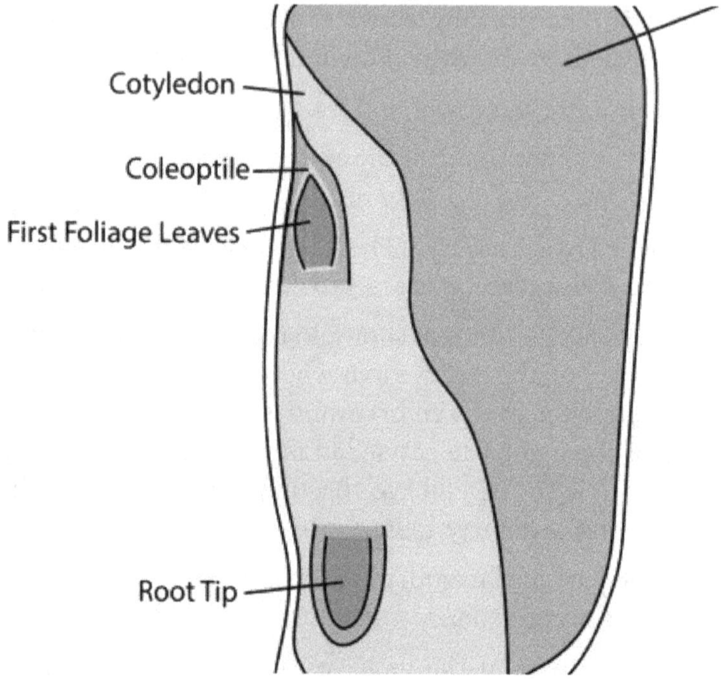

Cross section of corn seed. Sarah Greenwood. CC BY-SA 4.0

In flowering plants (angiosperms), there is a phenomenon called "double fertilization." The angiosperm pollen grain holds two sperm cells. One fertilizes the egg, and the resulting zygote grows to become the embryo. The other unites with two other maternal nuclei, called polar bodies, and these three nuclei together grow to become a tissue called endosperm (like the meat and milk of the coconut). This will be covered in more detail when we study meiosis and gametogenesis. For now, remember that in flowering plants there is a process called double fertilization that results in an embryo and an endosperm.

In grasses, like the corn in the illustration above, the endosperm is the major energy and nutrient storage tissue. This is different from the kidney bean, where the cotyledon is the storage organ. There is a cotyledon in the corn seed as well, but instead of storing energy and nutrients, it helps break down, absorb, and transfer the energy and nutrients stored in the endosperm to the embryo. Also in many monocots, there is a sheath covering the plumule and epicotyl that provides protection. This sheath is called the coleoptile. A similar sheath covers the radicle, and is called the coleorhiza. Coleoptile and coleorhiza are terms used specifically with plants in the grass family (Poaceae), and not in other families within monocots. Both function in providing protection to the emerging shoot and root. In the corn seed, the three functions of 1) propagation, 2) protection, and 3) nutrition are satisfied by the:

- Embryo (shown in this illustration extending from the shoot to the root) and that will propagate a new plant.
- Endosperm tissue that provides energy and nutrition for the embryo.
- Pericarp protecting everything inside. Remember that in corn, which produces a type of fruit called a caryopsis, the pericarp is fused right to the seed. The pericarp is the mature ovary wall of the female corn flower in the ear (labeled 'seed coat' in the illustration above).

This summary table contrasts corn and bean seed components:

	Cotyledon	Endosperm	Seed Coat	Pericarp
Corn – monocot	Energy absorption	Energy storage	Remnants (these tissues are absent or only fragments of tissue in the mature seed)	Protection (outermost layer of the corn kernel)
Bean – dicot	Energy storage	Remnants (these tissues are absent or only fragments of tissue in the mature seed)	Protection (outermost layer of the bean seed)	Protection (pod)

Second function: Storing energy and nutrients for embryo growth

Among flowering plants, energy and nutrients can be stored in the seed in the:

- Cotyledon
- Endosperm

What types of energy and nutrients are stored in these tissues? Think about the seeds you eat, and you can probably name many of these nutrients.

Carbohydrates

- Provide energy — complex molecules composed of carbon, hydrogen, oxygen.
- In plants, carbohydrates include starch and sugar.

Protein

- Sources of amino acids for production of enzymes and other nitrogen-rich compounds.
  - Amino acids are the building blocks of proteins.

- Protein is the solid material in tofu (pressed curds from coagulated soy milk). There are several different categories of protein in seeds based on the specific chemical structures of the molecules. Gluten is a type of protein found in wheat endosperm that confers elasticity to bread dough so that the stretchy dough traps the carbon dioxide given off by yeast in the bread-making process and forms a tender loaf full of air pockets.

Lipids

Plant oils, called triglycerides, are compact molecules for storing energy in a more compact way than in starch and sugar.

We are bombarded with a great deal of confusing nutritional information about lipids in our foods. When you encounter nutritional claims, keep this information in mind:

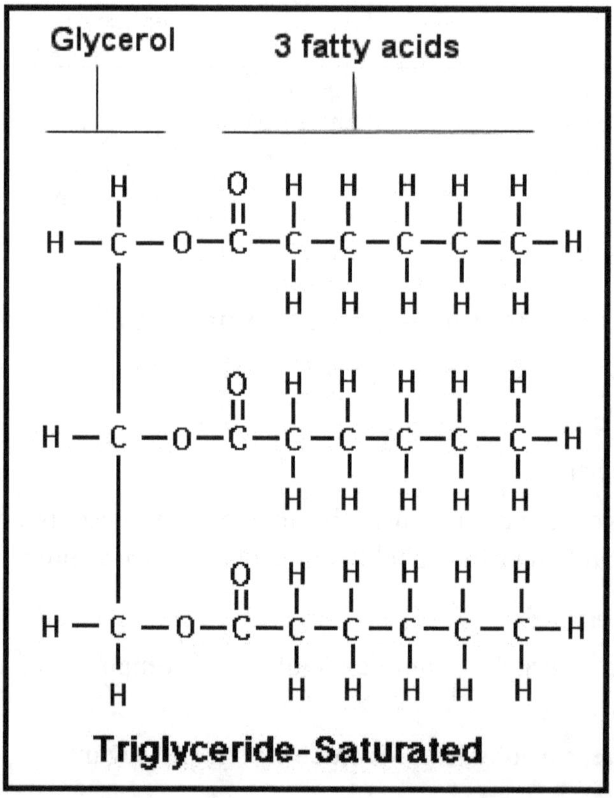

- The building blocks of a triglyceride are a Glycerol molecule plus 3 Fatty Acids. The illustration shows a triglyceride made up of glycerol linked to three saturated fatty acids.

- Fatty acids are long chains of carbon atoms with two hydrogen atoms attached to each carbon, except where there is a double bond between adjacent carbons on the chain. In the case of a double bond, each carbon involved in the double bond has only one attached hydrogen atom.
- Saturated fatty acids have no double bonds in the chain, and all carbon atoms in the interior of the chain have two attached hydrogen atoms – they are thus saturated with hydrogen atoms.
- Unsaturated fatty acids have one or more double bonds between one or more carbon atoms in the chain. These fatty acids lack some hydrogen atoms, and therefore the carbon atoms are not saturated with hydrogen – they are unsaturated.
- You have probably heard nutritional claims about saturated and unsaturated fats. The difference between the two is whether all the carbon atoms in the fatty acids are saturated with hydrogen atoms, or are missing hydrogen atoms as a result of double bonds in the carbon chain. Plant oils tend to be unsaturated, which results in them being liquid at room temperature – they have a low melting point. Animal fats tend to be saturated, which results in them being solid at room temperature – they have a higher melting point. If you are interested in more of the chemical nature of fatty acids, check out this link to The Fat Primer **(optional reading).**

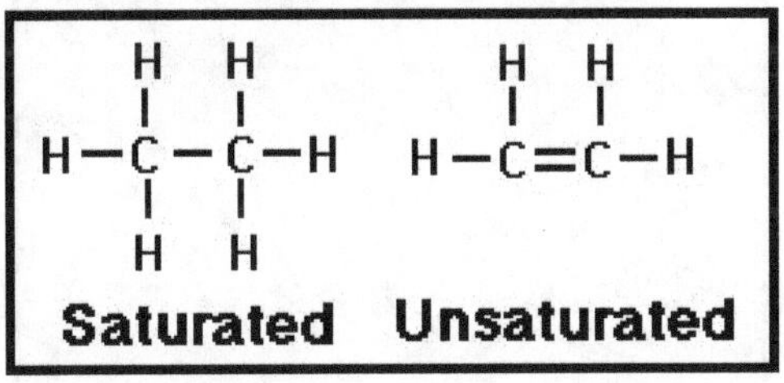

Nutrient differences between monocots and dicots (legumes in particular)

Dicots

As noted above, seeds with two cotyledons tend to have cotyledons whose function is storage. Legumes (Fabaceae family) like beans, peas, soybeans, and lentils are dicots that tend to store large amounts of protein in their cotyledons. Some legumes have high protein, high lipid, and low carbohydrates (like soybean and peanut); these are called oilseeds because they have high lipid content and oil can be squeezed or otherwise extracted from them. Others have high protein, low lipid, and high carbohydrate (like pea and bean), and these are called pulse crops. Pulse crops are hugely important foods because they are edible protein sources. They are a key source of protein in vegetarian diets.

Monocots

Monocots have a cotyledon too, but as noted earlier the cotyledon is primarily used for absorption. In cereals the endosperm stores the nutrients, which tend to be primarily starch and sugar and low amounts of protein and oil.

As a rule, legume seeds are high protein (and in some cases like soybean and peanut high oil), while cereal grains like corn, wheat, oats, barley, and rice are high in starch.

Review questions

- What seed structure(s) contains carbohydrates?
- What is the key characteristic of an unsaturated fatty acid?

Third function: Protecting the embryo and nutrients

Seeds have two layers of protection:

- The seed coat, which originates as ovule wall tissue.
- The pericarp, which originates as ovary wall tissue.

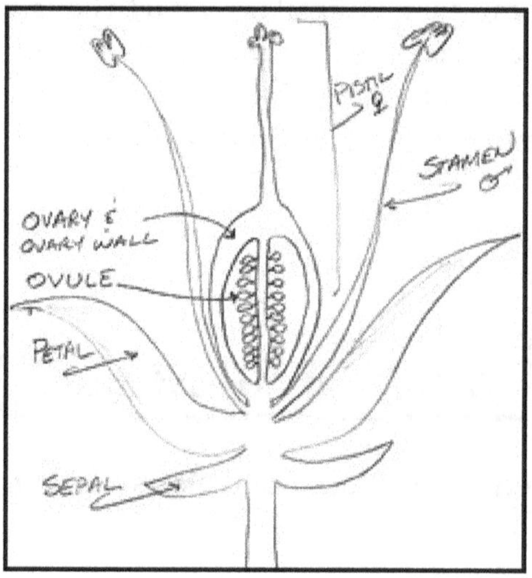

Cross section of a flower. Tom Michaels

The line drawing of a flower cross-section, right, shows sepals, petals, stamens (made up of a filament and anther), and carpel or pistil (made up of stigma, style, and ovary). Also identified are the ovary wall and the ovule wall. The ovary is at the base of the carpel and holds the ovules. The ovules are protected by the ovary and hold the egg. When the flower is fertilized, a pollen tube germinates from the pollen grain and grows into the stigma and down through the style. The sperm follows the pollen tube into the ovary. One sperm unites with the egg and the resulting zygote becomes the embryo. The other sperm unites with two polar nuclei to form the endosperm. This is the double fertilization noted earlier.

As the seed matures, the cells inside the ovule multiply and grow. The ovule wall, which is made up of maternal cells called integument tissue, matures to become the seed coat. An example of a seed coat is the red or tan "skin" on a peanut. The ovary wall (note the important difference between the words "ovule" and "ovary") matures into the protective cover called the pericarp. Again using peanut as an example, the pericarp is the "shell" of the peanut. So peanuts-in-the-shell are an example of a pericarp (shell, ovary wall tissue) and a seed coat (skin, ovule wall tissue) protecting the cotyledons and embryo within the seed.

In contrast, when corn matures, the ovule wall smashes up against the ovary wall as the cells inside multiply and enlarge. The ovule wall cells disintegrate, so at maturity there are only remnants of the seed coat. The ovary wall matures into the pericarp of the corn seed, which is the hard exterior of the seed.

These words belong together:

- Ovule wall — Seed coat
- Ovary wall — Pericarp

You might be interested to know that you are probably wearing integument/ seed coat tissue right now. The fine fibers attached to the outside of cotton seeds are made up of long chains of single cells of integument tissue and are extensions of the seed coat. These fibers are removed from cotton seeds, spun into a fiber, and woven into fabric.

9.2 SEED PHYSIOLOGY

Learning objectives

By the end of this lesson you will be able to:

- Describe the process of seed germination.
- List the external factors that are required for seed germination.
- Understand dormancy and the conditions needed for germination to occur.

Seed — a living plant in a quiescent state

In the ovary, within the ovule, after egg and sperm unite to form the zygote, the zygote cell repeatedly divides and develops into an embryo. The embryo differentiates into different structures — plumule, radicle, cotyledon(s), and the endosperm, seed coat, or pericarp develop. Once this development has occurred, the embryo's metabolism slows down to nearly zero and the maternal plant stops pumping energy into the seed. The nascent plant contained within the seed

— really an embryo surrounded by nutritive tissue and a protective covering — is now independent of the mother plant. The seed, which is the next generation of plant, enters a quiescent phase.

Recall that the seed contains:

- An embryo, which is the new plant,
- A nutrient source (typically endosperm and/or cotyledon), and
- A protective covering (typically a seed coat and/or pericarp)

Why do plants produce seeds? Because seeds allow plants to:

- Propagate the next generation
- Generate genetically variable offspring that can be sorted through natural selection
- Survive harsh conditions
- Disperse into new environments

Red chile pepper. Nir Sinay. CC BY-NC-ND 2.0

The first of these reasons is obvious, to propagate. One parent plant generates many seeds, and through these seeds potentially bears many offspring. A kidney bean plant, for instance, might average 4 seeds per pod and have 20 pods hanging on the plant, so one plant yields 80 seeds. A nice ear of field corn will have 16 rows of kernels with 40 kernels per row, for a yield of 640 seeds. The tiny hot pepper in your garden has over 50 seeds, and 20 peppers on a plant would yield 1,000 seeds.

The second reason, genetically variable offspring, results from cross-pollination (mentioned above under Seed Morphology), which is particularly common in wild, undomesticated plants. When plants cross-pollinate, an egg is formed within the maternal plant. The genetic constitution of the developing embryo within the seed is 50% from the paternal plant, 50% from the maternal plant. The particular combination of genes in the developing seed is different from that in either parent plant, and from the other seeds on the same maternal parent. The seeds share some of the same genes, but the specific combination of genes is different. That difference results in genetic variability, which can be expressed as differences in plant height, flower color, leaf shape, fruit size, or other morphological or physiological characteristics. In nature, this variability is the raw material on which natural selection operates. In plant breeding, it is the resource that sustains our efforts to select improved plant varieties.

Due to their protective coating and quiescent metabolism, seeds can survive harsh conditions that will kill the parent plant such, as freezing cold, protracted drought, and even fires. Once conditions are again favorable for plant growth, the seeds can then germinate.

Seed dispersal

Samaras are the flat-winged fruit produced by maple trees. Bob Larrick. CC BY-SA 2.0

Seeds disperse from the maternal parent plant "in space" through many wonderful and creative mechanisms for hitching rides on the wind, on animals, and sometimes in animals as they are eaten, pass through the gut, and are excreted.

You have probably seen fluffy cottonwood (poplar) seeds floating on the summer breeze, pulled burdock seeds off your sweater after bushwhacking through the woods in the fall, or washed bird guano full of mulberry seeds off your car's windshield. These seeds all used strategies for dispersal in space. Vanderbilt

University's Bioimages (optional) includes photos of mechanisms that help spatially disperse seeds.

Seeds also disperse from the maternal parent plant "in time." Some have dormancy mechanisms that delay germination until the next favorable growing season, which might be a year from now or even several years from now. Seeds are considered dormant if they are alive and don't germinate even if provided with favorable conditions for germination. There are many dormancy mechanisms; we'll address some of them later.

If you've ever dug up new sections of your yard for a garden, did you notice that, a few weeks after planting veggies or flowers in the spring, you saw a big flush of weeds? And if you pulled up all of those weeds and somehow prevented any new seeds from landing in the garden, you still got a big flush of weeds the next year? This is called the soil's seed bank, and it's due to dormant seeds that are resting in the soil. Every year a percentage of those seeds lose their dormancy and germinate, leaving you wondering in frustration whether the weeding will ever end. (It won't.)

Review questions

- What type of reproduction results in genetic diversity among seeds produced on the same plant?
- Distinguish between mechanisms causing dispersion in time vs. dispersion in space.
- What part(s) of the seed provide nutrition? Protection?

Germination

Germination is the reactivation of the seed's metabolism and the restoration of embryo growth. There are two main reasons why seeds don't germinate:

- They could still be quiescent because favorable external conditions do not yet exist. In particular, the environment could be too dry or too cold, or the oxygen levels could be too low to support embryo growth.
- They could be dormant – the characteristic that allows seeds to disperse in time.

External conditions

The external conditions required for germination to occur are:

- Water
- Oxygen
- Temperature

- Light (for some small or fine seeds, like lettuce)

Looking at these external conditions more closely:

Moist conditions allow the seed to imbibe water. Seeds are typically very dry — somewhere in the 8–15% relative humidity range — so they will readily take up moisture from damp soil. Water moves through the pericarp and seed coat into cells and leads to reactivation of the metabolic processes. The seed's nutritive reserves are metabolized for the embryonic cells to divide, enlarge, and differentiate.

The breakdown of nutrient reserves to form energy for plant growth is called respiration, and it requires oxygen. The seed must have oxygen to respire. If you keep seeds in an oxygen-depleted atmosphere, they will not germinate. One all-too-common type of oxygen-depleted environment in which seeds are sometimes placed is waterlogged soil. If you over-water newly planted seeds, the water will keep oxygen from reaching the seeds, and although the seeds will imbibe water and swell as if everything is going well, they will not germinate, and will likely rot because there is not enough oxygen available to sustain respiration. Waterlogged soil also encourages growth of bacterial and fungal organisms that can infect and decompose the seed.

Depending on the species, seeds have various temperature requirements for germination. Some spring flowers will germinate when soil temperatures are quite cool, even below 50ºF, while most of the seeds we plant in our gardens prefer temperatures in the 50–70ºF range. Again, respiration is the reason. Molecules move around faster when they are warmer, and this movement encourages the chemical reactions required for respiration. If the seeds are cold and the molecules aren't moving around, the seeds won't germinate.

If light is required for seed germination, the species is said to be positively photoblastic. This characteristic allows the seed to remain dormant when buried deep underground, but to germinate when brought to the surface. As you might imagine, weeds that are successful in annually-tilled soils may be positively photoblastic. They remain dormant until tillage brings them to the soil surface.

A note about respiration

Respiration refers to the set of reactions that take place in the plant cell to convert chemical energy stored in molecules into a form of energy that can be readily used by the cell to power other chemical reactions. Respiration converts the starch stored in the endosperm or cotyledon into ATP (Adenosine triphosphate) (optional reading), which is used in the apical meristems and the radical of the embryo to fuel cell division and the production of new cells.

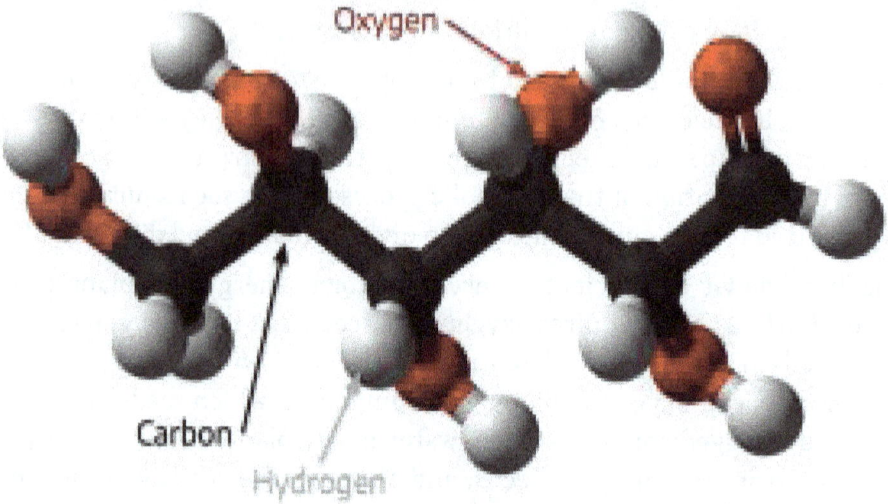

Glucose molecule. Public domain.

Starch is made up of a chain of glucose subunits. Glucose, shown in the three-dimensional model above, is a simple sugar that is made up of 6 carbon atoms, 6 oxygen atoms, and 12 hydrogen atoms. The shorthand formula for glucose is $C_6H_{12}O_6$. The shorthand formula for starch is $[C_6H_{12}O_6]n$, where the "n" indicates that there are "n" glucose molecules that, when linked together, make up a starch molecule.

When we put a quiescent, but not dormant, seed in the ground and it has access to appropriate moisture, warmth, and oxygen (and light if positively photoblastic), it begins to respire. Enzymes are secreted by the cotyledon and, depending on the species, by other specialized cells surrounding the cotyledon and endosperm, which break down the stored starch into its glucose subunits. It is important that the starch be broken down to glucose because glucose is a sugar that is physically small enough to pass through the semipermeable cell membrane; starch is too large to get through. Starch can't be moved from cell to cell, but glucose can. Starch can be stored in cotyledon or endosperm cells and be broken down to glucose, and that glucose then moves into the actively dividing meristem cells.

Once in the cell cytoplasm, the glucose is broken in half by a process called glycolysis **to form a -3carbon compound known as** pyruvate. Pyruvate first reacts with a carrier molecule and then moves into the mitochondria — the powerhouse organelles in the cell — where it is further metabolized to yield high-energy molecules of ATP. The processes in the mitochondria require the presence of oxygen. The ATP moves out of the mitochondria and to the parts of the cells where chemical reactions are taking place that need energy.

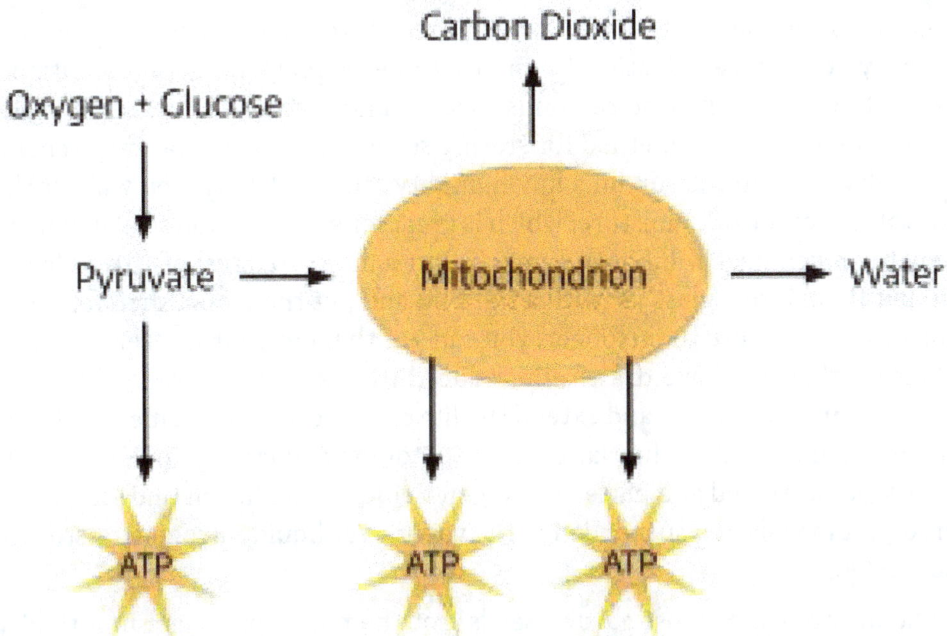

Cellular respiration. Emily Tepe

Starch stored in the seed is a form of stored energy composed of glucose. Glucose is a transportable form of chemical energy that can move through cell membranes, so it helps surround the seed with chemical energy. Pyruvate is a compound formed from glucose that can move into the mitochondria and be broken down to yield ATP. ATP leaves the mitochondria and provides the cell with the energy needed for a wide range of chemical reactions.

The inputs for respiration are glucose and oxygen. Respiration converts the energy stored in the glucose into ATP that will power reactions throughout the cell. Carbon dioxide and water are the two waste products.

Seeds also store lipids and protein in the cotyledons. These too can be broken down during respiration. The lipids are first biochemically deconstructed into their components: glycerol and fatty acids. The glycerol molecule is made up of carbon, hydrogen, and oxygen — so it can also be converted to pyruvate and heads into the mitochondria for conversion to ATP. The fatty acids take a different biochemical route, but still end up yielding ATP. Protein respiration is even more complicated, and yields nitrogen-containing building blocks of protein and cells called amino acids that are used in construction of other molecules in the cell. Stored protein in the seed is better at providing amino acid building blocks than it is in providing ATP to energize the cell. Protein can provide energy if necessary, but starch and lipid are more efficient energy storage molecules.

Storing seeds

Since germinating seeds require oxygen, moisture, and warmth, you can intentionally restrict germination by limiting one or more of these conditions. Why restrict germination? One reason is to store and save seeds for long periods of time. The most common method for storing seeds is to ensure that they remain dry. If you dry seeds in the sun on a low-humidity Minnesota day, you will get the seeds down to around 10% moisture, which is great for storage. (Don't bake them in the oven; too much heat will kill the embryo.) At a moisture level of 10% or lower you can put them into a glass jar with a tight lid and put them on a shelf for a few years of storage. To store them longer, you can put them in your freezer, which of course means that you have drastically reduced the heat in the seed, which will stall respiration even further and extend the life of the seed. In extreme situations, such as that maintained at the National Seed Storage Laboratory in Fort Collins, Colorado, seeds are dried and placed in oxygen-depleted conditions and stored in a freezer, or put in vials and suspended in the vapor over liquid nitrogen for storage at about -150ºC.

For home storage of most garden seeds, get them dry, put them in a tightly lidded glass jar, and, to make them to last 5–10 years, put the jar in the freezer. To dry small amounts of seeds, you can use a home food dehydrator set on a very low temperature. Don't put your seeds in the refrigerator unless you have them very tightly sealed in an air-proof container, because refrigerators are damp. A refrigerator is a great place to store popcorn, because the humidity ensures that the kernels have enough moisture to pop strongly, but it's a poor environment for seed storage.

Dormancy

The second reason seeds resist germination is dormancy. Dormancy is when the seeds do not germinate, even though conditions for germination are favorable. Something about the seed prevents germination. Barriers to germination could include:

Barriers in the protective covering:

- A seed coat, or pericarp, that is impermeable to water or oxygen.
- Compounds that act as germination inhibitors that are embedded within the seed coat.
- Barriers in the embryo:
  - Physiological immaturity of embryo — the embryo is initially immature and requires a period of cool temperatures or alternating warm and cool temperatures to fully mature.

- Endodormancy in temperate plants — internal biochemical processes must be met in the seed before germination can begin. Endo-dormancy is the first stage of dormancy for many seeds from plants grown in temperate environments like Minnesota. Once the internal biochemical processes are met, the seed usually goes into eco-dormancy.
- Ecodormancy — external factors are not optimal for germination. This is often due to temperatures being too cold, or to amounts of water not sufficient for germination.

Kentucky coffee tree seeds. Laura Irish

The ways in which horticulturists overcome seed dormancy depend on the type of dormancy. For seeds with impermeable seed coat, such as the Kentucky coffeetree (*Gymnocladus dioicus*) seen above, a technique called scarification is used. Sandpaper is used to break through the seed coat until the white cotyledon is visible. Under warm, moist conditions, these seeds will germinate. To germinate a large amount of seed, there are other ways to break the seed coat, such as using acid to "etch" holes in the seed coat to allow for water imbibition.

In the case of endodormancy, we usually have to be patient. Seeds are placed in a cool (38–42ºF) place under moist conditions. This process is called stratification. Depending on the plant species, stratification can take from 2 weeks to almost a year. It is as easy as putting seeds in a sealable plastic container or bag with moist media, and placing them into a refrigerator.

Dormancy is a great strategy for enhancing a plant's survival potential because germination is delayed until a later time when environmental conditions are more favorable. However, for horticulturists who prefer that a seed germinate as quickly as possible after being planted, dormancy is a nuisance. One of a horticulturist's important skills is to recognize dormancy, identify the dormancy mechanism, and take steps to overcome the inhibition so that plants can grow predictably from seed when planted.

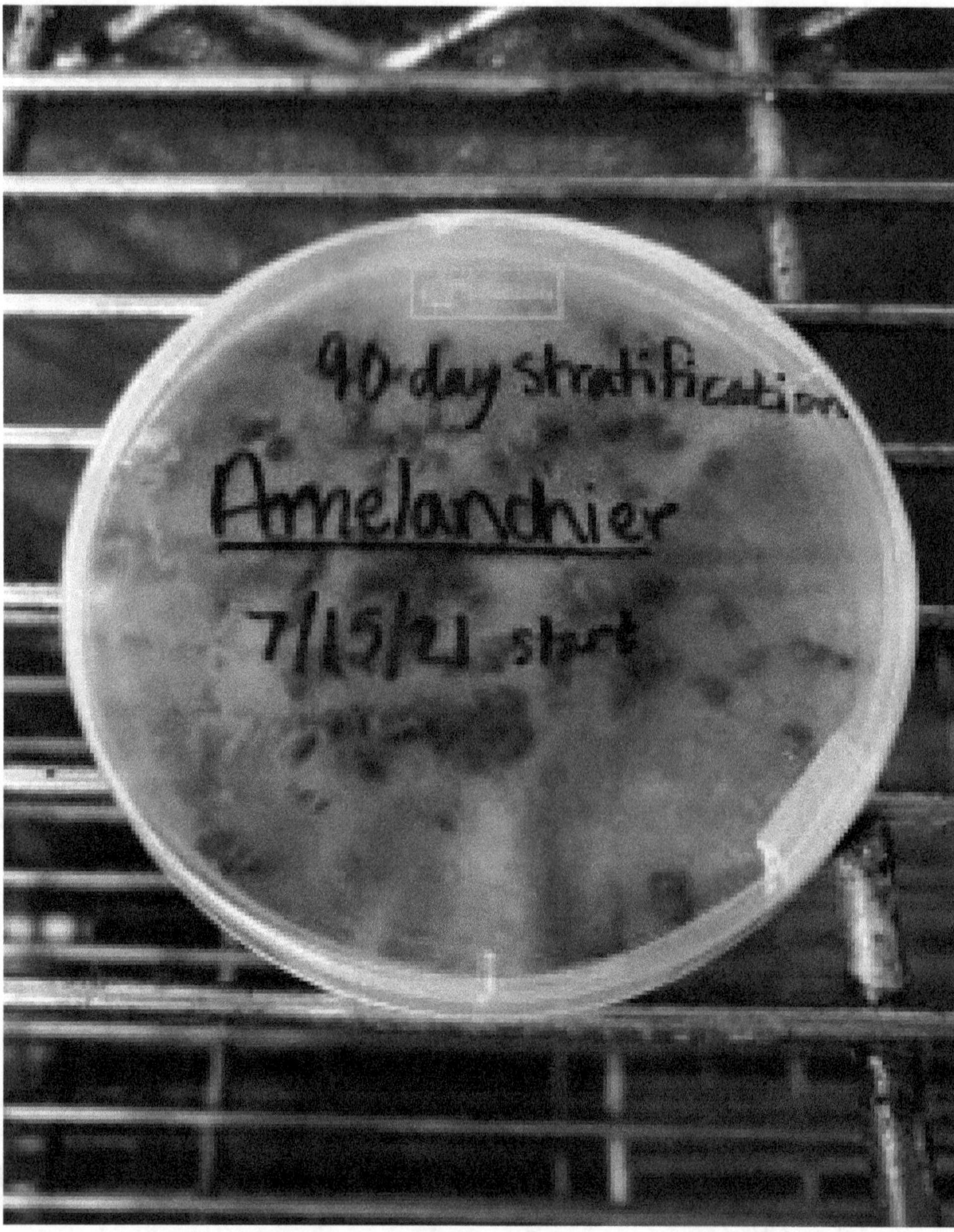

Stratifying Amalanchier seeds. Laura Irish

TERMS

Flashcards

Apomixis	A form of clonal reproduction where vegetative cells in the flower develop into zygotes to form seeds.
Carbohydrates	One of the three major types of nutrients found in seeds; provide energy in the form of starch and sugar.

Coleoptile	Protective sheath that covers in the plumule and epicotyl in the Poaceae family.
Coleorhiza	Protective sheath that covers the radicle in the Poaceae family.
Cotyledon	Food storage structure used in germination.
Dormant/dormancy	Term used when seeds are alive and don't germinate when provided with favorable conditions for germination.
Double fertilization	Where one haploid male sperm cell fuses with the female haploid egg cell to form the diploid zygote, and the second haploid male sperm cell fuses with two egg cells to form a triploid endosperm.
Ecodormancy	When external factors, usually environmental, prevent a seed from germinating.
Embryo	Nascent (new, young) plant resulting from the combination of genes from the male sperm transmitted by the pollen to the female egg held in an ovule in the ovary.
Embryo axis	Embryonic root and shoot.
Endodormancy	When internal factors within the seed prevent germination.
Endosperm	Tissue that results from the second haploid male sperm cell fusing with two egg cells during fertilization.
Genotype	Genetic composition of an organism.
Lipids	Compact plant oils that store energy; also called triglycerides.
Pericarp	Ripened ovary wall; made-up of three parts: exocarp, mesocarp, and endocarp.
Phenotype	Physical appearance of an organism.
Plumule	First true leaves of the plant; emerge from the seed, rise above the soil surface, and start to collect energy from the sun.
Proteins	Sources of amino acids for production of enzymes and other nitrogen-rich compounds in the seed.
Quiescent	When a seed does not germinate until given proper conditions for germination (oxygen, water, temperature, and sometimes light).
Saturated fatty acids	Fatty acids that have no double bonds in the chain with all carbon atoms in the interior of the chain having two attached hydrogen atoms.
Scarification	Process used to break a physical seed dormancy (hard seed coat).
Seed coat	Outer layer of the seed.
Stratification	Process used to break a physiological dormancy, such as embryonic or endo/eco-dormancies.
Triglycerides	Another name for lipids.
Unsaturated fatty acids	Fatty acids that have one or more double bonds between one or more carbon atoms in the chain, lack some hydrogen atoms, and therefore the carbon atoms are not saturated with hydrogen.

CHAPTER-10
GRAFTING

Some plants do not produce sufficient seed or lack competency to form adventitious roots by making cuttings, and we are left with grafting as a method of choice for asexual propagation. Grafting is the cloning of a scion or bud onto a rootstock, but adds the possibility of manipulating shoot properties through the choice of rootstock.

Other plants clone themselves naturally and may use special storage organs to help with perenniation. These can develop offsets, pups, and other clonal storage organs. The plant propagator can also induce these tissues to form more propagules. Excellent examples include bulbs, corms, and rhizomes.

In this research orchard, grafted apple trees are tested for suitability.
Photo by UMN Department of Horticultural Science.

Learning objectives

- Understand why plants are grafted or budded and the techniques used for specific outcomes.
- Describe how a graft union heals.
- Compare and contrast how a plant responds to wounding versus healing a graft union.
- Characterize the differences between bulbs and other storage organs.
- Describe techniques for propagating plants with different clonal strategies from storage organs

10.1 GRAFTS AND WOUNDS

Learning objectives

- Explain why plants are grafted or budded and the techniques employed for specific uses.
- Describe how a graft union heals.
- Compare and contrast how a plant responds to wounding versus healing a graft union.

Grafting

Grafting is the art and science of connecting two pieces of living plant tissue together in such a manner that they will unite and subsequently grow and develop into one composite plant. The union of these two different plant materials via grafting creates a chimera, — two different plant genotypes growing together in the same plant. The roots of an apple tree in a commercial orchard, for instance, likely came from a plant with a genotype that induces dwarfing of the tree. The picture below shows the swelling of the graft union a few inches above the soil surface.

The scions are exact genetic copies, and are sometimes called clones or clonal propagation. Although the graft can combine two different plant types, species, or even genera between the scion and rootstock, the scion material being propagated is an exact genetic copy of the parent plant that donates the vegetative parts used in the graft. Usually the scion or bud and rootstock are the same species, because this favors compatibility and successful grafting. The rootstock can thus be from mixed types, but the scion is from the same plant type for clonal propagation.

The shoot of a grafted apple tree, for example, comes from a plant with great-tasting apples, while the rootstock may not produce great apples. The root and shoot were grafted together and the tree grows as one plant, but with two genotypes: the roots have one genotype — one set of genes — and the shoot another, yet they are all growing as one plant. That's a chimera. Grafting is means of asexual reproduction, which is vegetative propagation.

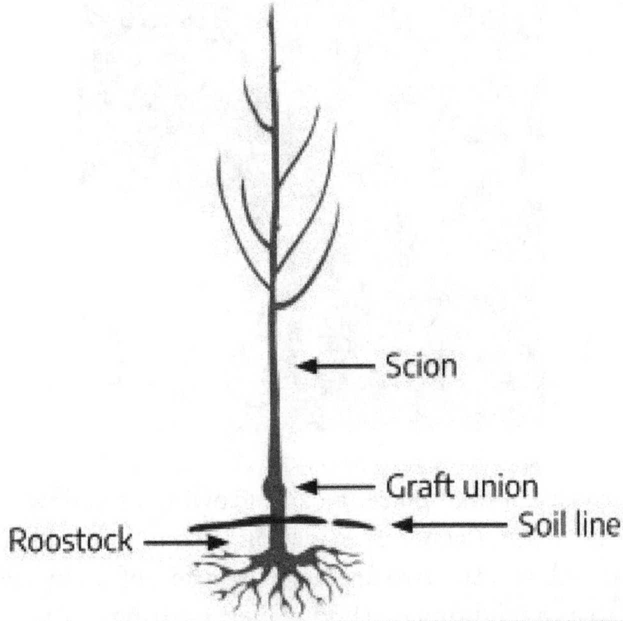

Grafted apple tree. Emily Tepe

When a horticulturist makes a graft, as in the apple trees above, one genotype of the tree species is typically used for the above-ground part of the plant, called the scion, and another for the below-ground portion, called the rootstock. A grafted plant has a scion growing on a rootstock, and the scion and rootstocks have different genotypes. In many grafting situations the scion and rootstock are of similar diameter, uniting the two genotypes where the graft union is made.

Grafting is a very old horticultural technique; there are historic records of horticulturists grafting olives 2,000 years ago. They may not have known how the cells divided and healed, but they knew how and why to graft.

Reasons to graft

While grafting is a valuable vegetative propagation technique, not all plants are easily grafted. For those that do respond well to grafting, there are different reasons for employing the technique.

Create unique, commercially desirable ornamentals

Grafted cactus. hmerinomx. CC BY-NC-SA 2.0

For photosynthesis to take place, a cacti growing as yellow or orange cannot live without the rootstock. Cactus propagation is commonly done by grafting; the cambiums are aligned, and the two are held together, often by rubber bands. This is often referred to as a modified method of cleft grafting.

Perpetuate genotypes that do not root from cuttings

Japanese maple tree. Brian Holsclaw. CC BY-ND 2.0

Japanese maples, above, have highly desirable leaf and plant forms. Genotypes don't root successfully from cuttings, but can be grafted onto a rootstock grown from seeds of less desirable types. The seedlings provide great rootstocks, and the shoot that has been grafted onto this rootstock expresses the highly desirable plant type. In this process horticulturists usually use T-budding, addressed later in the course.

Colorado blue spruce. Ali Eminov. CC BY-NC 2.0

Conifers: Colorado blue spruce cuttings don't root easily either, but blue spruce scions can be grafted onto Norway spruce rootstocks grown from seed to form a strong plant with the highly desirable blue spruce color and form.

Change cultivars

Most fruit trees, including apple trees, can survive for many years. In commercial apple orchards, some trees might be of cultivars that are no longer popular. Instead of pulling out these trees and planting young trees of a new cultivar, which difficult and expensive, new scions can be grafted from popular cultivars onto older trees after cutting those trees back to stumps. This is called cleft grafting or, more commonly, "topworking." By putting the new scion on an established rootstock, an orchard will come back into production much sooner than if new, young trees were planted.

Produce trees with specialized forms

Weeping flowering tree. Janine and Jim Eden. CC BY 2.0

In the case of some flowering trees, like cherry and dogwood, propagators graft a weeping-type scion (one that looks like an umbrella rather than growing upright) onto a 3′ to 5′ standard rootstock. The "weeping" form has branches that arch downward rather than upright. The graft union is far above the ground rather than down by the soil, as is typically the case with grafted maple, spruce, and fruit trees. These weeping-habit trees are more prone to poor wound healing and regrowth of the rootstock at the point of the graft union.

Repair damaged plants

The idea here is to create a graft that re-unites the cambium in the tree trunk after being severed by a chewing rodent or mechanical damage such as might be dished out by an indiscriminately piloted weed whipper. The graft is inserted into a point below the damage and runs to a point above the damage. This grafting operation is particularly important when the tree is girdled or nearly girdled (the cambium is damaged all the way around the trunk) because otherwise the tree will die. This technique is called bridge grafting and can be quite successful in saving a tree. The picture is from Maple Valley Orchards where they bridged grafted several trees to save them from dying. You can see how the cambium and phloem have both been disrupted.

Bridge graft. John Kring. Used with permission.

Take advantage of rootstock characteristics

All fruit trees grown in Minnesota are grafted, primarily in order to make the trees shorter. In apples, there are numerous rootstocks to choose from. If you are interested in growing Honeycrisp apples, for instance, you can grow a tree that will be anywhere from 6 feet tall at maturity to 25 feet tall at maturity, depending on which rootstock you choose. Same cultivar, "Honeycrisp," but a different root system.

In a home landscape, many of us can think of a place for a tree that might grow to 12 feet high, but not many of us have places to grow trees that will get to 25 feet. In commercial orchards, almost all the trees that have been planted in the last 30 years are on dwarfing rootstocks. These trees are far easier to harvest, prune, and maintain.

Some rootstocks offer improvements in disease and insect resistance, or environmental stress such as cold or drought tolerance.

Grafting challenges

There must be a good reason to graft, and no easier or cheaper way to produce the desired plant. For most plant material that is grafted commercially, such as apple or peach trees, there is no other way to preserve the desired scion cultivar.

Grafting is an art. People who have grafted a lot of plant material understand aspects of their plants that will increase success, such as choosing compatible genotypes and matching stem diameters. Experienced grafters also know the timing of growth stages to increase grafting success. For example, whip and tongue grafts are done when both the rootstock and scion are dormant. In contrast, t-budding is done in August when the bark is "slipping" and the buds themselves are newly formed.

So why don't grafts heal and produce new plants? Some of the reasons include:

- The type of plant is not conducive to grafting. Dicots and gymnosperms both have cambium encircling the stems, so aligning is possible. Monocot stems have cambium scattered around the stem, so aligning cambia between two plants is basically impossible.
- While the same species are often compatible and a successful graft results, as the relatedness decreases between the scion and rootstock, for example with different genera, grafting success is significantly reduced.
- Time of year and growing season both make a difference in graft success. Grafting success is more likely when the cambium is actively growing, but leaf growth is minimal, because water loss is at a minimum.
- The environment under which the grafts are healing, including temperature and humidity, is important. Light is less important.
- Technique is VERY important. If there is poor cambial contact, the callus bridge does not form, no vascular tissue is produced, and the grafted plant does not grow.

While most reasons to graft relate to desirable characteristics of the scion, there are situations in which the rootstock genotype or characteristics are also important. For example, apple and grape rootstocks are critical for correct plant form and disease resistance. Outside of those and perhaps a few other exceptions, the point of grafting is to propagate the scion.

Again, because of the expertise, time, and expense required for grafting, grafting is used only if it's not possible to propagate a plant by other easier and cheaper methods, like planting seeds or taking cuttings. If a plant roots easily, and doesn't require special rootstock characteristics like dwarfing or resistance to diseases and insects, there is no need for a graft.

Types of grafting and budding

Techniques for grafting have different names and are based on the type of plant material available and its stage of growth. Here are a few of the grafting or budding methods and when they are used.

Bridge grafting

Bridge grafting is used to repair damage to tree trunks. Smaller-size woody branches are used to reconnect the vascular tissues in the trunk so that water can flow up the trunk and sugars can flow down. These branches are the conduits.

Whip and tongue grafting

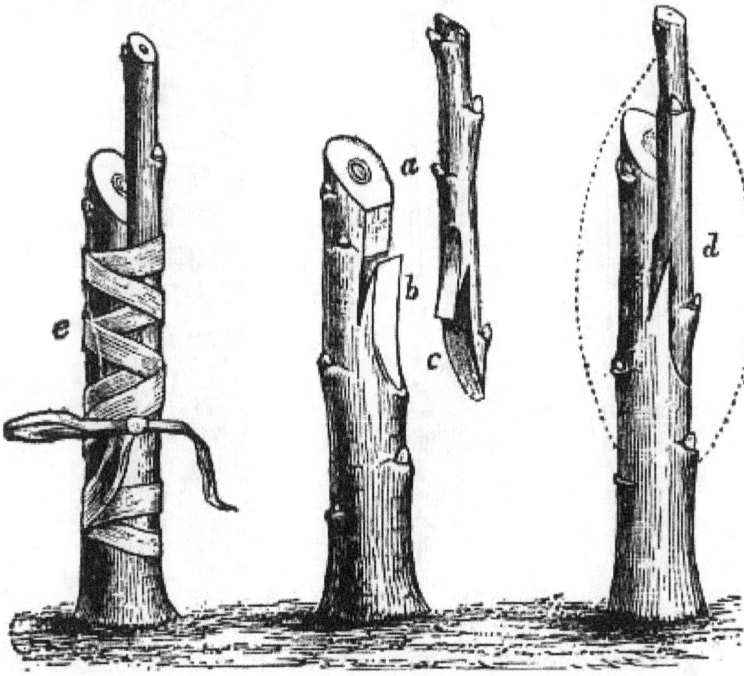

Whip-and-tongue grafting. Anonymous. Public domain.

The whip and tongue graft is used when the rootstock and scion are the same size. If you have two pieces of woody plant material that are the same size in diameter, it is fairly easy to line up the cambium of each piece and slip the two pieces together. The video linked above shows the steps of this type of grafting, including the cuts that need to be made, how to tie the two pieces together, and the finishing step of reducing water loss at the graft union by applying a wax. For additional pictures and an explanation of the entire method, Texas A&M has an excellent short article with more information.

Approach graft

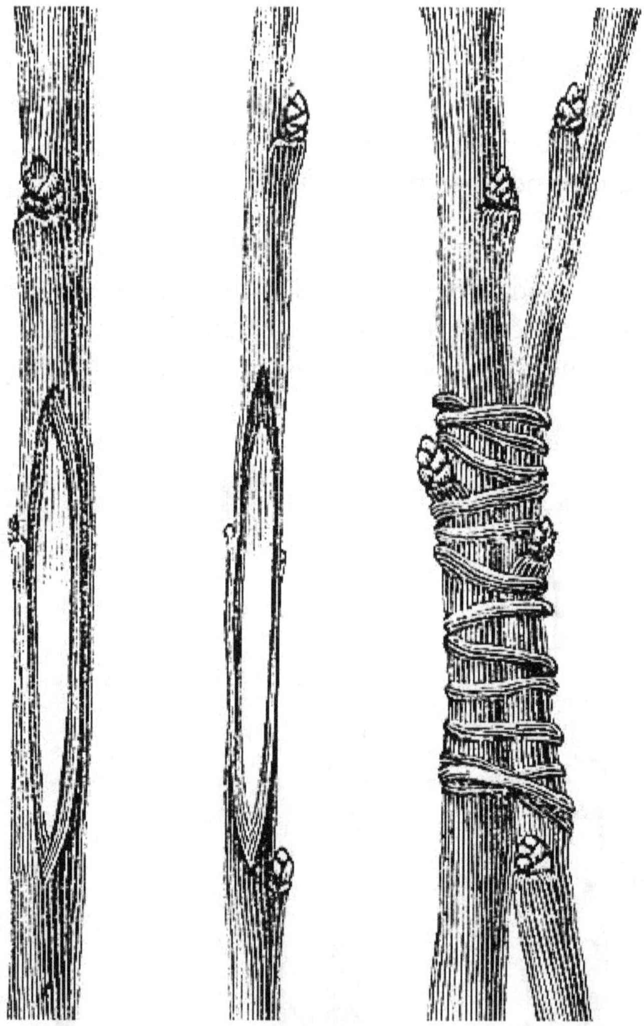

Approach graft. Gmihail. CC BY-SA 3.0 RS

In approach grafting, the scion is often smaller than the rootstock. As the name indicates, the two plant materials are brought close together, the cambiums of both are aligned, and they are allowed to grow together. Often, in plant propagation, we have done approach grafting by using potato as the rootstock and tomato as the scion (remember that these two plants are closely related). Using the same methodology described above, the tomato and potato plants are potted into the same pot, a thin slice is taken off each stem to reveal the cambium, and the stems are tied together. When the graft heals, the plant on the top will produce tomatoes, and the one on the bottom will produce potato tubers that can be harvested at the end of the season. For a bit more reading, and a few more pictures, see the Texas A&M article.

Cleft grafting

Cleft graft photo. London Permaculture. CC BY-NC-SA 2.0

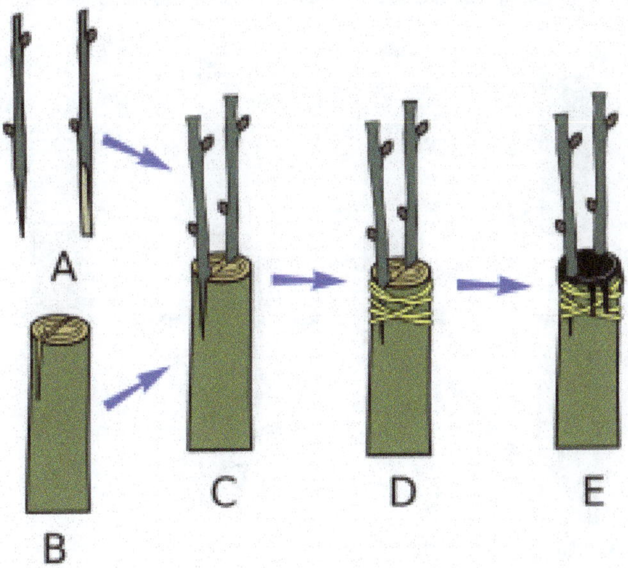

Cleft graft diagram. Giancarlodessi. CC BY-SA 3.0

Cleft grafting is a technique which allows the union of a rootstock that is much larger in size than the scion. It is conducted in late winter, when both the rootstock and the scion are dormant. Common applications for cleft grafting include changing the variety of an existing orchard (also called topworking), adding a branch of an untested scion cultivar to an existing tree for observation, or repairing a tree that may have had a branch broken off by storm damage or fruit overloading. The technique can also be used for producing one tree with multiple cultivars on it. Have you ever read about an apple tree with 10 different types of apples on it? This is one method that can be used to make that happen.

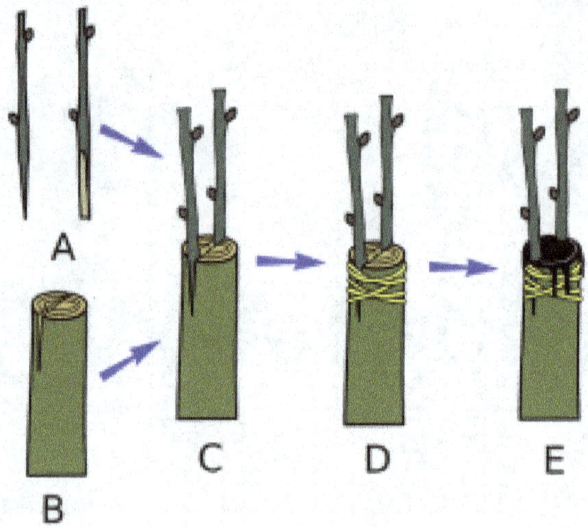

Cleft graft diagram. Giancarlodessi. CC BY-SA 3.0

T-budding

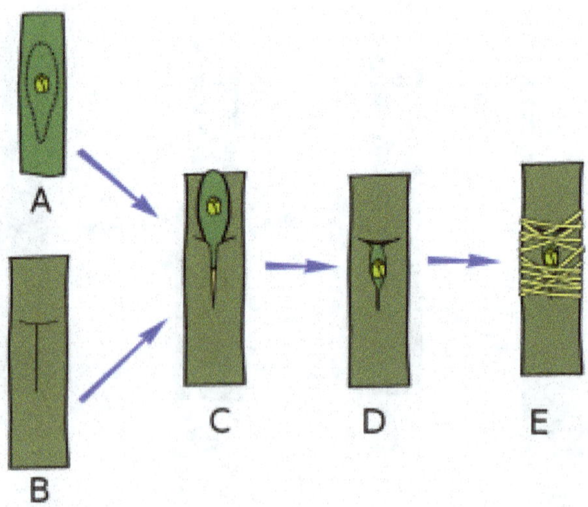

T-budding. Giancarlodessi. CC BY-SA 3.0

T-budding is done when bark is "slipping," meaning the plant is actively growing. The technique is used frequently because it does not require as much plant material of the scion. This is particularly true with new plants, such as a new apple genotype, from which you want to propagate as many trees as possible. T-budding requires only one bud to make one tree, while whip and tongue grafting uses more buds per tree.

Flowering tree with multiple species. Mojo0306. CC BY-SA 4.0

The grafting site at University of Missouri Extension is a great place to visit for more information.

Another interesting video highlights the work of a Syracuse art professor and his development of a single tree with 40 different types of plums, peaches, and apricots on it. What type of grafting did he practice?

Review questions

- How does grafting result in a chimera?
- What is the difference between grafting and budding? Why would you use one over the other?
- What types of plant habit can be produced through grafting onto special rootstock? Why are these habits desirable?

- Provide three good reasons to use grafting and three reasons not to use it.
- Provide two reasons why grafts are not successful.
- What type of grafting is recommended to repair trunk damage?

Wound healing

What happens when you graft?

First, recall what happens when you do a stem cutting:

- The cells along the cut surface are sliced open, die, and become necrotic (dead) tissue.
- The surviving cells one layer in from the cut (parenchyma cells in the cortex) respond to the wound. There are two responses:
    - The cells rapidly exude compounds like suberin (gummy substance) to protect the plant from excess water loss and invasion by diseases and insects.
    - The cells are stimulated to divide and produce a mass of new cells to cover and protect the wound. The mass of new cells is called callus. This response takes quite a bit more time (like years) compared to the time required to exude suberin (hours).

From the plant's point of view, the slicing that happens when making a graft is similar to the slicing that happens when you make an incision in the stem for a cutting. You are still slicing open cells, but because the rootstock and scion are in close contact, the environment around the cut ends is quite different from a cutting in which the wound is exposed to the environment. With a graft you have two stems that are held in close proximity and therefore are more protected than the open wound of a cutting.

- Within a few days, cell division starts and a callus of undifferentiated cells forms. The callus cells will continue to differentiate by developing cells with specialized form and function maturation of new cambium, xylem and phloem.
    - The callus originates from parenchyma cells in the cortex, pith, or vascular bundles. The origin depends on the species.
    - The callus grows from both the scion and rootstock.
- Callus from the rootstock and scion grow together to form a callus bridge.
- The parenchyma cells in the callus bridge that lie between the cambium of the rootstock and the scion differentiate into cambium cells.

- Additional parenchyma cells on either side of the cambium may differentiate into xylem and phloem until a connection among xylem, cambium, and phloem has been formed across the callus bridge.

- Once these links are formed, the cambium begins to initiate new xylem and phloem.
- Differentiation of the xylem and phloem is faster where cambium layers are closely aligned.

Again, it is important when grafting to get good cambium-to-cambium alignment between scion and rootstock. This will promote development of cambium cells that will differentiate into the callus bridge. The callus bridge is what joins the scion and rootstock. If your technique has poor cambial alignment, or gaps between scion and rootstock at the graft junction, the cell-to-cell linkage will not happen, or it will only happen weakly, and the graft will fail because there is no new xylem and phloem production, and therefore no water or sugar transport pathways. In addition to this being a linkage of plumbing so that water and sugars flow, it is also a structural linkage that gives the plant strength and rigidity across the graft union.

T-budding. Chrizz. CC BY-SA 3.0

Five requirements for making a successful graft

- The rootstock and scion must be compatible. Even if they are from the same species, some scions won't graft on to some rootstocks.
- The cambial layers of the rootstock and scion must be closely aligned and in contact.
- The graft must be done at the appropriate time of year. The buds, whether grafting whole stems or just the buds themselves, must be dormant. In temperate latitudes, grafting is typically done in winter when rootstock is also dormant, but budding is done late in the growing season when buds are formed, but dormant, and the bark is "slipping" so that a T-shaped cut in the bark can be opened into a pocket for the bud (image on the right).
- Grafts must be protected from drying.
- The grafted plant must receive proper post-graft care such as removing sprouts and suckers that emerge from the rootstock and might be mistaken for scion, and pruning or training the scion so that it develops the appropriate plant form.

10.2 UNIQUE STORAGE ORGANS

Learning objectives

- Characterize the differences between bulbs and other storage organs.
- Describe techniques for propagating plants with different clonal strategies from storage organs.
- Demonstrate other ways in which plants colonize through natural clonal propagation.

Geophytes

Plants called geophytes have evolved to store carbohydrates and nutrients underground in special structures which allow them to regenerate during the growing season.

Unlike trees and shrubs that keep their dormant buds above ground during the winter, some plants keep their dormant nodes underground. Other plants may grow and store their nodes under water in a pond or stream. Many orchid species live in trees and are called epiphytes. The specialized storage organs have several roles which allow them to perenniate and colonize.

These organs store carbohydrates that can be used for rapid growth when favorable conditions return. They protect the nodes from injury from herbivory and injury (frost and freeze), and are a natural form of cloning by producing new daughter plants away from the mother plant. The plant propagator uses techniques that take advantage of these natural processes and can therefore rapidly increase the number of plants.

Bulbs

The onion and lily family (Liliaceae; monocots) are typical examples of bulb-producing herbaceous perennials. The bulbs are subterranean, which prevents them from predation and weather conditions. Bulbs consist of a highly condensed stem (with nodes and internodes) and adventitious roots that form along the basal plate. Two main types of bulbs include imbricate and tunicate. Imbricate bulbs, typical of some lily species, consist of scales, as shown in the photo below. Scales are specialized leaves that form radially around the meristem. During the growing season, leaves and a stem emerge from the soil from the center of the plant, where the apical meristem is located. These above-ground leaves are able to photosynthesize. The central meristem converts to a floral meristem where flowers are produced. New lateral bulblets form from the meristem in axils underground, allowing the plant to colonize and perenniate. In some cases, the main bulb may not continue to grow as the meristem has converted to flowering. The new bulbelts can be removed to produce new daughter plants. The plant propagator can remove scales from the bulb and create a new daughter plant from each. This scaling technique is highly effective, though it may take several years for the new plantlet to flower again.

The common Easter lily (Lilium longiflorum) has an imbricate bulb structure. Gardening Solutions. CC BY-NC 2.0

Tunicate bulbs (below), or those with paper coverings, are exemplified by onions, tulips, and garlic. Specialized leaves make up the mass of the storage organ for the plant and form concentric rings around the meristem. Typically, when we cut an onion from the tip through the root end, we can see the layers of leaves (the parts that we eat) and then stem at the bottom. We know it is a stem because there are nodes and internodes. When cooking, we typically discard the stem, basal plate, and adventitious roots. To view the different structures of a bulb during the growing season, look in particular at green onions, scallions, and leeks. Newer leaves form in the interior of the bulb (the apical meristem is in the middle) and the papery skin is the oldest layer formed. From the middle of the bulb, leaves emerge to photosynthesize when the conditions are right, and the floral meristem emerges from the middle. Lateral tunicate bulblets may form inside the main bulb, or along the exterior at the basal plate, especially if there is an injury. Scooping the basal plate is one way to induce bulblet formation. The basal plate can also be scored to have a similar effect.

Lateral tunicate bulblets on Albuca bracteata. CloveMill. CC BY-SA 4.0

Garlic is an example of a tunicate bulb. Scouse Smurf. CC BY-ND 2.0

To propagate a tunicate bulb, it can be sectioned in many pieces by cutting from top to bottom through the basal plate. The sections are allowed to dry for a few days, treated with a fungicide, and then planted in a cold frame or other suitable system to induce bulb formation and plant regeneration. It may take several seasons for the bulbs to develop a suitable size for flowering. When doing any sectioning, scaling, scoring, or scooping the tools should be cleaned between cuts, or at the very least between bulbs, to prevent the spread of disease. Alcohol is suitable for killing microbes and for cleaning the cutting surface and fingers.

Bulbils on Allium vineale v. vineale. Dr Mary Gillham Archive Project.

Liliaceae is unique in that bulb-like organs may also form above ground. In this family, bulbils, shown in the photo above, are formed in the leaf axils above ground or in some flowers. Bulbils are clonal (asexual) propagation and not formed through pollination. When the lily's flower stalk dies, it falls to the ground and the bulbils are scattered away from the mother plant. This colonization strategy is highly effective and can be very weedy for the gardener. Mature or "ripe" bulbils can be removed from the plant and potted as soon as they easily detach. Alternatively they can "self sow" and be moved the following spring.

Some bulbs develop contractile roots that pull the bulbs into the soil. As bulbs grow each year, they would logically move closer to the surface, putting them at risk. The contractile roots are able to get the bulb to the proper depth after a growing season.

Corms

Giant elephant ear corm. Matthew Clark

Like bulbs, corms are underground storage structures that have evolved in some plants. Corms are a storage unit made of compressed stems, unlike the leaves that provide the storage function in bulbs. Because they are stems, corms have nodes, internodes, and meristems just like above-ground stems. The roots of corms are adventitious and develop from a basal plate and nodes. Corms may be short-lived in herbaceous perennials like Gladiolus, where a new corm is replacing itself regularly; in tropical climates, taro (Alocasia) may grow for several years with the same corm structure. The giant elephant ear plant above shows a corm structure. Corms reproduce asexually through cormels, akin to the bulblets produced by bulbs. Watch this short video (0:47) to see a closeup of this corm.

Corm with 2 cormels formed. Matthew Clark

Corms can be induced to produce cormels by removing the apical bud, which removes dominance and stimulates new growth at the nodes. This is generally not necessary, as many corm-producing plants naturally increase with new corms annually. In the photo above, the corm on the left shows the aggressive nature of colonization. The oldest corm at the bottom is being replaced by the actively growing corm attached above it. We can also see two cormels that have formed (red arrows). Eventually the lower, older corm will dessicate and decompose. This is a great example of why some geophytes have contractile roots, for in one growing season the newest cormel could grow above the soil surface.

Rhizomes

Rhizomes are also underground stems. The Canna lily (below) produces large rhizomes each year. Notice how it looks a lot like ginger "root;" the two plants are in the same family. As stems, they have nodes that produce shoots and they have adventitious roots. Rhizomes typically grow horizontally at or near the soil surface. Some grasses use rhizomes to help with rapid colonization from the crown. Rhizomes can be divided into many pieces as a way to produce multiple clonal propagules.

Canna lily rhizome. Matthew Clark

Tubers

Potato (Solanum tuberosum L.) is the typical example of a common plant that uses a tuber for storage and propagation. Tubers are another example of an underground stem, with nodes and internodes. The nodes may not be obvious when the potato is fresh from the grocery store, but in storage the potato may grow "eyes" which are new shoots that are growing from the once dormant nodes. Most people have seen this in their pantry or on the kitchen counter. To plant potatoes, the farmer cuts up pieces of the last year's potatoes so there are enough carbohydrates and nodes for a new plant to emerge from the soil.

Roots

We've already learned that roots play an important role in storing nutrients for plants. Biennial plants like carrots, parsnip, and beets have fleshy roots which help them overwinter. These are not easy to propagate into new plants and typically are grown from seed. Horseradish and some poppy plants, however, can easily be propagated from root cuttings. A common example of a tuberous root is the sweet potato (looks like a tuber, functions like a root). The common garden radish (Raphanus sativus) may seem like a root, but close investigation reveals that it is mostly the swollen hypocotyl (image below).

Radish hypocotyls in 4 stages of growth. Matthew Clark

Other methods of colonization

Crowns

Hosta plants are common examples of herbaceous perennials that are excellent at colonizing through an underground crown. The underground stem grows radially each year. New shoots can be removed, as shown below. Notice the adventitious roots. Many perennial grasses, asparagus, and other herbaceous perennial plants can be divided into multiple new plants, taking advantage of this colonization strategy. Offshoots (sometimes called pups) can be removed from the mother plant and replanted elsewhere.

Agave plants – which are used to produce tequila – are clonally reproduced this way, as seed production results in new genetic recombinations that are different from the mother plants. Clonal propagation ensures reproducibility for uniform forms and consistent tequila production. Gardeners have been using this method to propagate plants for millenia.

Hosta crowns. Matthew Clark

Stolons

Perennial grasses and strawberry (Fragaria spp., photo below) are excellent examples of plants that produce runners or stolons that extend the reach of the mother plant for colonization. These above-ground stems (there are underground examples as well) are not for storage, as in the previous examples, but are adaptations to reduce local competition with the mother plant while also spreading her genetic material. A strawberry stolon may root adventitiously at a node and produce a new plant from the bud (remember, buds are located at nodes and form into new shoots). The stolon likely will continue to grow and do the same at each node thereafter. A gardener can peg the stolon into the soil to help the plant root, then remove it later and replant it.

This 'Pink Panda' ornamental strawberry has produced many stolons. Frank Vincentz. CC BY-SA 3.0

Stem layering

Plants with stems that creep along the soil or whose shoot tips bend and touch the soil take advantage of being able to form adventitious roots for colonization. Some raspberry species (Rubus occidentalis), for example, grow long canes after fruiting, which extend beyond the mother plant and push their shoot tips into the soil. In the photo below, the plant grows roots and reorganizes the developing shoot (see how the leaves face upward). Now anchored into place, a new crown can form, often several feet from the mother plant. Vining and climbing plants often can root at each node, which makes them able to grow along the forest floor until they find a structure to climb. This adaptation can allow clones of the same plant to "move" over time to conditions that are more favorable. This may be more evident in tropical regions, where plants grow faster and local conditions can change suddenly. Vining plants, like grape (Vitis), are often easy to root from hardwood cuttings because they have evolved to root quickly at a node.

Raspberry stem with roots from tip-layering. Matthew Clark

TERMS

Approach graft	A type of grafting where two independent plants are grafted together and severed only once the graft has "taken."
Bridge graft	A type of repair graft used when a plant has been girdled; scion pieces are inserted above and below the girdled site and act to repair the disruption of the cambium.
Budding	A form of grafting where a single scion is used rather than an entire stem.
Callus	Growing mass of unorganized parenchyma cells produced in response to wounding.
Callus bridge	Parenchyma cells that lie between the cambium of the rootstock and the scion and differentiate into cambium cells.

Chimera	When two different genotypes are growing on a single plant.
Cleft grafting	A form of grafting where the rootstock is much larger than the scion; both are dormant.
Corm	A condensed stem and storage organ; typically growing underground and covered in scale leaves.
Differentiation	Process by which cells or tissues undergo a change toward a more specialized form or function.
Genotype	Genetic composition of an organism.
Geophyte	New growth begins underground and the function of the underground growth is storage of food, nutrients, and water during adverse environmental conditions.
Graft union	Location where the rootstock and scion meet.
Grafting	Art and science of connecting two pieces of living plant tissue together in such a manner that they will unite and subsequently grow and develop into one composite plant.
Imbricate bulb	Underground storage organ formed primarily of modified leaves (scales) without a papery covering. Individual scales do not encircle the entire bulb.
Pruning	Cutting away dead, overgrown, or unwanted branches or stems to improve safety, aesthetics, or productivity.
Rhizome	Stem that grows horizontally underground and is a swollen storage organ for the plant.
Rootstock	Portion of a graft that contains the root system.
Scion	Portion of a graft that contains the shoot system and all above-ground parts.
Stolon	Creeping horizontal stem, sometimes called a runner, that roots and forms plantlets at nodes that extend away from the mother plant.
Suberin	Impermeable (to water and gases), waxy substance present in the cell walls of corky tissues.
T-budding	A type of budding performed using dormant scion buds on actively growing rootstocks; typically done outdoors in late summer.
Topworking	A type of grafting performed on established orchard trees.
Tuber	A thickened underground stem used as a storage organ for many plants to allow for perennation.
Tunicate bulb	An underground storage organ formed primarily of modified leaves formed in concentric circles around the active meristem. The bulb is covered with a papery covering.
Whip and tongue graft	A type of graft where both scion and rootstock are dormant and the same diameter; much more secure than other types of bench grafts.

CHAPTER-11

WATER AND LIGHT

Plants are sessile and cannot move to locations that might be more suitable for growth and reproduction when environmental conditions become less than favorable. To maintain growth, plants monitor signals of light, temperature, humidity, wind, and soil water availability; these signals inform the plant on how to modify water movement and gas exchange, which also affect photosynthesis. These lessons review how these signals impact physiology and the role of photosynthesis in growth.

Learning objectives

- Understand the meaning of photoautotroph and where in the plant the various photosynthetic reactions take place.
- Explain how the energy from light is converted into carbon-based chemical energy and building blocks in plants
- Map the movement of water from the roots to leaves and carbon-based building blocks from source to sink.

11.1 PLANTS AND WATER

Learning objectives

- Summarize the mechanics of evapotranspiration.
- Describe how leaves adapt to lack of soil moisture.

Evapotranspiration

Most of the water molecules taken up by a plant's roots move up the stem into the leaves, out the stomata in the leaves, and then evaporate into the atmosphere. The stomata open to allow oxygen (as a waste product of photosynthesis) to escape the leaf, and carbon dioxide (donor of the carbon atoms that are the building blocks of the sugar molecules assembled during photosynthesis) to enter. When these stomata are open, water vapor exits. We often refer to stomata as associated with gas exchange in the leaves because of the movement of these three gasses: oxygen (out), carbon dioxide (in), and water vapor (out).

Evapotranspiration (often just called transpiration) refers to the movement of water in the plant from root to stem to leaf and out through the stomata to the atmosphere. This isn't just a dribble of water. An acre of corn will transpire about 3,000–4,000 gallons of water each day, and a large oak tree can transpire 40,000 gallons each year.

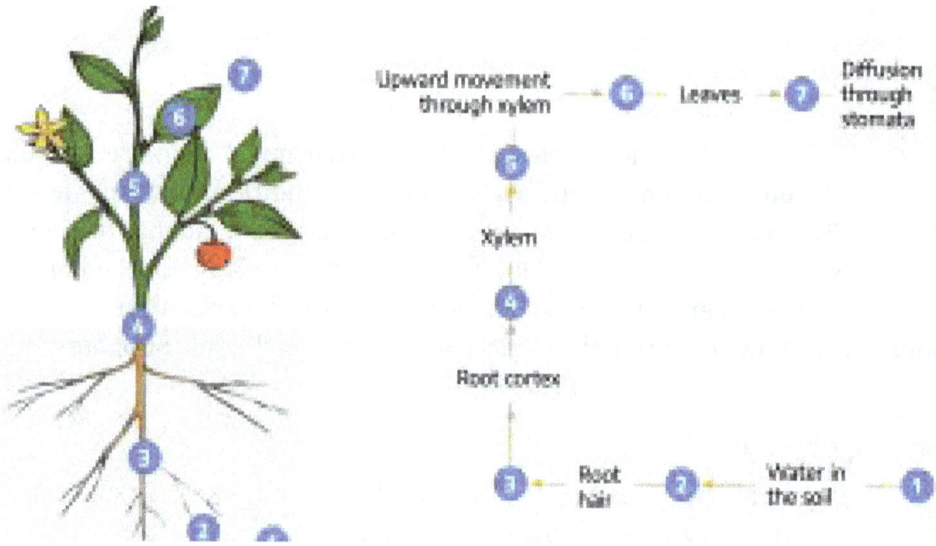

Diagram of evapotranspiration. Sheng-Yang He. CC BY-SA 4.0

As illustrated above, a stream of water is constantly moving up from the roots and out of the plant. Note the tissues and cells that are involved, and recall that water moves from the soil through the epidermis and cortex toward the xylem in the vascular bundle in one of two ways, symplastically or apoplastically. Symplastic means that water and minerals move interior to the cell membrane, or through cells, while apoplastic water moves around the cell membrane in the space outside the cell. Symplastic movement starts with water entering the epidermis cells through root hairs and then continuing from cell to cell through the cortex to the xylem in the vascular bundle. Entry of symplastic water into the root is regulated by the cell membrane of the root hair. Apoplastic movement of water occurs between the cells. This movement is unregulated until the water hits the cutin barrier formed by the Casparian Strip around the innermost layer of cortex cells in roots The Casparian strip blocks apoplastic water movement. The apoplastic water must then move symplastically into the cortex cells through the cell membrane, which controls the entry of water and minerals.

From here, the water moves from cell to cell to the xylem, and then is pulled up the plant as described below.

The rate of evapotranspiration depends on environmental factors such as:

Light — Due to the occurrence of photosynthesis, plants transpire more rapidly in the light than in the dark. The guard cells, part of the stomata, are stimulated to swell, opening the stomata in the light of the day.

Temperature — As temperatures rise, water evaporates out of the leaves more readily. On hot summer days, leaves thus have a tendency to wilt due to lack of water in the soil and to the increased rate of transpiration.

Humidity — When the air around the leaf is drier, there is greater movement of water vapor out of the leaf than if the air around the leaf is saturated with water.

Wind — A breeze will clear water vapor away from the surface of the leaf, leaving the humidity on the leaf surface low and increasing the rate of transpiration.

Soil water availability — The water that is transpired must come from somewhere, and that somewhere is the soil. When the roots can't absorb enough water to keep up with the evapotranspiration demand, the leaves lose more water than they can replace. Water pressure inside the cells, called turgor pressure, is reduced because some water is pulled out of the cells to satisfy the demand from evaporation. This loss of turgor pressure relaxes the guard cells, causing the closure of the stomata, which shuts off a major avenue for gas exchange and the main channel for evaporation. This is a key strategy used by plants for managing stress from insufficient water. If the loss of turgor is severe, the plants will temporarily wilt. When the evapotranspiration demand is reduced through a change in environmental conditions, or when water supply increases, the cells again fill with water, turgor is reestablished, the stomata reopen, and the plant leaves recover from their temporary wilting. You have likely seen this happen when you have forgotten to water a house plant. So long as you water it soon enough, the plant regains turgor and survives the neglect.

Review questions

- What is turgor pressure and how do leaves compensate when cells begin to lose turgor?
- How can wind result in low turgor pressure?
- What are three gasses that move through the leaf stomata? What is their involvement in plant function?

Mechanisms of water movement in plants

How does water move from the soil to root to stem to leaf and out to the atmosphere? This is a more complex question than it may first appear. Unlike animals, plants do not have a heart to pump water from roots to leaves. There is a push explanation and a pull explanation.

Guttation. Noah Elhardt. Public domain

Push explanation

Water pressure (turgor) in the root cells during the night or during cloudy days can push water and dissolved materials up into the stem. This root pressure is the cause of guttation, the dew-like drops of water that are forced out of leaves. This same pressure is the force driving sap up the trunk of sugar maples in the spring. One problem with this mechanism is that, at most, root pressure can move water upwards only about 60 feet, and this only happens at night and when it is cloudy, and it only happens in some plants, but not in them all. So the push explanation has many limitations that make it unsatisfactory as a general theory for water movement up the xylem. How does water get to the top of a plant when it is sunny? And how does it make its way to the top of tall plants?

Pull explanation

The cohesion – adhesion – tension theory

Water is a polar molecule — like a magnet, it has positive (+) and negative (-) regions. When water molecules are near one other, the negative region of one molecule is attracted to the positive region of another. This attraction is called a hydrogen bond. This type of bond is weak compared to covalent bonds, where molecules share electrons, but when there are lots of hydrogen bonds holding these water molecules together, this type of chemical bonding is quite tenacious.

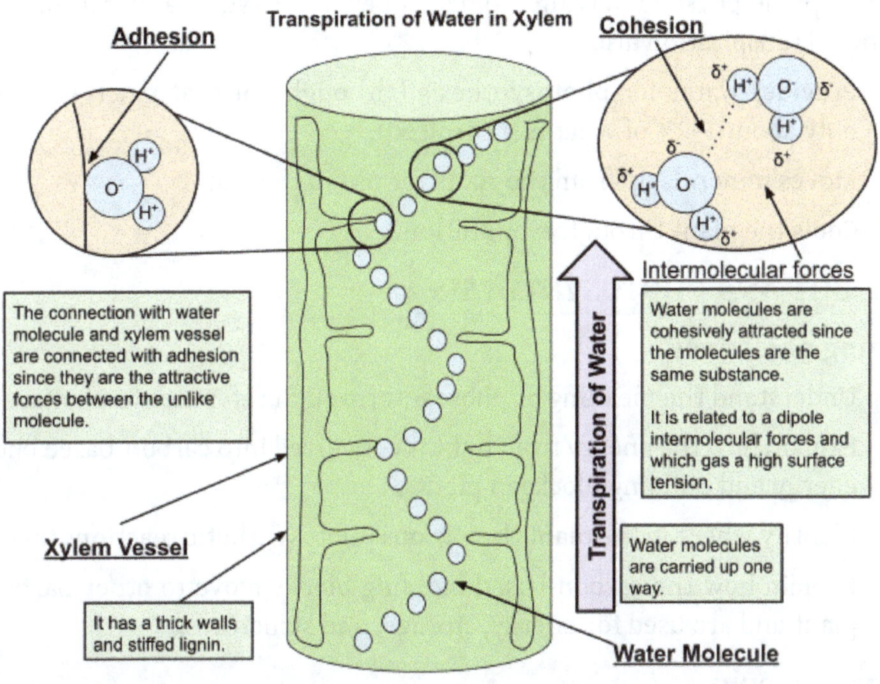

Cohesion/adhesion. FeltyRacketeer6. CC BY-SA 4.0

When water is held in a very small tube, such as an xylem vessel (above), the cohesion among water molecules due to the hydrogen bonds is very strong – strong enough to hold the column of water together very tightly over long distances, like from the root through the stem and into the leaf. Although an individual bond is weak, there are so many that a column has enormous tensile strength.

Water is also attracted to the walls of small tubes like xylem vessels. This force of adhesion between the water and walls of the xylem helps hold the water in the xylem against the downward force of gravity.

As a water molecule moves out of the leaf xylem into the air spaces among spongy mesophyll cells, out the stomate, and into the atmosphere through evaporation, it creates a void or empty space in the xylem, which is filled by the next water molecule in line. As this water molecule moves forward, it exerts tension (pulls) on the cohesive column of water that extends all the way back down to the root. As one water molecule leaves, the next takes its spot, and as it moves forward in line, it pulls upward the molecules behind it.

This force of cohesion-adhesion-tension is sufficient to pull water up to the top of the tallest tree, and is very effective while the sun is shining, when the stomata are open and transpiration is active.

The enormous flow of water through the plant isn't simply waste and the price the plant pays for having stomata open for oxygen and carbon dioxide exchange. Transpiration also:

- Provides water for photosynthesis (although not that much is needed — only about 1–2% of what is transpired).
- Moves minerals up from the roots for use in the leaf.
- Cools the plant through evaporation.

11.2 LIGHT AND PHOTOSYNTHESIS

Learning objectives

- Understand the meaning of photoautotroph in reference to plants.
- Explain how the energy from light is converted into carbon-based chemical energy and building blocks in plants.
- Identify where in the plant the various photosynthetic reactions take place.
- Explain how the carbon-based building blocks move to other parts of the plant and are used for energy, storage, and structures.

Photoautotrophs

Plants are autotrophs, meaning that they are self-nourishing (Greek autos = self and trophe = nutrition). Specifically, plants are photoautotrophs, because they use the energy from light to produce organic molecules with which they build their cells and store energy.

Organic molecules are compounds associated with living organisms that contain carbon atoms. It was once thought that organic molecules could only be synthesized in nature by living organisms through the intervention of a "life force." This hypothesis was disproved in 1828 when urea, a simple organic compound, was synthesized in a laboratory. Since that time, a major branch of chemistry, organic chemistry, has arisen to study and synthesize organic molecules. In contrast to organic compounds, inorganic compounds were historically defined as those lifeless minerals that are dug up from the ground.

Note that this chemical definition of organic (containing carbon atoms) has little or no relationship to the contemporary use of the word to describe a method of producing food. Organic food production, by regulation, relies strictly on inputs of organic molecules that come from life (like manure) and also on inorganic compounds like minerals, and eschews the use of organic molecules that have been synthesized by humans.

The organic molecules that a plant produces must be:

- Storable within the plant.
- Metabolized by the plant to yield energy for use in growth, maintenance, and producing other required organic molecules.
- Reasonably compact so that enough energy can be stored for growth.
- Transportable within the plant.
- Stable and non-toxic to the plant.

Since plants are photoautotrophs, they must have a mechanism for capturing energy from the sun or other sources of light and using that energy to produce organic molecules with the characteristics noted above. Photosynthesis is the process on which photoautotrophs rely to capture that light energy and to produce carbon-based organic molecules. The carbon used to make these molecules comes from the carbon dioxide (CO_2) in the atmosphere. Because photosynthesis removes carbon from the atmosphere and incorporates it into organic molecules which eventually become the plant's leaves, stems, roots, and fruits, photosynthesis is sometimes said to fix carbon. Fix, in this sense, means to secure or sequester rather than to repair.

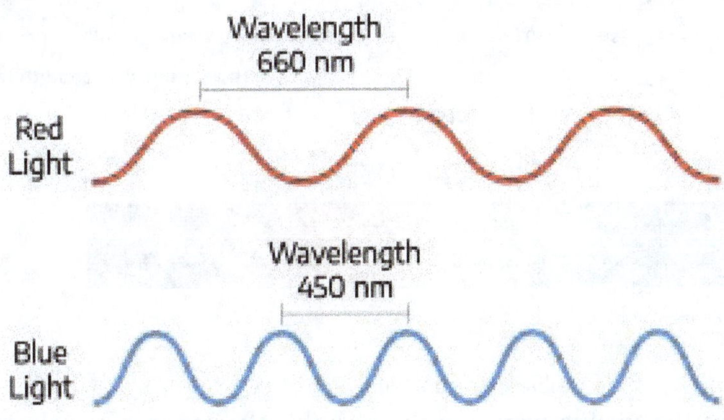

Red light/blue light wavelengths. Emily Tepe

If you follow the public discourse on climate change, you are aware that global warming is accelerated by the accumulation of greenhouse gasses in the atmosphere which trap and re-radiate sunlight and heat back to the earth. CO2 is one of these greenhouse gasses. Removal of CO2 from the atmosphere, for instance by planting trees that photosynthesize, fix carbon, and store the carbon-rich product as wood, is one method of carbon sequestration. An emerging and increasingly popular strategy for remediating greenhouse gas emissions is through the buying and selling of carbon credits, where industries that discharge CO2 into the atmosphere purchase credits from organizations whose activities (such as tree planting) sequester carbon. Photosynthesis and sequestration of carbon by trees is one tool used to offset the industrial release of CO2.

Review questions

- In what sense does photosynthesis fix carbon?
- Where does the carbon come from that is used by photosynthesis, and where does it go within the plant?

Light reaction

Let's start with light, because that's where the plant gets the energy for photosynthesis. Here are some characteristics of light:

- Light travels in waves.
- The length of the wave is measured from one peak to the next and is called the wavelength, which differs for different colors of light.

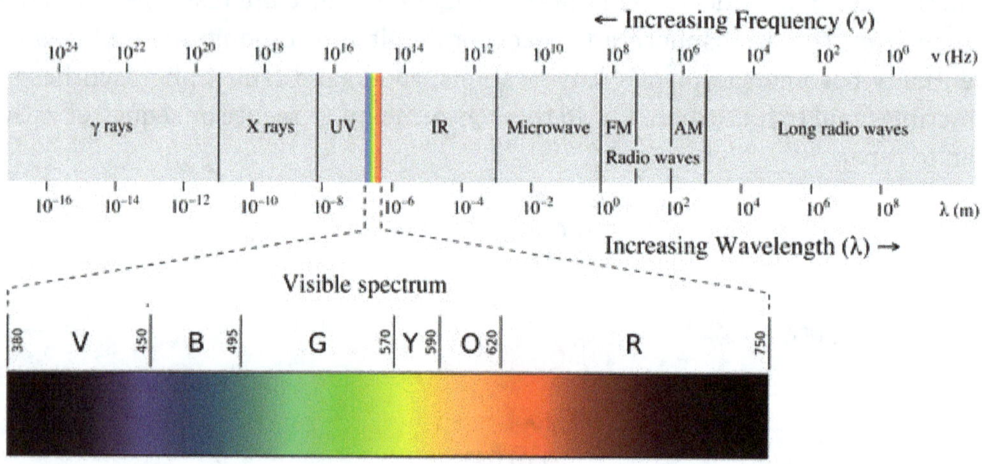

Visible spectrum. Philip Ronan. CC BY-SA 3.0

- Within the visible wavelengths of light, the longest wavelengths are red light; outside the visible range of wavelengths, even longer wavelengths include infrared radiation, microwaves, and radio waves.
- Shorter visible wavelengths include blue and purple light, and beyond the visible range even shorter wavelengths include UV light, X-rays, and Gamma rays
- Light also has a particulate nature, and those particles are called photons.

The photons in light provide the energy that drives photosynthesis. This energy is used to incorporate carbon found in CO2 from the atmosphere into organic molecules and, in particular, into simple sugars used by the plant. The chemical formula is the same for the two types of simple sugars produced by photosynthesis: glucose and fructose: $C_6H_{12}O_6$. The equation that summarizes photosynthesis is:

water + carbon dioxide -> oxygen, water, and simple sugars

$$12H_20 + 6CO_2 \rightarrow 6O_2 + 6H_2O + C_6H_{12}O_6$$

This balanced equation tells us that 12 molecules of water plus 6 molecules of carbon dioxide, in the presence of chlorophyll, accessory pigments, and light, produces 6 molecules of oxygen gas, returns 6 molecules of water back to the cell, and produces one molecule of a simple sugar like glucose or fructose.

Two reactions make up photosynthesis: the Light Reaction (abbreviated LR) and the Light Independent Reaction (abbreviated LIR). As the names suggest, the LR requires light while the LIR does not. The LR uses light energy to split water, which transforms the energy from the sun into hydrogen ions and electrons. The LIR uses that energy to grab the carbon from carbon dioxide and use the carbon to build simple sugars.

Let's start with the light reaction. You've heard of chlorophyll, and may recognize this molecule as a green pigment that captures light for photosynthesis. There are two chlorophyll pigments in plants that are critical for absorbing light: Chlorophyll a and Chlorophyll b.

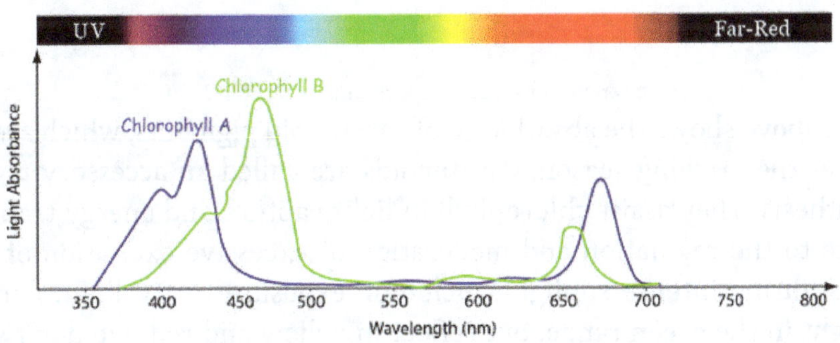

Chlorophyll A & B Absorption Spectrum. byr7. CC BY 2.0

The graph above shows % absorbance of different wavelengths by these two chlorophylls. The Y axis (the vertical one) shows the percentage of the light that is absorbed (rather than reflected). High levels of absorption mean that the chlorophyll molecule uses that wavelength of light for energy. Low absorption means that the molecule does not use that wavelength, and is thus reflected away. The X axis indicates the wavelength of light in nanometers (nm) (the wavelength of green light, roughly 500 nm). The bar at the top represents the color of the light at the wavelength shown. The blue line is a typical absorption curve for chlorophyll a, while the green line shows chlorophyll b.

High absorbance at a particular wavelength means that pigment is collecting that light at that wavelength to harvest energy. Low absorbance means that the plant is reflecting that light back. Both chlorophyll a and b absorb blue and red light wavelengths and reflect green. Chlorophyll a has a peak in the violet and red regions and chlorophyll b in the blue and orange regions. Notice how their absorbance is very low in the green region. That's why we think of chlorophyll as green, and why we perceive leaves, which have chlorophyll as the predominant pigment, as green. Also notice that chlorophyll reflects some yellow wavelengths, but when the yellow and deep green wavelengths are mixed, we see the green leaf color.

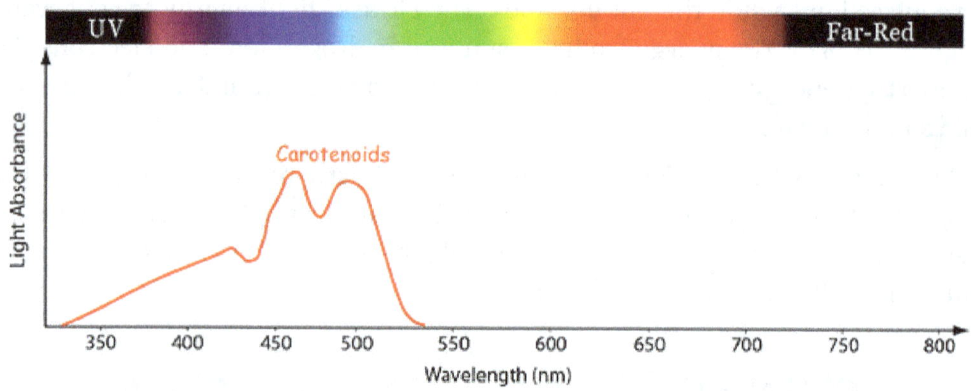

Carotenoids absorption spectrum. byr7. CC BY 2.0

The graph above shows the absorbance of carotenoid pigments, which are present throughout the growing season. Carotenoids are called an accessory pigment in photosynthesis. They assist chlorophyll in light capture and energy transfer, and contribute to the regulation and moderation of excessive excitation of pigment molecules during intense sunlight, including exposure to UV light. Carotenoids absorb light in the green range, but reflect in yellow and red. We don't see these pigments during the growing season because they are much lower in concentration than the chlorophylls, so the green reflected light overwhelms the orange, and we see green. But when the chlorophyll fades in the fall, due to decomposition of chlorophyll, the orange can be seen in beautiful fall leaf colors.

Chlorophyll *a* and *b*, as well as the accessory pigments, are found in the chloroplasts, which are membrane-bound organelles within cells. The highest concentration of chloroplasts is most commonly found in the palisade mesophyll cells of the leaf.

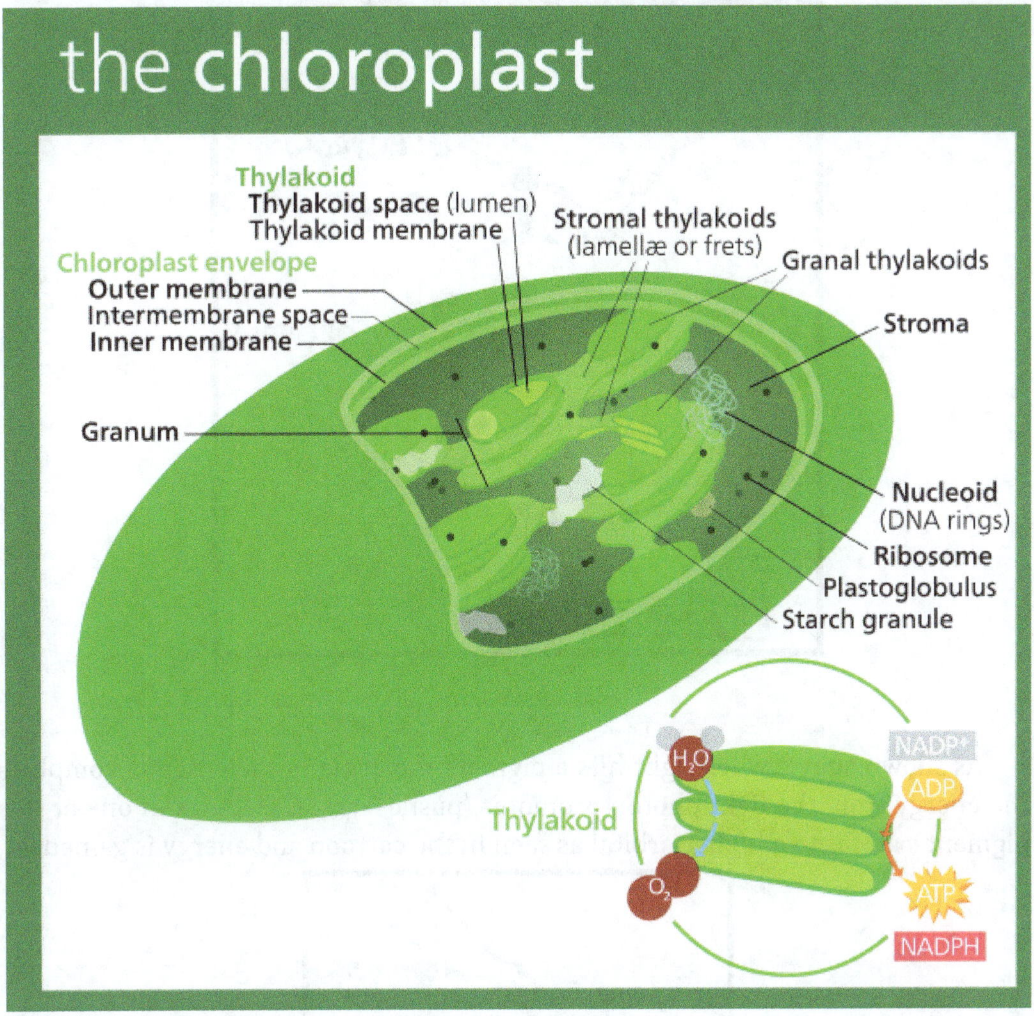

Chloroplast. Kelvinsong. CC BY 3.0

The above illustration of a chloroplast labels the internal structures. The chloroplast has a double membrane. The interior of the chloroplast is called the stroma. Within the stroma are coin-like thylakoids. The stacks of thylakoids are called grana. The thylakoids are also surrounded by a membrane, called the thylakoid membrane. The green chlorophyll pigment that you associate with photosynthesis, as well as the accessory pigments, are embedded in the thylakoid membranes and arranged in a structure called the antenna complex — given this name because it captures and routes the energy from sunlight to a collector called a reaction center.

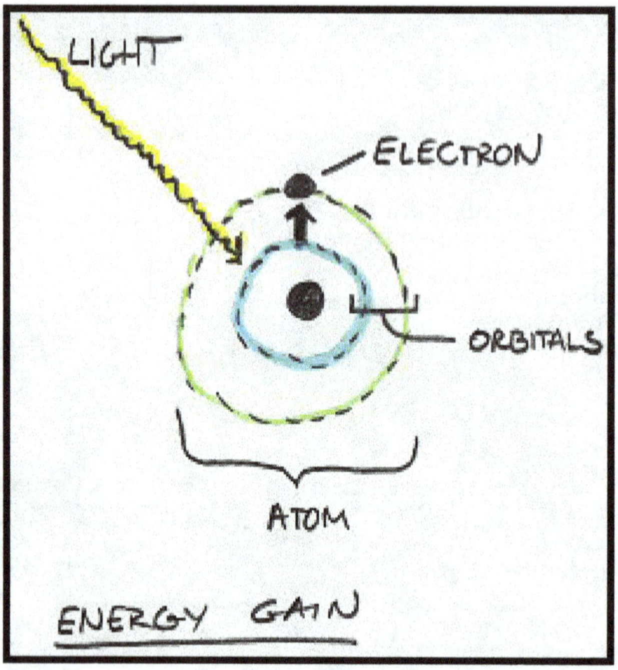

Energy gain. Tom Michaels

As shown above, when light hits a pigment molecule in the antenna complex, the energy from the light photon promotes (pushes up) an electron in one of the pigment's atoms to a higher orbital as seen in the cartoon and energy is gained.

Energy loss. Tom Michaels

The electron can't stay in that higher orbital indefinitely, and when it drops back to its home orbital it releases the energy it absorbed from light, denoted as energy loss. This released energy can be passed to another pigment molecule.

This process of one pigment capturing the photon's energy and passing that energy onto adjacent pigment molecules is the crucial step in energy transformation that takes place in photosynthesis. This is the step that takes light energy and converts it into chemical energy — one of the only known biological processes that allows this type of energy transformation.

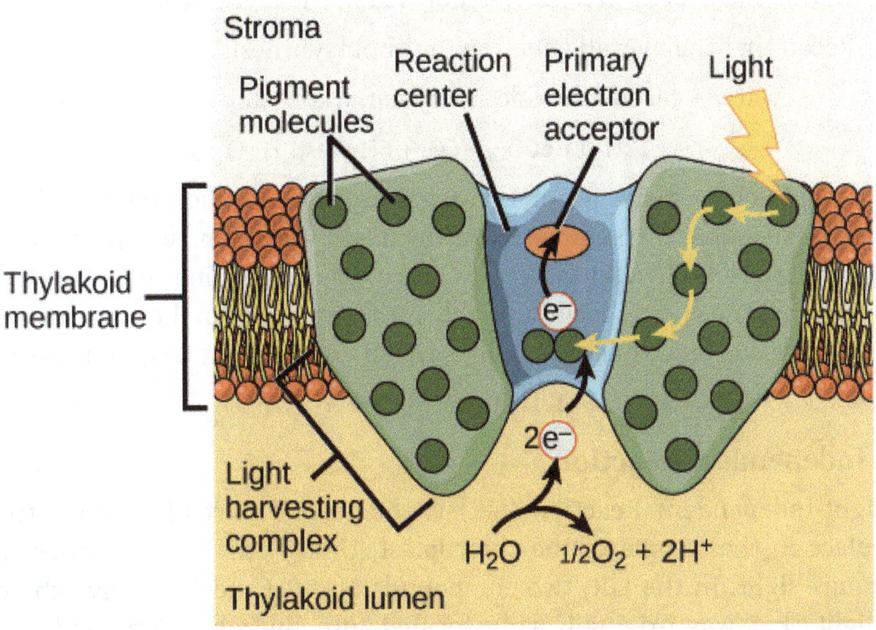

Light harvesting complex. OpenStax. CC BY 4.0

A light photon excites an electron of one pigment molecule in the antenna complex, or light harvesting complex, and by resonance this energy is transferred from pigment molecule to pigment molecule The energy transfer makes its way to the reaction center, where the first major chemical reaction in photosynthesis — splitting water — takes place. This reaction is called the light reaction or light-dependent reaction because it requires light. Water is split when the reaction center grabs electrons from water, which separates water into oxygen gas (O_2), hydrogen ions (H+), and electrons (e–).

To reiterate, the light is captured by the light harvesting complex (antenna complex) where electrons in the chlorophyll atoms are excited and jump up to a higher orbital. When the electron drops back, the energy is transferred to an adjacent pigment atom. This resonance energy travels down the antenna complex to the reaction center, where the captured energy pulls electrons out of water molecules, and water is split into oxygen gas, hydrogen ions, and electrons. The energy that was present in the photons of light has been transferred to the hydrogen ions and the electrons. We'll see more of how that energy is used in the next section.

Review questions

- What wavelength(s) of light does chlorophyll a absorb? Chlorophyll b? What wavelengths do these two molecules reflect?
- What pigments make up the antenna complex?
- How is the energy in light transformed in the Light Reaction?
- Recall that the overall equation for photosynthesis is:

water + carbon dioxide -> oxygen, water, and simple sugars

$$12H_20 + 6CO_2 \text{ -> } 60_2 + 6H_2O + C_6H_{12}O_6$$

This equation is made up of two parts called half-reactions. The first half-reaction is an equation summarizing the Light Reaction, where energy from sunlight is used to split water molecules into oxygen gas, some electrons, and some hydrogen ions. The energy from sunlight is transferred from the pigments to these hydrogen ions and electrons. The half-reaction for the Light Reaction is as follows:

$$12H_2O \text{ -> } 60_2 + 24e^- + 24H^+$$

Light independent reaction

The Light-Independent Reaction (LIR) is the second part of photosynthesis. It takes place in the stroma of the chloroplast. Unlike the Light Reaction, it does not require light. In the LIR, two compounds, NADPH and ATP, carry the energy from light that was originally transformed into hydrogen ions and electrons through the splitting of water. The NADPH and ATP, along with carbon dioxide from the atmosphere, enter a process called the Calvin Cycle, where the energy is used to fix carbon into a molecule abbreviated G3P. This process requires the help of an important protein abbreviated RuBisCO (Ribulose-1,5-bisphosphate carboxylase/oxygenase) that catalyzes the step in the process where the carbon from atmospheric CO2 is incorporated into an organic molecule. RuBisCO is the most abundant protein in leaves and, given the number of leaves in the world, likely the most abundant protein on the planet. The G3P produced by the carbon fixation process is called a triose phosphate, meaning it is a 3-carbon sugar (triose) with phosphorus and oxygen atoms (phosphate) attached. Triose phosphate moves out of the chloroplast into the mesophyll cell's cytoplasm, where two of these three-carbon molecules are combined to produce the 6-carbon molecules glucose and fructose. The glucose and fructose molecules then combine to form sucrose, a 12-carbon organic molecule. Sucrose is important because it is the sugar that is transported by the phloem throughout the plant to provide energy and building blocks for other organic molecules like starch and cellulose.

The half-reaction for the LIR is:

$$24H^+ + 24e^- + 6CO_2 \text{ -> } C_6H_{12}0_6 + 6H_2O$$

Review questions

- Does the Light Independent Reaction require darkness?
- What sugar is moved throughout the plant through the phloem?

Photosynthesis summary

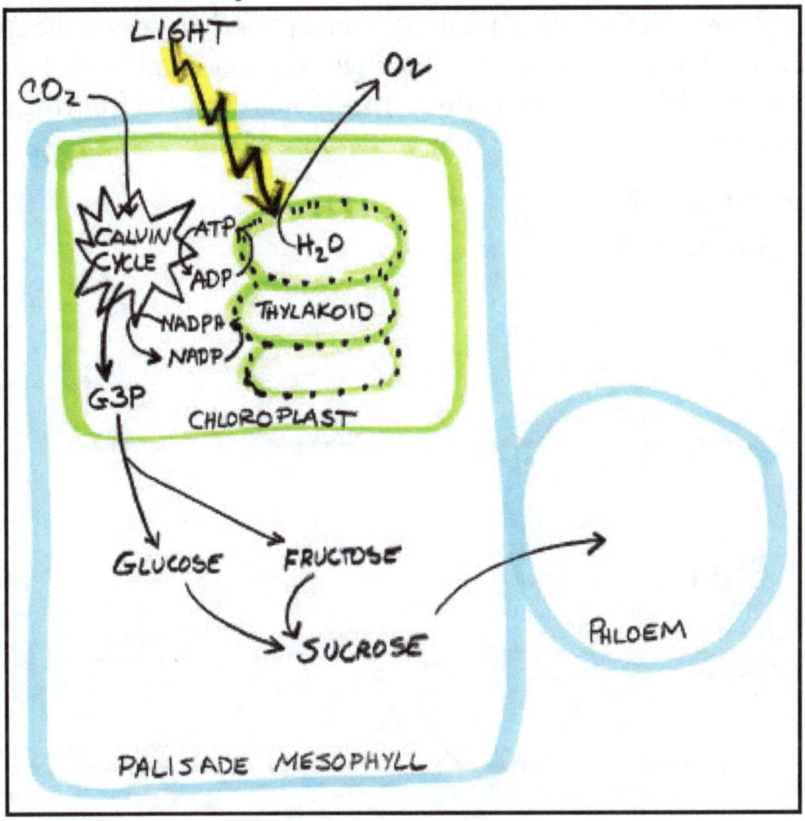

Mesophyll cell. Tom Michaels

When we add the two half-reaction equations for LR and LIR together, we get back to the summary equation for photosynthesis:

$$12H_2O + 6CO_2 \text{ -> } 6O_2 + 6H_2O + C_6H_{12}O_6$$

The illustration above is a summary of what happens in a mesophyll cell. The rectangular blue outline represents a palisade mesophyll cell in a leaf. Inside the cell is a green rectangle, representing a chloroplast. Inside the chloroplast is a stack of green ovals with black dots. These ovals are the thylakoids, and the stacks are grana. The black dots in the green thylakoid membrane represent the antenna complexes. Light hits the antenna complex and transfers its energy to pigments, and the energy is funneled to the reaction center where water (H_2O) is split in the light reaction to form the energy carriers ATP and NADPH. This is the Light

Reaction. The waste product formed at this stage is oxygen, which might be waste for the plant, but is quite useful for us.

In the Light Independent Reaction the energy is carried to the Calvin Cycle, represented by the multi-pointed star in the chloroplast, which uses the energy in ATP, the NADPH, and CO2 from the atmosphere to form the three-carbon G3P triose phosphate with the help of RuBisCO. Triose phosphate leaves the chloroplast and passes into the cytoplasm of the mesophyll cell to be transformed into glucose and fructose, which are combined into sucrose that is exported from the mesophyll cell to the phloem.

Cellulose and starch

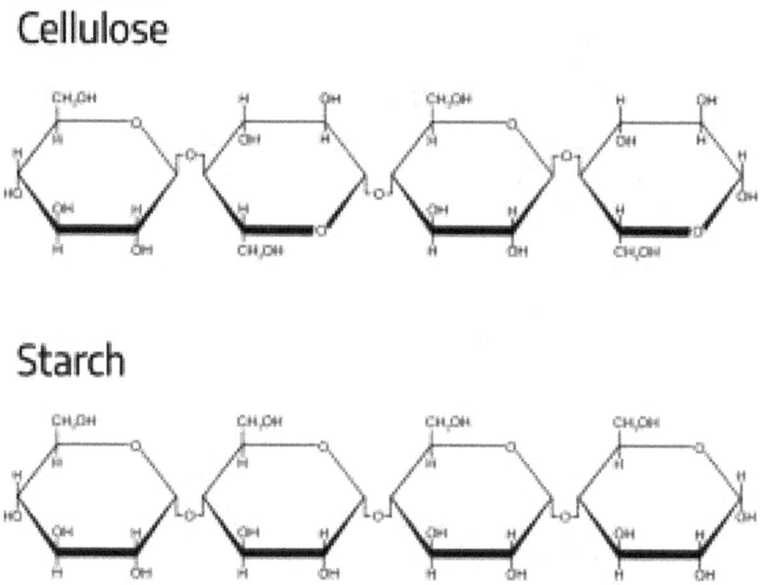

Cellulose and starch molecules. FreeSVG. CC0 1.0

Within a plant, the regions of photosynthesis and sugar production are called the source. Leaves are typically the main source within the plant, since that is where most photosynthesis takes place. Those regions that do not support photosynthesis (like roots), but that still need organic molecules to survive, are called sinks. Movement of solutes (molecules dissolved in water) like sucrose from source to sink through the phloem is called translocation. Translocation of sucrose through the phloem to the sink provides cells with a source of stored energy, and also building blocks for organic molecules, as noted earlier. Sucrose can be broken down to glucose and fructose, building blocks used to form other extremely useful organic compounds. Two particularly useful compounds result from the production of long glucose chains: starch, a key energy storage compound in plant cells, and cellulose, the main constituent of the cell wall and key to a plant's structural integrity. Wood, for instance, is primarily made up of the cellulose-rich cell walls of dead xylem. Both starch and cellulose are long chains of glucose, but they differ in the way the glucose molecules are linked together.

Cellulose is the molecule into which carbon extracted from atmospheric CO2 is sequestered for long-term storage. Starch sequesters carbon for a much shorter period of time because it is either eaten, used by the plant for new growth, or decomposed by bacteria and fungi that can utilize starch for energy.

To review

- In the light reaction, pigments in the thylakoid membrane capture energy from sunlight.
- The energy is used to split water, which releases oxygen to the atmosphere.
- The energy used to split water is transferred into electrons and hydrogen atoms, and eventually to ATP and NADPH.
- In the light independent reaction, the ATP and NADPH power the Calvin cycle that captures carbon from atmospheric CO_2 and incorporates it into simple sugar molecules.
- These simple sugars can be translocated to sinks, where they are used for energy, converted into energy storage compounds, or converted into structural molecules.

Some of the explanations and pictures used in this lecture came from Introduction to photosynthesis and other pages at that McDaniel College site. For additional explanations, an overview of the chemical reactions of photosynthesis, and another description of the properties of light and pigments, navigate to the pages listed at the top of that web page.

TERMS

Flashcards

Accessory pigments	Light-absorbing pigments, other than chlorophyll, that are found in chloroplasts.
Adhesion	A force where dissimilar molecules stick together; in plants this force of adhesion between water and the walls of the xylem helps hold the water in the xylem against the downward force of gravity.
Antenna complex	Structure of chlorophyll and accessory pigments that are embedded in the thylakoid membranes; captures and routes energy from sunlight to a collector called a reaction center.
Apoplast	Space outside the cell membrane where water and minerals can move freely; interrupted by the casparian strip in roots.

ATP	A principle molecule for storing and transferring energy in cells; created in the LR.
Carotenoid pigments	Accessory pigments that absorb green light and reflect yellow and red light; overtaken by chlorophyll during the growing season, so we do not see the yellow-red reflection.
Cellulose	A long chain of glucose that is a main constituent of the cell wall and key to a plant's structural integrity; sequesters atmospheric CO2 for long-term storage.
Chlorophyll	Green photosynthetic pigment found in plants, algae, and cyanobacteria that captures light for photosynthesis.
Chlorophyll a	Type of chlorophyll; mainly absorbs violet and red light while reflecting green light.
Chlorophyll b	Type of chlorophyll; mainly absorbs blue and orange light while reflecting green light.
Chloroplasts	Membrane-bound organelles found within cells that house chlorophylls and accessory pigments.
Cohesion	A force where similar molecules stick together; in plants this occurs with water molecules bonding together.
Evapotranspiration	Movement of water in the plant from the root through the stem to the leaf and out the stomata to the atmosphere; also called transpiration.
Fructose	Simple sugar; can be produced via photosynthesis.
Gas exchange	Movement of oxygen and carbon dioxide through stomata in the plant.
Glucose	Simple sugar; can be produced via photosynthesis.
Grana	Stacks of thylakoids.
Guttation	Dew-like drops of water that are forced out of the leaves of some plants due to root pressure.
Hydrogen bond	When water molecules are near each other and the negative region of one molecule is attracted to the positive region of another; a weaker bond than covalent bonds.
Light absorption	Process in which light is absorbed and converted to energy.
Light Independent Reaction (LIR)	Second half-reaction in photosynthesis; occurs without the presence of light and uses the energy produced in the Light Reaction to grab the carbon from carbon dioxide and use the carbon to build simple sugars.
Light Reaction (LR)	First half-reaction in photosynthesis; occurs with the presence of light and uses light energy to split water, which transforms the energy from the sun into hydrogen ions and electrons.
Light reflectance	Light wavelengths that are not absorbed, but are reflected back.
Light wavelength	Length of the wave from one peak to the next; measured in nanometers.
NADPH	Energy created in the LR; used to drive the LIR.
Palisade mesophyll	Densely packed, columnar-shaped, elongated cells full of chloroplasts; analogous to cortex parenchyma cells in the stem, but in the leaf they are specialized for light energy capture.
Photoautotrophs	Name given to living things, namely plants, that use energy from light to produce organic molecules with which they build their cells and store energy; self-nourishing.

Photon	Particle representing a quantum of light; provides the energy that drives photosynthesis.
Photosynthesis	Process of capturing light energy and producing carbon-based organic molecules.
Reaction center	Complex of pigments, proteins, and other factors that execute the primary energy conversion reactions of photosynthesis, primarily where water is split in the LR to form the energy carriers ATP and NADPH.
Resonance	Energy that is passed from one molecule to the next.
Rubisco	One of the most abundant proteins on earth; catalyzes the step in the process where carbon from atmospheric CO2 is incorporated into an organic molecule; full name: Ribulose-1,5-bisphosphate carboxylase/ oxygenase.
Simple sugars	Monosaccharides; examples include glucose and fructose.
Starch	Key energy storage compound in plant cells; a long glucose chain that sequesters atmospheric carbon for short-term use.
Stroma	Interior of the chloroplast; site of the LIR.
Sucrose	Sugar that is transported by the phloem throughout the plant to provide energy and building blocks for other organic molecules like starch and cellulose.
Symplast	Interior to the cell membrane, where water and minerals are transported through cells.
Tension	Differential pressure; in plants this occurs as water molecules are pulled through the plant via transpiration.
Thylakoid membrane	Membrane that surrounds the thylakoid.
Thylakoids	Membrane-bound compartments inside chloroplasts and cyanobacteria; the site of the light-dependent reactions of photosynthesis.
Transpiration	Movement of water in the plant from the root to stem to leaf and out through the stomata to the atmosphere; also called evapotranspiration.
Triose phosphate	A 3-carbon sugar (triose) with phosphorus and oxygen atoms (phosphate); G3P is an example.
Turgor pressure	Water pressure inside of cells.

CHAPTER-12

SOILS, FERTILITY, AND PLANT GROWTH

When a plant grows in soil or potting mix it removes nutrients as well as water from the soil. Although a plant produces (fixes) its own carbon-based molecules from photosynthesis, all other nutrients are taken up by the roots from soil. As we harvest plant material and dead plant material decomposes, these nutrients are depleted. This lesson discusses how this process changes soil structure, texture, and fertility.

Learning objectives

- Understand how the texture, structure, and fertility of soil affect plant growth.
- Appreciate the different types of soil and manufactured soil-less media for growing plants.

12.1 SOILS, FERTILITY, AND PLANT GROWTH

Learning objectives

- Describe how soil texture and soil structure affect plant growth.
- Use simple tests to determine the texture of soil.
- Determine the meaning and impact of the three numbers typically displayed on fertilizer labels.

This is a course about plant propagation, but propagation is only useful if you can successfully grow the plants you propagate. Soil, light, and water are key to growing healthy plants. Here you'll have a brief introduction to soils and soil fertility — a huge area of knowledge and study.

Soil texture

Soil has two key properties: texture and structure. Soil texture refers to the relative proportion of sand, silt, and clay particles in the soil. Sand, silt and clay are the three sizes of mineral particles (originating from rock rather than from previously living material) that make up soil. Sand is the largest particle, silt is intermediate, and clay is very small. In relative terms, if sand is a 55-gallon barrel, silt is the size of a plate, and clay is the size of a dime.

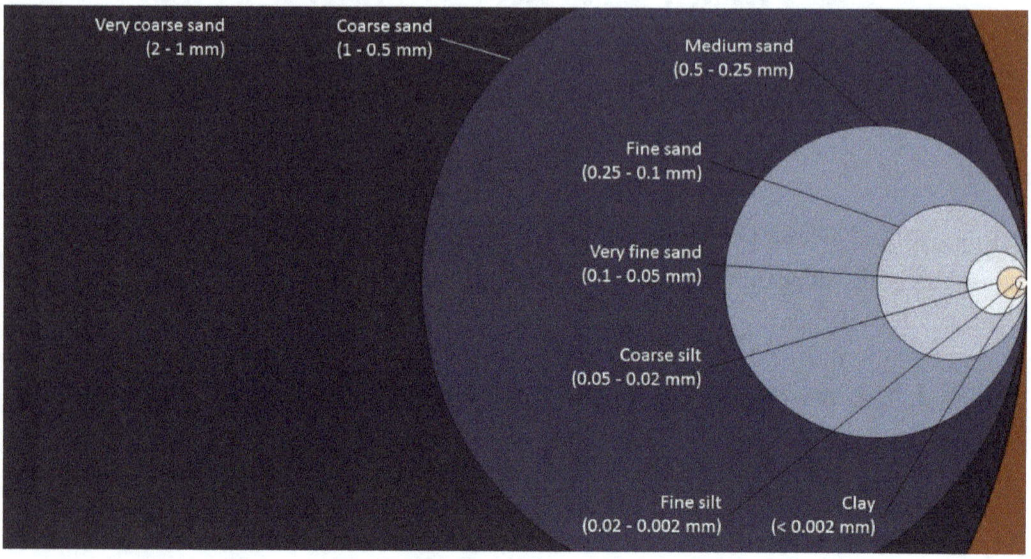

Soil particle size. Antonio Jordán. CC BY-SA 3.0

This mixture of different-sized particles is called texture because of how different combinations of these particle types make soil feel when you rub a sample between your thumb and forefinger. High amounts of sand makes a soil sample feel gritty, more silt makes it feel floury, and lots of clay makes it feel like velvet when dry and sticky when wet.

You can get a good idea of the texture of a field soil by doing a simple "jar test:" put soil in a jar, add water, shake the jar, then wait a few days to see the layers of different size particles settle.

The USDA Soil Texture Triangle, above, indicates the type of soil for different percentages of sand, silt, and clay. Notice that there are lines running through the triangle; these are to help you associate the percentages on the margins of the triangle with locations in the interior. The numbers on the margins are angled so that they are roughly parallel to the associated index lines. For example, the 60% clay index line is a horizontal line extending to the right of the 60 on the percent clay margin. The 20% silt index line runs from the upper right to the lower left of the triangle. And the 20% sand index line runs up from the lower right to upper left; all three lines are marked with a red arrow. These lines intersect at the red dot in the middle of the Clay area, indicating that a soil with 60% clay, 20% silt, and 20% sand is classified as a clay soil. When farmers talk about their field soil they often use the terms in the texture triangle rather than the percentages of sand, silt, and clay. This approach is much less common when talking about the potting mix used in greenhouses, as there is very little real mineral soil (sand, silt, clay) in these mixes.

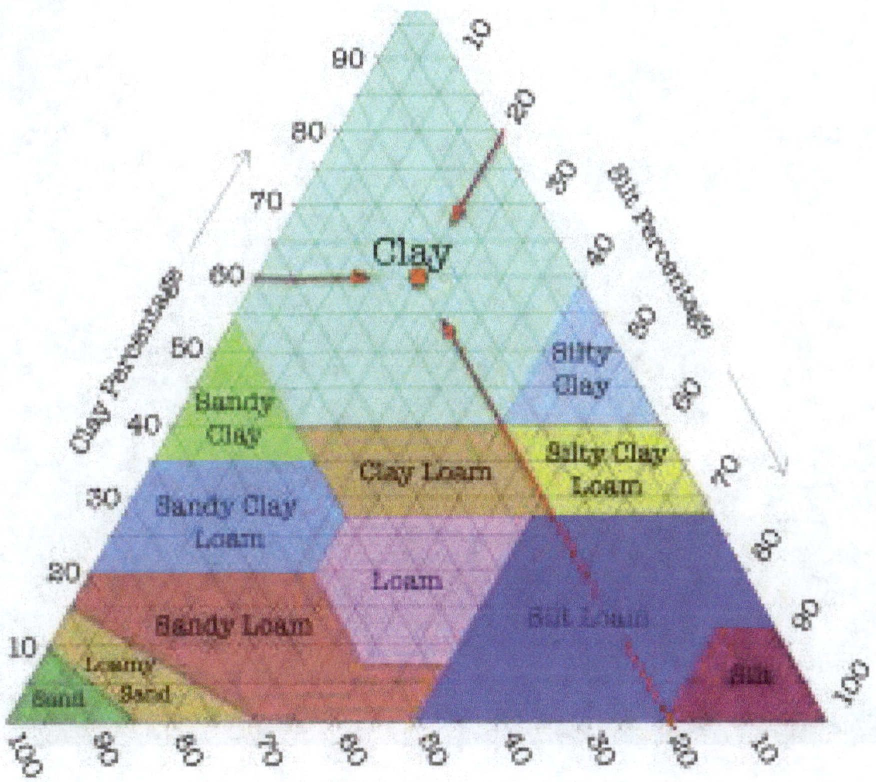

Soil texture triangle. Lewi1224. CC BY-SA 4.0

Soils high in sand have great drainage and aeration so that roots are exposed to air in the soil and don't rot as easily. Roots can penetrate sandy soil easily. But sandy soils are poor at holding moisture when the weather turns dry, and sands don't hold nutrients well. Nutrients and moisture hang on to a soil best when the soil particles have a lot of surface area, and sand has the least surface area (relative to particle volume) of the three particle types.

Clay, in contrast, holds on to water so tenaciously that it is tough for the plant to get the water for itself. Wet clay is sticky, and clay packs together so tightly that when it dries it clumps together and turns into hard clods. Roots have difficulty penetrating a dry, clay soil. But clay does have a lot of surface area for its volume, and holds nutrients better than other particles. Clay soils tend to be fertile.

Silt has intermediate properties between sand and clay, as you might expect. An ideal soil has some of each type. A silt loam soil with 60% silt and 20% each of clay and sand is perfect for growing corn, wheat, and soybeans. But crops whose economically valuable part is in the ground, like potatoes and peanuts, do well in a sandy soil, because the tubers and pods come out of the ground cleaner and with less mechanical effort than they would from a soil with higher clay or silt.

Soil. Natural Resources Conservation Service Soil Health Campaign. CC BY 2.0

All soil textures have advantages and disadvantages, depending on climate, topography, and crop. The soil texture of large areas, like fields, can't really be modified to suit a particular crop, so a crop must be chosen that does well in the available soil. For example, if you have a field with sandy soils you aren't going to truck in tons of clay and silt to make the soil suitable for corn. Instead, you'll grow a crop like potato that does reasonably well on a lighter, sandy soil. For crops grown in greenhouses or containers, however, you can choose the soil texture to suit the crop you want to grow.

Review questions

- What are the three particles that make up soil texture?
- Which is smallest? Largest?

- Is a loam soil high or low in clay relative to the other particles?
- How do you determine soil texture using a soil jar?

Soil structure

Soil structure. Soil Science. CC BY 2.0

Soil structure refers to the way in which the soil particles and other materials like the organic matter in the soil bind together into clumps. These clumps are called aggregates. Pure sand does not clump together into aggregates at all (think about how hard it is to get sand at a beach to stick together for a sand castle). When sand, silt, clay, and organic matter interact to form small aggregates, like the ones shown below, they create what is called a granular structure. Large holes in the aggregates provide spaces for gasses and water to pass through, while smaller holes hold water. The need for water is obvious, but the need for gas exchange may not be. As you know, root cells are growing, which means they require oxygen and give off carbon dioxide as a waste product. Oxygen needs to be available in the root zone, and carbon dioxide needs to be vented. If soils are waterlogged, plants die because too much carbon dioxide builds up around the roots and the roots are starved of oxygen. It is therefore important for soils to have these holes in the aggregates for gas exchange. This is called the aeration-porosity of the soil. Organic matter, which in this case refers to decaying bits of formerly living material, helps build the aggregates by gently sticking the soil particles together. The space between and within aggregates provides aeration-porosity.

The illustration below includes a cross section of soil, showing several soil aggregates packed together. Each aggregate is built from sand, silt, clay, and organic matter (also called humus). Note the micro- and macropores for water and gas exchange.

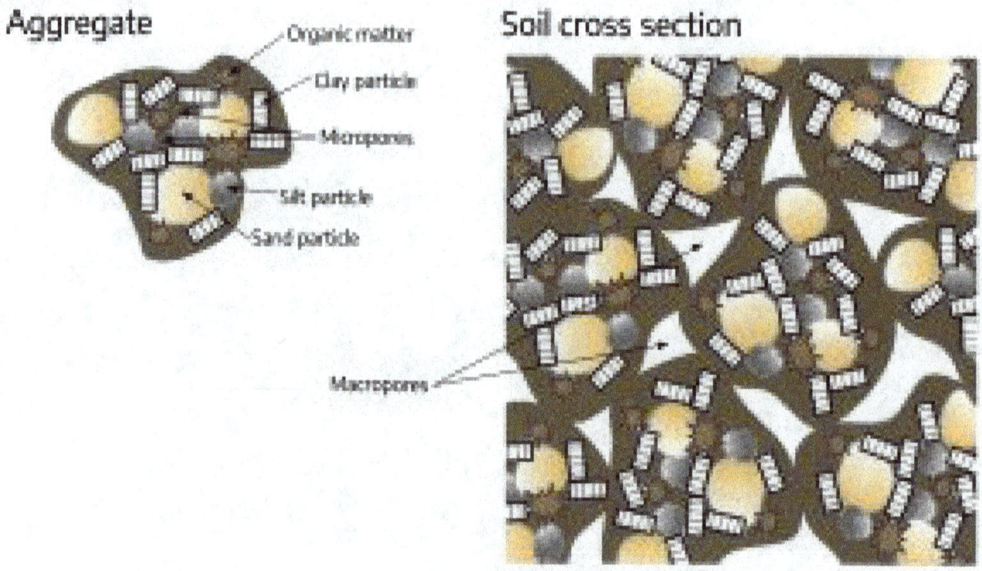

Soil aggregation. Queensland Government. CC BY 4.0

Soil with granular aggregation that favors plant growth by holding water and nutrients, yet allows for drainage and gas exchange, is said to have good tilth. The soil hangs together (unlike sand), doesn't form hard clumps (unlike clay), and breaks apart into crumbly moist chunks when you turn over a spade of earth. While gardeners are usually stuck with whatever soil texture they might have in their gardens., one of the most important and readily accomplished tasks a gardener can take on to improve garden soil is to improve the soil structure by:

- Increasing the soil organic matter, and
- Reducing **soil compaction.**

Increasing soil organic matter will improve and stabilize soil aggregation. Reducing compaction, like foot traffic through the garden, will maintain the macro-and micropores in the soil to promote drainage, moisture retention, and gas exchange.

Review questions

- What makes up the glue that holds the soil particles into aggregates?
- Why is gas exchange in soils important for plant growth?

Soil organic matter

Soil organic matter refers to carbon-based material in the soil that was originally a living organism, whether plant, animal, or microbe. Sometimes, soil organic matter also refers to organisms such as bacteria, fungi, insects, and worms that are still living in the soil, but this discussion refers to the materials that were once alive and are now dead and decomposing. Leaves, stems, and roots eventually die, are incorporated into the soil, and decompose. Soil organisms decompose the former living material and transform it into material called humus. Humus is sticky, and helps bind soil particles together into aggregates, as noted above. Humus also can absorb and hold up to six times its weight in water, so it is very important in improving light (sandy) soils. The decomposing organic matter also releases nitrogen and other nutrients that the plant can take up for growth. And finally, humus, like clay, holds nutrients in the soil through electrochemical charge; organic matter is negatively charged, so it holds positively charged cations like calcium that are important for plant growth.

In summary, organic matter is formerly living matter that is transformed in the soil into humus. Humus helps stick soil particles together to improve soil structure, holds water in droughty soils, and holds plant nutrients. Decomposing organic matter makes nutrients such as nitrogen available to plants.

Organic matter is added to soils in several forms:

Compost

For gardeners, this may be the most familiar form of organic matter. Leaves, weeds, grass clippings, and other organic material are mixed together and occasionally turned to promote decomposition. This results in humus that, when added to the soil, builds soil structure. Most of the nutrients have been used by the organisms that are decomposing the organic matter, are lost to the air, or are leached away by rain, so compost isn't very effective as a nutrient source. Its main purpose is to build soil structure and assist in retaining available moisture and nutrients.

Compost pile. nancybeetoo. CC BY 2.0

Green manure, or cover cropping

A crop grown with the sole purpose of tilling the crop into the land to increase the organic matter is called green manure. Green manure crops are used to change soil structure by incorporating organic matter directly into the soil. This technique is also used extensively in horticultural crop production to reduce soil-borne pathogens, and these crops serve a very useful purpose of smothering weeds.

Mixed species green manure crop. UGA CAES/Extension. CC BY-NC 2.0

Incorporating crop residues

After a crop is harvested, it is good agricultural and horticultural practice to incorporate the remaining plant material into the soil. Sometimes this is done with a moldboard plow to completely bury the residue, but the more modern method is to use the bare minimum of tillage, or to leave the residue on the top of the soil and plant over the dead material the next spring. The latter method, called no-till, is particularly useful for minimizing soil erosion caused by soil particles blowing away with the wind or moving with flowing water.

The addition of too much organic matter that has too much carbon and not enough nitrogen can deplete the soil of nitrogen and harm plant growth. For instance, if you try to improve the organic matter of your soil by tilling in bales of straw or sawdust (both of which are almost all cellulose, which is very high in carbon), when the microbes begin to break the straw down they need to absorb nitrogen from the soil just for their own growth. If instead you add manure to the soil, which is a blend of straw (high carbon) and animal waste (high nitrogen), the microbes can use the nitrogen from the manure for their own growth as they decompose the organic matter and make more nitrogen available to plants.

Review questions

- Why add organic matter to the soil?
- Is all organic matter of the same value when added to soil, or are some types of organic matter better than others? Why?

Containers and raised beds

Garden soil cannot be used for container gardens, sa it compacts too tightly in pots and has terrible drainage. Instead, it is best to a soil-less mix like those available at nurseries, or to make a mix that is high in an organic matter like peat moss or rice hulls, to increase aeration porosity.

Salad table before planting. Tom Michaels

Author Dr. Tom Michaels developed this salad table, above, which has great potential for use in urban areas with smaller areas for growing greens, including apartment patios. This table is made with 2×4 lumber for the sides and legs, with hardware cloth and landscape fabric for the bottom of the table. The growth medium is normally a peat-based potting mix. A table this size supplies enough salad greens throughout the summer for two adults. You could modify it to have deeper soil so that you can raise a tomato or pepper plant.

Salad table with plants. Tom Michaels

Since the growing medium is potting mix, it dries out quickly. You can see a few modifications in this Hydroponic Salad Table, also created by Dr. Michaels. It's about 2′ x 4′ x 7.5″ deep and made with lumber, a plastic liner, and a styrofoam lid. About 30 gallons of nutrient solution is added to the box, the box is covered with a lid, and salad green seedlings like lettuce, spinach, chard, and kale are placed in holes in the lid. The plants yield greens for most of the summer and little or no water needs to be added.

The big three on fertilizer bags

Fertilizer bags and containers display a series of three numbers separated by dashes. This is called the fertilizer's analysis. The numbers represent the percentage of the fertilizer that is nitrogen (N), phosphorus (P), and potassium (K) — always in that order. N, P, and K are the elements needed by plants in the greatest quantities. Nitrogen is a key element found in protein, phosphorus is an important component in energy transfer molecules like ATP and as part of the DNA backbone, and potassium is an essential part of the mechanism for moving nutrients into and out of cells. Other elements can also be important in small quantities and in special circumstances, but N, P, and K are the most common plant nutrients.

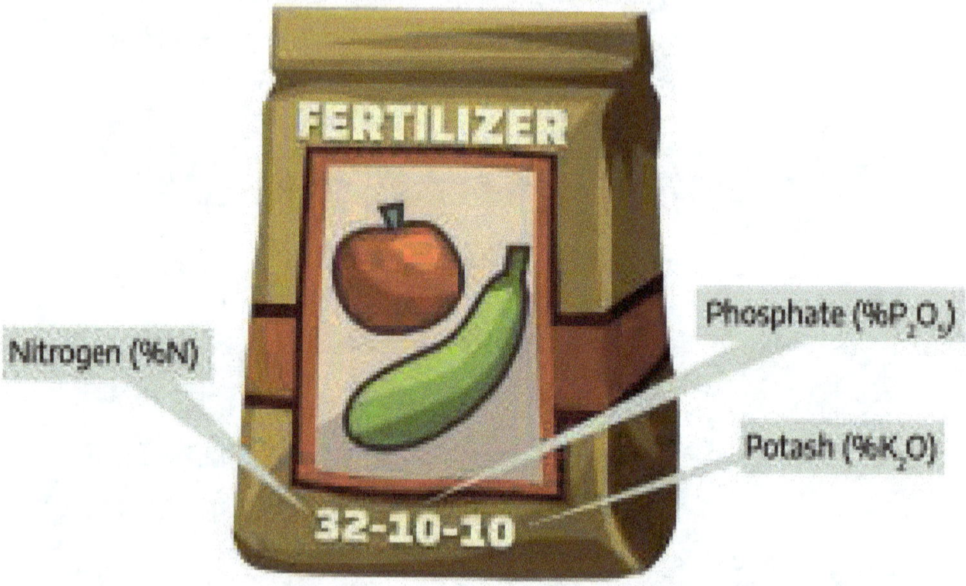

Fertilizer bag. Pixabay. Pixabay license

A 10-10-10 general purpose garden fertilizer has 10% nitrogen, 10% phosphorus, and 10% potassium. The rest is filler, like sand or fine gravel. In Minnesota, fertilizers available to homeowners typically have no phosphorus because of legislation aimed at reducing phosphorus runoff into our lakes. Phosphorus is considered to be the limiting factor in algae growth, so if phosphorus runs off yards and gardens into lakes it causes algae blooms. In addition, our garden soils normally have sufficient phosphorus. A general fertilizer analysis without phosphorus would be 10-0-10. Nitrogen is usually the nutrient most limiting for plant growth, so it is worth it to read through the labels.

A caution: seeking the best value per pound of N isn't always the right strategy. Sometimes the form of the nutrient is important. If you are interested in growing a hydroponic salad table, the plants need a particular form of nitrogen called nitrate, which is not usually found in big, cheap bags of fertilizer; it's more likely to be found at a hydroponic shop, and costs more per pound of N than other forms.

TERMS

Aggregates	"Clumps" in the soil; see soil structure definition.
Clay	Smallest particle in soil; has high nutrient holding capacity.
Compost	A type of organic matter that builds soil structure and assists in retaining moisture and nutrients.
Cover cropping	Crop used to benefit the soil rather than the main crop species.
Fertilizer analysis	N-P-K content of a bag of fertilizer; shown in percentages by weight.
Granular aggregation	Interaction of small soil aggregates; it is important to have a mixture of large and small holes between the aggregates to allow for water and gas exchange.
Green manure	Crop grown to purposefully be tilled back into the soil to increase the organic matter (and thus change the soil structure); can also smother weeds.
Humus	Sticky material made from organic matter that helps bind soil particles together into aggregates; can absorb and hold up to 6x its weight in water, releases nitrogen, and holds positively charged cations for plant growth.
Nitrogen (N)	One of the most important elements for plant growth (by quantity); a key element found in protein.
Organic molecule	Chemical compound associated with living organisms that contain carbon atoms.
Organic material/ matter	Material that has come from a recently living organism (such as plants) that may be partially or fully decomposed.
Phosphorus (P)	One of the most important elements for plant growth (by quantity); a key component in energy transfer molecules like ATP and as part of the DNA backbone.
Potassium (K)	One of the most important elements for plant growth (by quantity); a key part of the mechanism for moving nutrients into and out of cells.
Sand	Largest particle in soil; helps increase aeration.
Silt	Particle of intermediate size in soil.
Soil compaction	When the pore spaces between soil aggregates are compressed.
Soil organic matter	Carbon-based plant, animal, and/or microbe tissues that are in the process of breaking down; increasing soil organic matter improves and stabilizes soil aggregation.
Soil structure	The way in which the soil particles and other materials, like the organic matter in the soil, bind together into clumps.
Soil texture	Relative proportion of sand, silt, and clay particles in the soil.
Tillage	Process of incorporating the residue from the top of the soil into the soil; there are many types.

CHAPTER-13

SEXUAL REPRODUCTION

Mitosis is essential to all plant propagation. It produces the mighty oak from a single zygote cell and makes all forms of propagation possible. The adventitious rooting process of any cutting begins with mitosis and continues through many mitotic divisions to produce a root meristem. Meiosis is also fundamental to plant propagation for the production of gametes that fuse, producing diversity for evolution and plant breeding.

Learning objectives

- Understand the structure of DNA.
- Compare the different states of ploidy and which process of cell division – mitosis or meiosis – they are produced from.
- Recognize what happens to the chromosomes, cell wall, cell membrane, and nuclear membrane in each of the stages of mitosis and meiosis.
- Compare the similarities and differences in the outcomes of mitosis and meiosis.

13.1 DNA

Learning objectives

- Discuss how ribose, phosphate, purine, and pyrimidine molecules are combined to make up a strand of DNA.
- Summarize how the strand of DNA is coiled and packed to become a chromosome.
- List the steps in the cell cycle and describe where in the cycle you would find particular types of cells.

Introduction to chromosomes

You have learned about two types of meristems: apical meristems, that result in primary growth, and lateral meristems, like vascular cambium and cork cambium, that result in secondary growth. Meristems are sites of cell division in plants. These new cells may themselves divide once or twice, but then they begin to enlarge and differentiate into cells with specialized function depending upon the type of tissue in which they occur. (Parenchyma cells are an exception because they remain undifferentiated. They can become meristematic and divide to form plant parts like lateral roots, to form interfascicular cambium, and to respond to wounds by filling in space with new cells called callus tissue.)

For a cell to divide into two identical cells, it is critical that all the components of the original cell are duplicated prior to cell division and then distributed between the two sister cells before they separate. It is also critical that the hereditary material, the DNA in the nuclear chromosomes, be exactly duplicated and equally distributed between sister cells. Exact duplication and equal distribution ensure that each cell in the organism has all the genetic instructions it needs to carry on its metabolic processes, and that every cell in the organism has the same instructions.

The type of cell division that takes place in meristems, which is the type where one cell divides into two identical sister cells, is called mitosis.

Mitotic cell division was first observed under light microscopes well over 150 years ago. Microscopists, scientists who use microscopes, noticed dark-staining cell bodies lined up in the middle of a cell that was about to divide. They called these cells chromosomes (based on the Latin for dark staining, "khrôma," and bodies, "sôma"). Once lined up in the middle of the cell, the chromosomes divided and moved to opposite poles in the cell just prior to the actual division of the cell into two sister cells. It was clear to those scientists that the process of cell division ensured that the chromosomes were specially and carefully handled during cell division. In 1902, Water Sutton, studying grasshoppers, provided proof (optional reading) that chromosomes contained the hereditary material for the organism. That's why cell division includes such careful division of the chromosomes — the organism needs to ensure that every cell has an exact copy of all of the hereditary material associated with that organism.

What makes up these chromosomes? The short answer is that chromosomes found in the nucleus of plant cells are composed of chromatin (optional reading). Chromatin is made up of DNA wrapped around proteins, called histones. These proteins around which the DNA wraps are called histones. We'll start with the structure of DNA and build up to a chromosome.

DNA structure

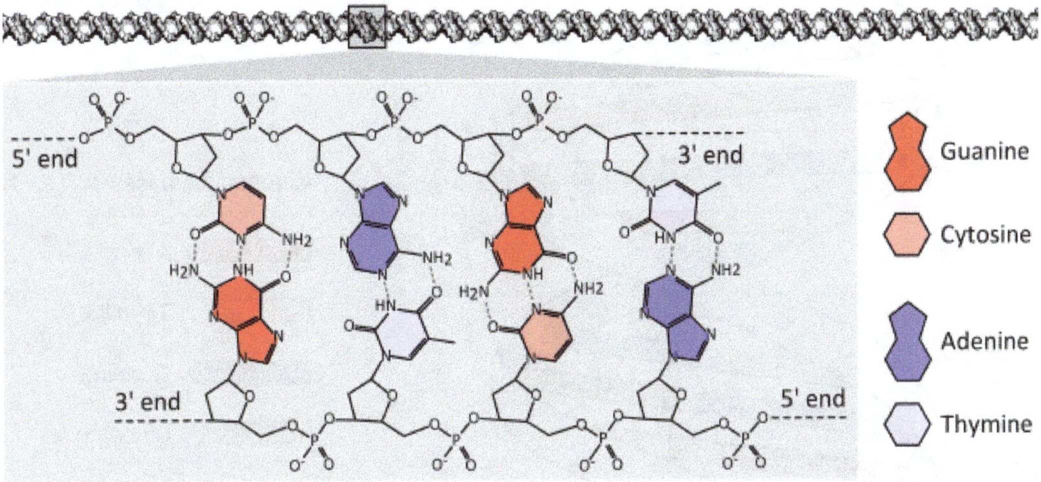

DNA molecular diagram. Thomas Shafee. CC BY 4.0

DNA is a double-stranded chemical polymer (a polymer is several types of molecules bonded together) that looks like a flexible ladder twisted on its long axis. Each side, or single strand of the ladder, is made up of a chain of alternating ribose and phosphate molecules. Ribose is a sugar molecule composed of a ring of five carbons. These molecules are linked to each other by a phosphate molecule made up of one phosphorus atom and two oxygen atoms. This sequence of ribose and phosphate molecules constitutes what is called the DNA's ribose-phosphate backbone.

The steps, or rungs, of the ladder are constructed from pairs of bases. Four types of these bases compose both strands of DNA. Two of the bases, collectively called purines — Adenine (A) and Guanine (G) — contain two rings of carbon atoms. The other two, collectively called pyrimidines — Cytosine (C) and Thymine (T) — have only one ring of carbon atoms. There is enough space between the sides of the ladder (or ribose-phosphate backbone) to fit two bases with a total of three carbon rings, so each rung of the ladder always has one purine bonded to one pyrimidine.

There isn't enough space for two purines, and there is too much space for two pyrimidines. And because of the molecular structures of the bases, Adenine (A) always bonds with Thymine (T), because they each can share two hydrogen bonds, and Cytosine (C) always bonds with Guanine (G), because they each can share three hydrogen bonds. As a consequence, the rungs on the ladder are always made up of AT, TA, GC, or CG. And the sequence or order of As, Ts, Gs, and Cs making up the rungs along one strand of the ladder is the genetic code carried in the DNA. With only rare exceptions, the code is read three-bases at a time and encodes all of the information for a plant's life cycle.

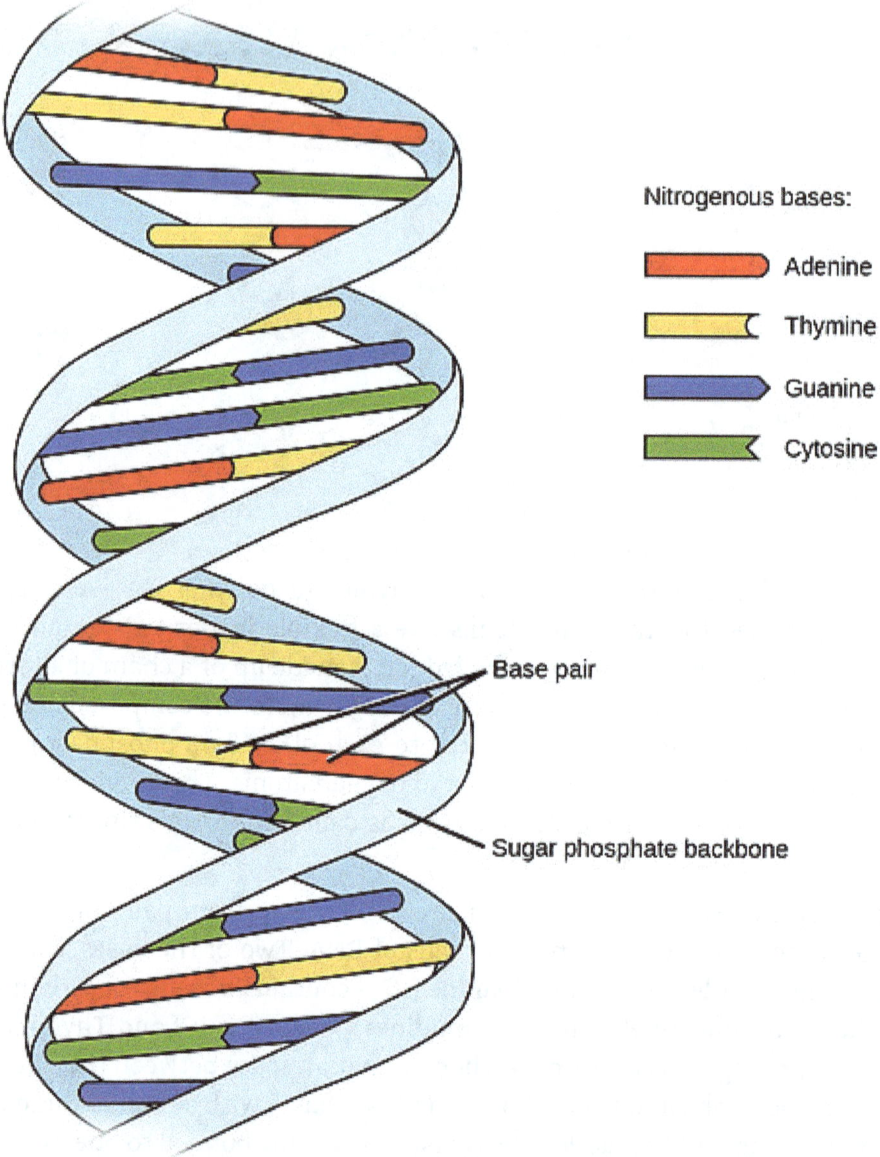

Structure of DNA. OpenStax. CC BY 3.0

Review questions

- If you know that a rung of the DNA ladder has the base Adenine, then the other base to which it is bonded is Thymine. Why not Guanine? Why not Cytosine?
- We take for granted now that chromosomes are the hereditary material in plants. When was this proven? A) about 15 years ago when the first

genetically modified foods were developed, B) around the time the atom bomb was developed, C) just before WWI, or D) back when Mendel was doing experiments with peas and discovering his principles of inheritance.

- Why is the shape of DNA called a double helix?

Nucleosomes

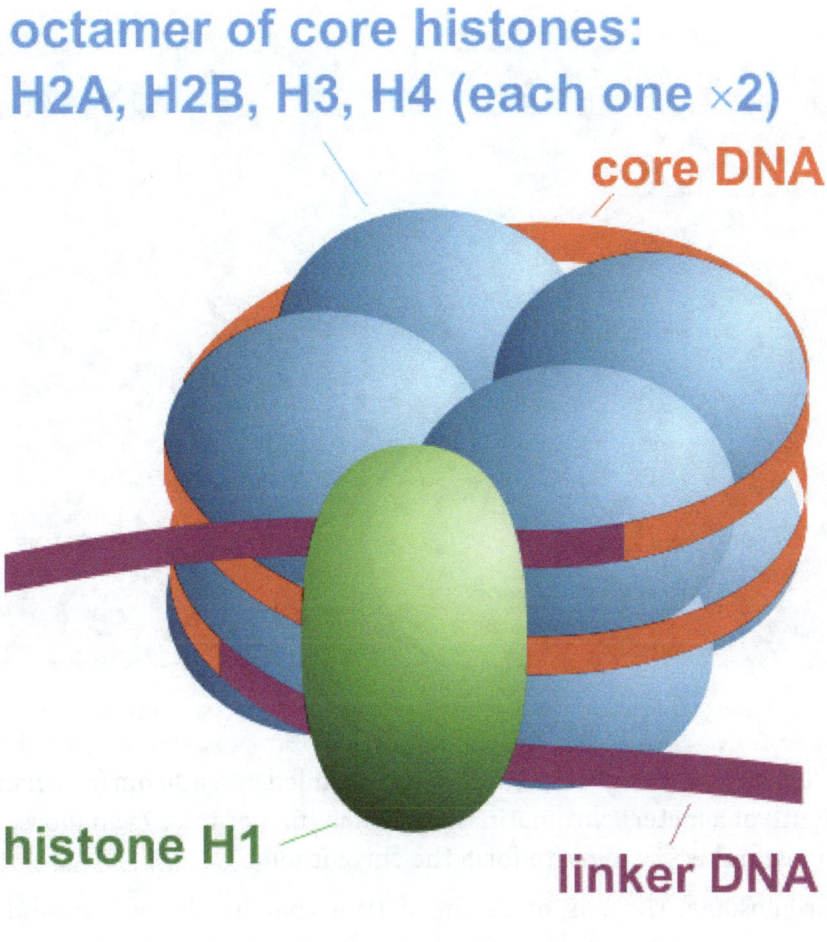

Nucleosome. Darekk2. CC BY-SA 3.0

We now have a DNA helix made up of alternating ribose and phosphate rails with rungs containing a purine and a pyrimidine. The DNA helix strand next loops around nucleosomes (optional reading) which are made up of histone proteins (illustration above). The histone proteins are essentially the spools around which the DNA thread wraps, except that DNA doesn't wrap continuously around one histone protein the way thread wraps around one spool. It makes a couple of loops and then moves on to the next histone protein.

The loops of DNA around these nucleosomes are separated by portions of naked DNA called linker DNA, so the effect is called "beads on a string" (see micrograph below). where the beads are the DNA loops and histone protein, and the string is the strand of linker DNA. This description dates back to the early days of exploring DNA using light microscopy. Chromosomes could be observed with the aid of the microscope and described, but their composition was unknown.

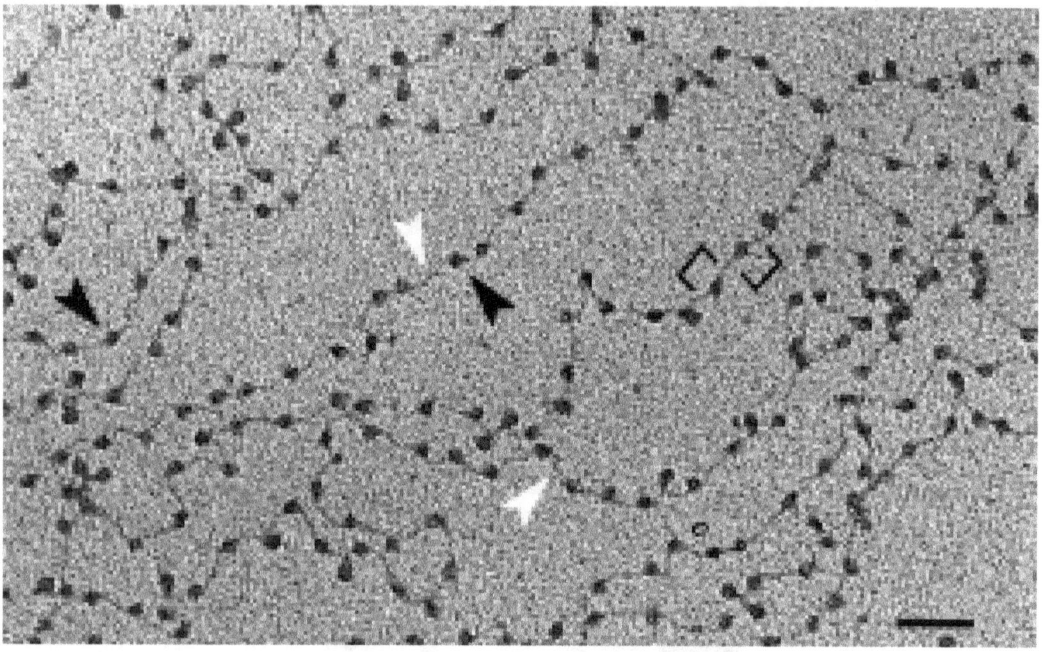

Black brackets highlight individual nucleosomes; black arrowheads point to nucleosome core particles (the beads), white arrowheads to linker DNA (the string). Scale bar: 50 nm. Chris Woodcock. CC BY-SA 3.0

The "beads on a string" structure is further coiled into a 30 nm (nm=nanometer, one billionth of a meter) chromatin fiber, which further folds even more, with the association of other proteins, to form the chromosome.

A chromosome, then, is made up of DNA that has looped around histone proteins, then coiled, then folded. Think of a chromosome as a tightly and carefully packed long thread of DNA that is associated with histone proteins to help with the packing.

Here's a diagram of this coiling:

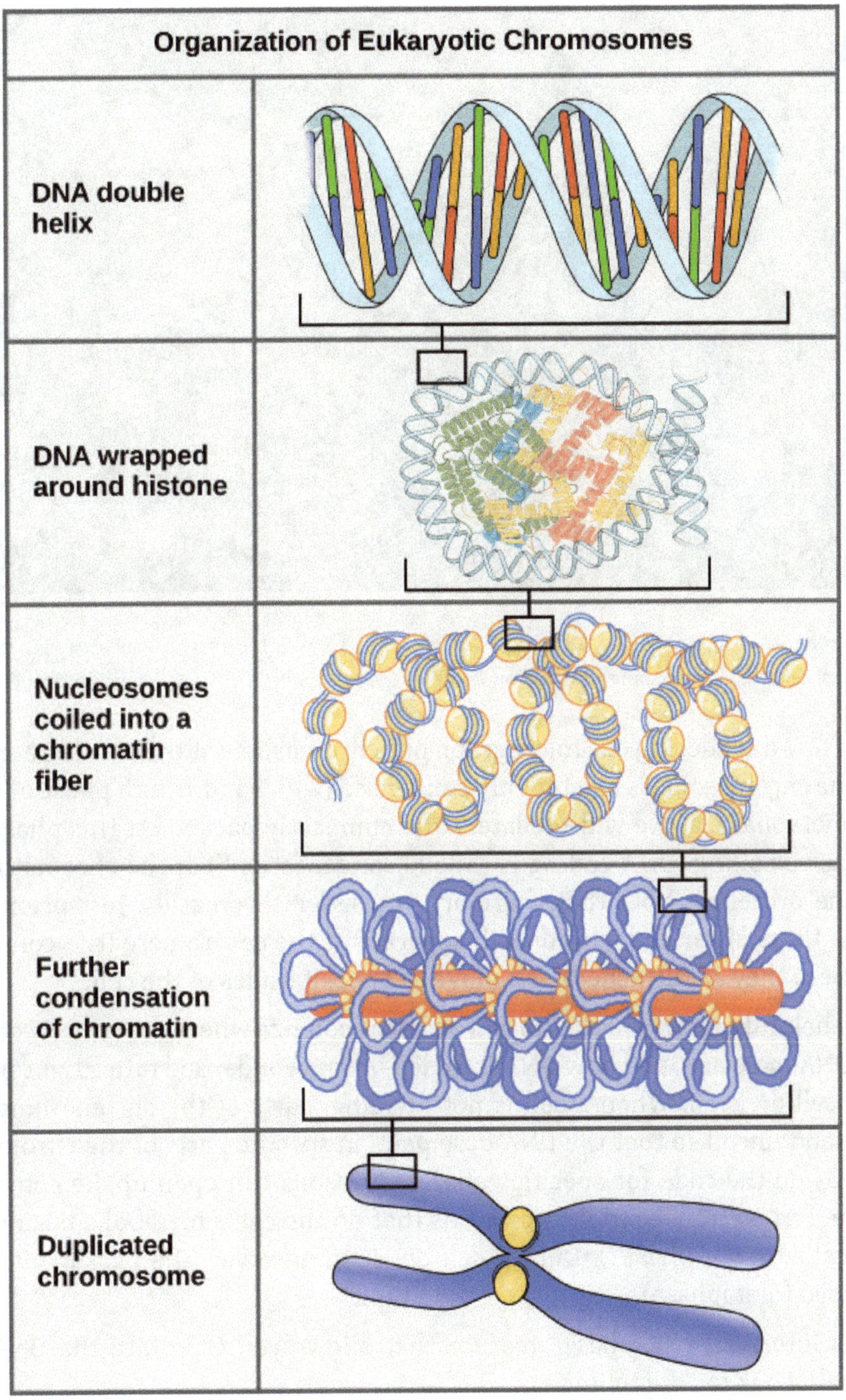

Chromosome coiling. OpenStax.

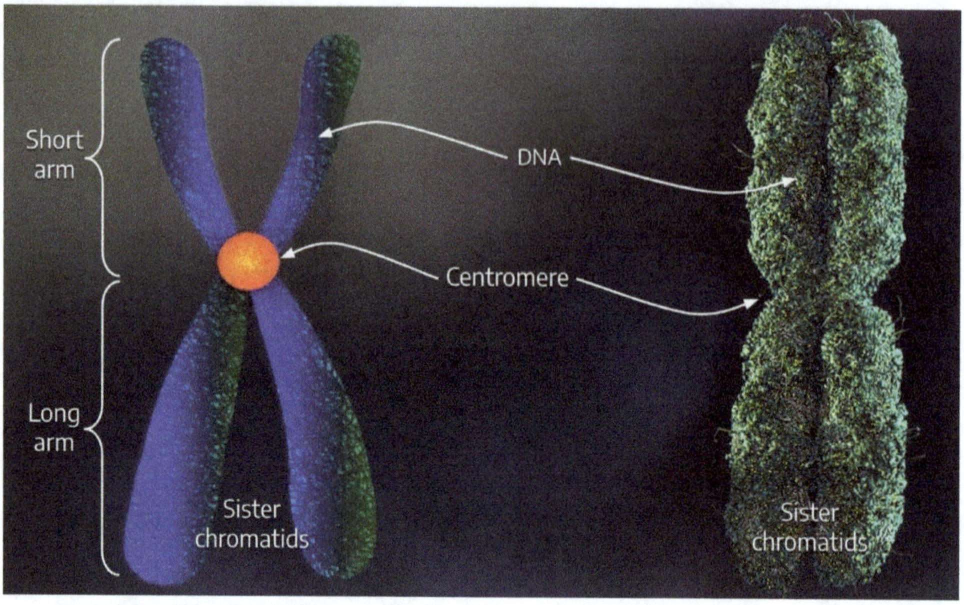

Sister chromatids. 3D Rendering X and Y Chromosome with Telomeres © Shutterstock, used with permission. Simple chromosome illustration by Prateek Pattanaik. CC BY-SA. Images adapted by Emily Tepe.

Above is a rendering of a micrograph of a chromosome at the cell division stage, when the chromosome is most highly condensed (which is during a phase of mitosis called metaphase, as we will see later) and compactly packed. At this phase there is extensive coiling (DNA coiling is called "condensation") of the chromatin (DNA + histone proteins). Note that there are two sister chromatids. Just prior to cell division these sister chromatids will separate at the centromere (the constricted spot where they are attached) and move to opposite sides of the cell.

It is helpful to have this detail so you can recognize where the genetic code sits in the DNA molecule, and how DNA is folded up and condensed into a tight package during cell division. When a cell is not dividing, parts of the chromosome relax, unfold, and uncoil so that the DNA base pairs in specific parts of the chromosome that provide the code for specific cellular functions can open up, be copied, and the code can be translated into proteins that do the cell's metabolic business — a phase called "interphase"). During cell division, however, the package is tightly condensed (metaphase).

This interactive site has a feature that allows you to rotate the DNA helix (optional). Look for the following:

- 5-carbon ribose ring making up part of the backbone
- phosphate linking the ribose rings completing the backbone
- 2-ring purine base (one per rung)

- 1-ring pyrimidine base (one per rung)
- hydrogen bonds (two or three per rung depending on the composition)

Review questions

- Why do you think the chromosome is so highly condensed (meaning wrapped and folded) during metaphase, which is in the middle of cell division? Think about why a chromosome might need to be very tightly packed during cell division.
- What is the connection among nucleosomes, linker DNA, and beads on a string?

The cell cycle

Below is a diagrammatic summary of the cell cycle, the cycle a cell goes through during its lifetime. You'll see that about 3/4 of the cycle is called Interphase and the other 1/4 is called M (for mitosis). As noted, during interphase the cell isn't dividing (but may be preparing to divide); the cell divides during M. Interphase is divided into three parts: G1 (the "G" is short for "Gap" or "Growth"), S (S for DNA "Synthesis"), and G2.

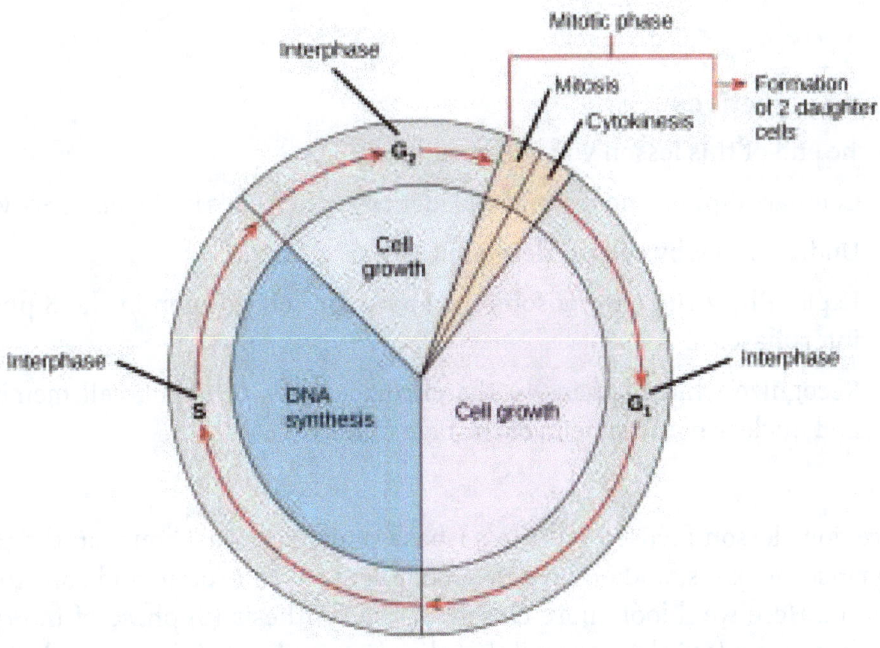

Cell cycle. OpenStax. CC BY 4.0

Consult this page on the cell cycle (optional reading) for additional information about the cell cycle.

The stages in the cell cycle are:

- G1 – cells are enlarging, and some are differentiating into specialized cells like epidermis, collenchyma, sclerenchyma, and cells in the xylem and phloem tissue that will no longer divide. These differentiated cells stay in G1 until death.
- S – chromosomes in cells like apical meristem cells, vascular and cork cambium, or undifferentiated parenchyma cells prepare to divide by replicating their chromosomes. To prepare for division, the cell makes a complete copy of all of its DNA, so in the process of copying the DNA there is DNA Synthesis. Once a chromosome has been replicated, the two copies are called sister chromatids and they are held together at a spot called the centromere. The sister chromatids are exactly alike, down to the exact order of AT, TA, CG, and GC bases in the rungs of the DNA backbone.
- G2 – the cell builds up the chemical machinery that it needs for division.
- M – finally, the cell heads into mitosis (M), addressed next.

Note that the size of the slices of pie in the diagram above are not indicative of the actual time duration.

13.2 MITOSIS

Learning objectives

By the end of this lesson you will be able to:

- Compare diploid and haploid and identify which cells in the plant are which.
- Understand why cells undergo mitosis.
- Explain how the chromosomes prepare for cell division in the S phase of interphase.
- Recognize what happens to the chromosomes, cell wall, cell membrane, and nuclear membrane in each stage of mitosis.

Ploidy

The previous lesson focused on DNA's ribose-phosphate backbone, on the purine and pyrimidine bases, and on how DNA complexes with protein and coils to form chromatin. Here we'll look more closely at the synthesis (S) phase of interphase and at the mitosis (M) phase. Recall that the mitosis phase of the cell cycle "pie" is divided into four stages; we'll look now at what happens in each of those stages and how it contributes to the outcome of mitosis, the equal division of chromosomes into two daughter cells.

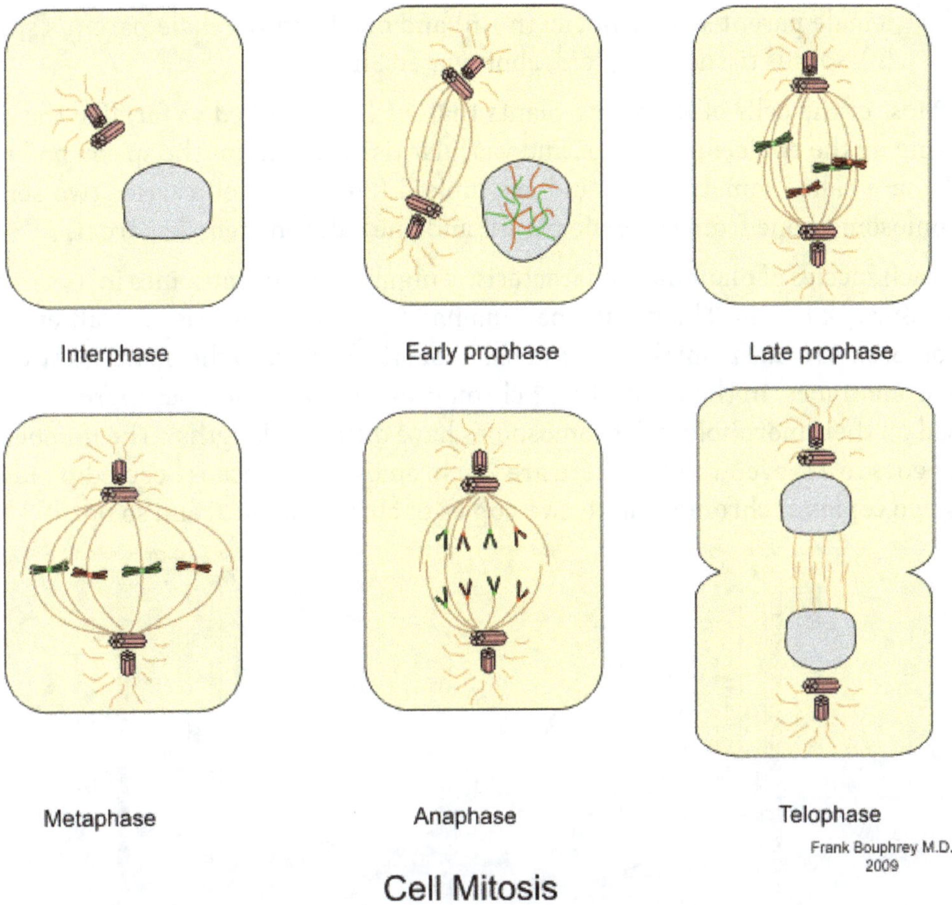

Phases of mitosis. Boumphreyfr. CC BY-SA 3.0

Ploidy refers to the number of sets of homologous (identical) chromosomes in a cell.

- In higher organisms like plants (and animals, including humans), gamete cells (egg and sperm) typically each contain one set of each of the chromosomes found in that particular species. When cells contain one set of chromosomes characteristic of the species, this state is called haploid and is abbreviated n.
- When the sperm and egg, each of which are n, unite to form a zygote, the zygote cell now has two sets of chromosomes, one from the male parent's sperm and one from the female parent's egg. When cells contain two sets of chromosomes, they are described as diploid, abbreviated 2n.
- Recall that one result of double fertilization in plants is that one sperm cell unites with two female polar bodies to create the endosperm found in seeds. Endosperm cells have three sets of chromosomes, two from the

female parent's polar nuclei (n + n) and one from the male parent's sperm (n), so this tissue is triploid, abbreviated 3n.

Most of the cells of flowering plants that we have studied so far, like the cells making up the epidermis, cortex, and vascular tissues (but not the sperm and eggs cells), are called somatic cells, and are diploid (2n). Each cell carries two sets of chromosomes: one from the male parent and one from the female parent.

Each species of plant has a characteristic number of chromosomes in its somatic cells. Bur oak has 24. The garden petunia has 14. Green bean has 22. Half of those chromosomes came from the egg and half from the sperm, so the plant has two sets of chromosomes. In the bean, the 22 chromosomes can be numbered from 1 to 11 based on their morphology (chromosomes have different lengths). The numbering only goes to 11, even though there are 22 chromosomes, because each diploid cell has two copies of chromosome 1, two copies of chromosome 2, and so on.

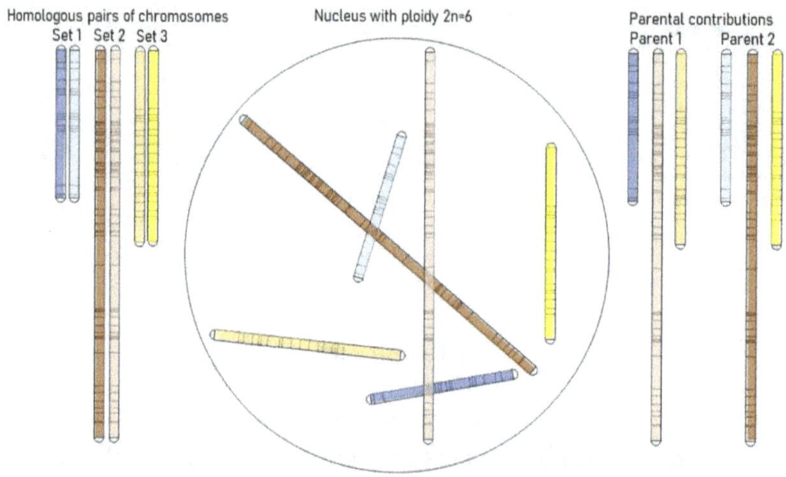

Illustration by Dr. Matt Clark, University of Minnesota

The illustration above shows this for a hypothetical plant's somatic cell's nucleus containing 6 chromosomes. On the left side, the chromosomes are rearranged into three pairs of homologs. The matching chromosomes from the two different sets (for instance, the two copies of chromosome 1) are called homologous chromosomes **or** homologs. Homologs carry, at the same location on the chromosome, the genetic information that affects the same characteristic or function. The version of the information can be different between the homologous chromosomes — that is, the sequence of base pairs may be somewhat different because one homolog came from the female and the other from the male.

The homologs look identical and carry genetic information about particular cell functions at identical places on the chromosome (shown using dark bands at specific locations on the chromosome), but the exact base pair sequences at those locations may differ, resulting in different alleles and gene function. The parental combinations are shown at the right, and are the haploid contribution that resulted from meiosis. The 50% reduction in the sex cells ensures that offspring have the proper diploid chromosome number and matching homologs that are the full compliment of the plants genome.

A cell in the plant's apical meristem that is preparing to divide is a somatic cell, so it is diploid, and contains two sets of chromosomes. When it undergoes mitosis, the outcome will be two identical diploid sister cells. Each of these sister cells will also be diploid, and will contain exact copies of the two sets of chromosomes that were in the original cell. "Daughter" and "sister" cells refer to the same thing—the new cells that arise as the result of mitosis.

From our study of meristems, you know that growth is the result of the formation of new cells, and the subsequent elongation of those cells. Mitosis is the process that results in the formation of new cells. Cells undergo mitosis, therefore, as part of plant growth.

Review questions

- Somatic cells of beans have 22 chromosomes. How many chromosomes in a bean sperm cell?
- Corn egg cells have 10 chromosomes. How many chromosomes are found in a corn seed's endosperm cells?
- Why do cells undergo mitosis? Is it important?

Synthesis

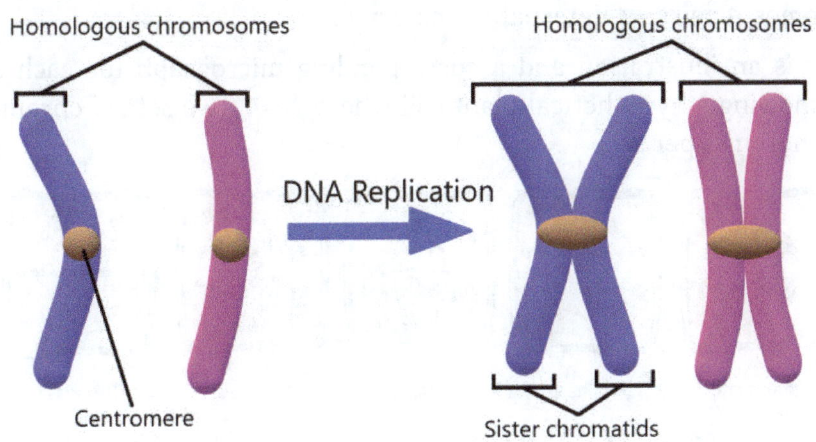

Outcome of S phase. Christinelmiller. CC BY-SA 4.0

If a cell that undergoes mitosis divides into two cells, how can both of these new cells be identical to each other and to the original cell? Won't the chromosomes in the original parent cell be divided in half during division? Won't the resulting cells be haploid instead of diploid? No. The number of chromosomes isn't reduced during mitotic cell division because, prior to division, each of the chromosomes replicates (duplicates), meaning that the cell makes an exact copy of each chromosome. This replication process happens during the synthesis (S) phase of the cell cycle.

Remember that G1, S, and G2 phases of the cell cycle are collectively called interphase. Most cells in the plant go about their business in the G1 phase. Only those cells called upon to divide make the next step, which is to replicate their chromosomes in the S phase. Once the chromosomes are replicated, the cell moves into the G2 phase of interphase and awaits mitosis. The S phase is called synthesis because making a copy of the chromosome requires new DNA production, or synthesis.

The two chromosomes that are exact copies are called sister chromatids and remain connected at one spot along their length; this spot is called the centromere, as shown in the illustration.

Review questions

- If a diploid cell enters S phase with 2n=20 chromosomes, how many sister chromatids are in the cell when it enters G2?
- Are the replicated sister chromatids independent or are they connected in some physical way?

Stages of mitosis

This video provides a view of the fluidity of mitosis in a cell where 2N = 8 chromosomes, 4 pairs = 4 paternal + 4 maternal.

Below is an illustration and a corresponding micrograph for each stage in mitosis, showing a hypothetical plant cell where 2n=4 (two sets of chromosomes, two chromosomes per set).

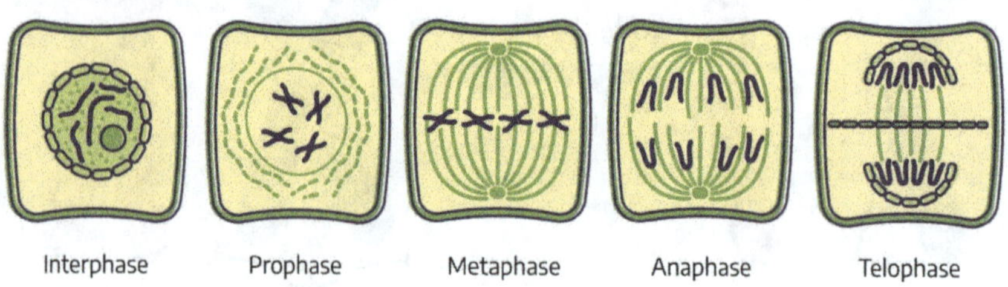

Stages of mitosis. © VectorMine | Dreamstime.com. Used with permission.

The micrographs below are onion (*Allium cepa*) root tip cells. Onion has 2n=16 chromosomes. Each of the cells has two sets of chromosomes where each set is made up of eight chromosomes. The micrographs are real examples of the illustrations above.

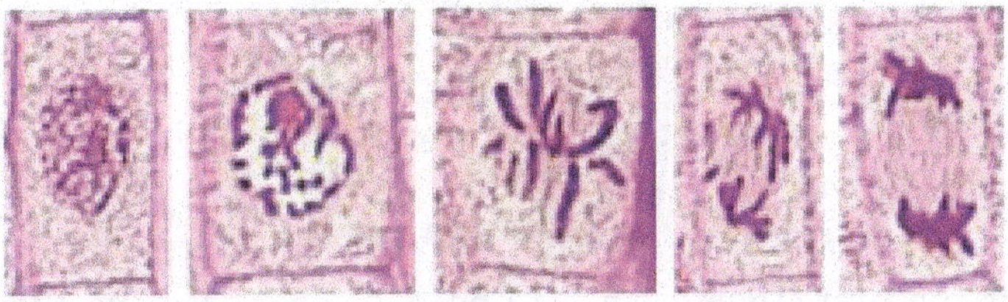

Stages of mitosis in onion root tip. Melissa Ha. CC BY-NC-SA 2.0

Interphase

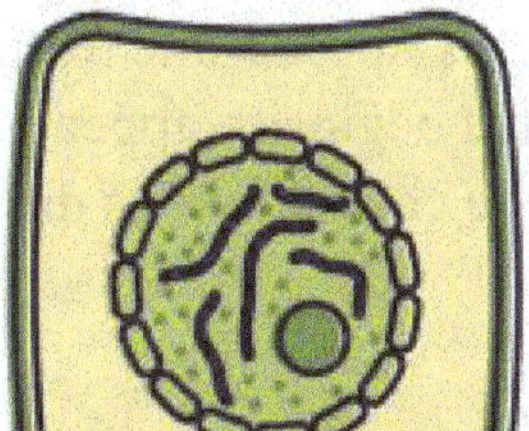

The lefthand frame of the illustration shows interphase cells. The deep red stained structures in the center of the onion cell micrograph are the chromosomes. They are corralled together within the nuclear membrane. Recall that during interphase the chromosomes are relaxed rather than highly condensed (that is, not extensively coiled or folded), and during the S phase of interphase each chromosome replicates. It makes sense that the chromosomes are relaxed because they can't go through the replication process if they are tightly coiled, and because chromosomes only need to be coiled so that they can withstand movement and not break. They aren't moving, just replicating, so being in a relaxed state is perfect. The two identical copies are called sister chromatids and they are held together at a site called the centromere. Note that sister chromatids are not the same as homologs. Homologs are corresponding chromosomes, one contributed through the sperm, the other through the egg. Sister chromatids are chromosomes that have replicated, are identical to each other, and are held together at centromeres. You can't distinguish individual chromosomes in the picture because they are relaxed rather than tightly coiled and folded, making them so fine that they are difficult to see.

Prophase

Prophase is the first stage of the M phase. In prophase the chromatin begins to coil and condense to form chromosomes. They are coiling because they are preparing to move around. You can begin to notice that each chromosome appears to have two strands (sister chromatids) and that these sister chromatids are attached to each other at a centromere. In prophase the nuclear membrane disappears and the chromosomes spread out to fill up much of the cell. During this phase, the spindle apparatus begins to appear. The spindles are microtubules associated with movement of the chromosomes during division.

Metaphase

In metastage the spindle grows and forms attachments to the pairs of sister chromatids at the centromere that connects the sister chromatids. This point of attachment is called the kinetochore. The sister chromatids move to an imaginary equatorial plate (called the metaphase plate), which is formed along the midline of the cell between the poles. The sister chromatids are in their most condensed state at metaphase.

Anaphase

The sister chromatids begin to separate at anaphase. When the sister chromatids separate, the centromeres divide so that one sister chromatid migrates to one pole, and the other migrates to the opposite pole. The chromatids that formed back in the S phase of interphase, when the chromosome replicated, now separate, and the spindle fibers shorten. With the sister chromatids separated, we can return to calling them chromosomes. Anaphase is the stage where the chromosomes carrying the DNA code are divided precisely so that each of the resulting cells has exactly the same chromosomes that were in the mother cell prior to division. One complete diploid complement of chromosomes (two sets) is delivered to each daughter cell.

Telophase

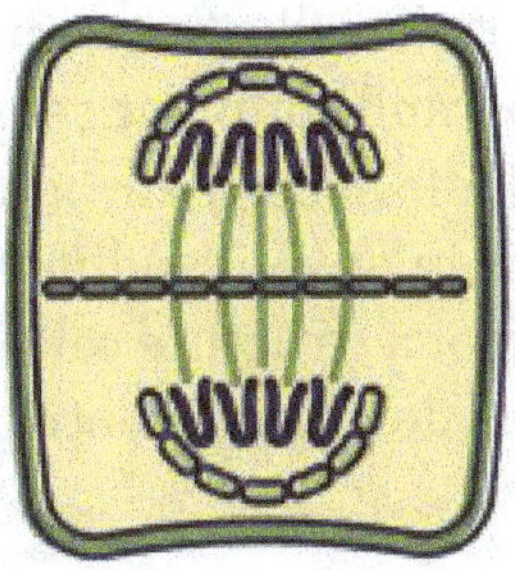

In telophase, the nuclear membrane forms around the chromosomes in each of the daughter cells, a cell plate forms between these cells, and cell walls separate the newly formed cells in a process called cytokinesis. The chromosomes decondense and again become relaxed chromatin. Telophase is the last stage of the M phase. After telophase and cytokinesis, the cells return to G1 of interphase.

Review questions

- When do the sister chromatids separate from each other?
- Do the chromosomes replicate during mitosis or during interphase?
- Why are the chromosomes in their most condensed state during metaphase and retain this condensed state through chromatid migration in anaphase?

Review

Recall that the outcome of mitosis is two cells with DNA identical to that in the original cell. There are three keys to understanding how two cells are formed from one, both with the same DNA as the original cell:

- The DNA is completely replicated during the S stage of interphase. This replication results in twice as many sister chromatids as there were chromosomes, and once these sister chromatids separate and are evenly allocated to the two new sister cells, both sister cells have the diploid number of chromosomes, just like the original cell prior to division.

- As the cell prepares to divide, the DNA condenses. This packaging helps keep the very thin DNA helices from being broken, and keeps the DNA organized into a tight package so that the cell can keep track of it and move it around.
- The process is very organized. For instance, the sister chromatids all line up in the middle of the cell at metaphase, split at the centromere, and half the chromatids go to one side of the cell, half to the other. This orderly separation of the sister chromatids ensures that the right number of chromosomes is packaged into each of the new sister cells.

There are many sites online that illustrate mitosis, but particularly relevant here are ones that show micrographs of plant cells. You may discover that there are some details about the spindles and their apparent site of origin that differ between descriptions of mitosis in animal and plant cells; not everything online pertains to plants. Any mention of a structure called a "centriole" refers to animal cell mitosis, not plants (as plants don't have centrioles).

13.3 MEIOSIS

Learning objectives

By the end of this lesson you will:

- Understand how meiosis starts with one diploid cell and results in four haploid cells.
- Know how meiosis produces gametes that are genetically diverse.
- Be able to model the stages of meiosis.
- Be able tompare the similarities and differences in the mechanics of mitosis and meiosis.

Review of sexual and asexual propagation

As seen earlier, there are two broad categories of plant propagation: sexual and asexual. When new plants are produced from existing plant parts, like pieces of leaf, stem, or root, reproduction is asexual and the only type of cell division that has taken place is mitosis, where one diploid cell produces two identical diploid cells.

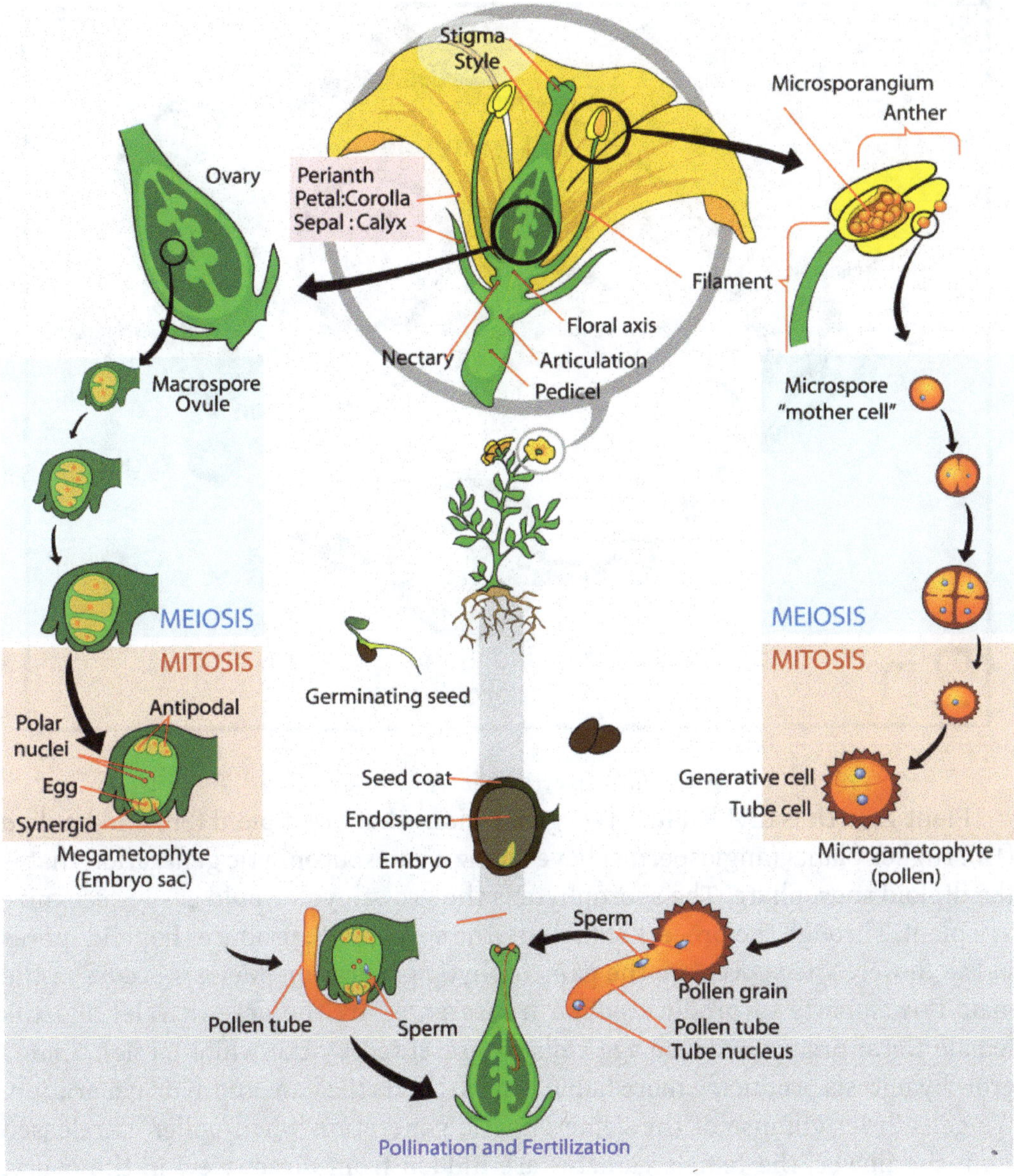

Sexual reproduction of an angiosperm. LadyofHats, Public domain, via Wikimedia

If new plants are instead produced from seeds, this is a strong indication (but not a certainty...optionally read about apomixis) that reproduction was sexual. Plants that practice sexual reproduction use mitotic cell division when increasing the diploid vegetative parts of the plant like stem, leaf, and root, but use meiotic cell division to initiate the haploid stage of the plant that ultimately results in production of egg and sperm cells central to sexual reproduction. Instead of two diploid cells from one diploid cell (the outcome of mitosis), the outcome of meiosis is four haploid cells from one diploid cell.Commons.

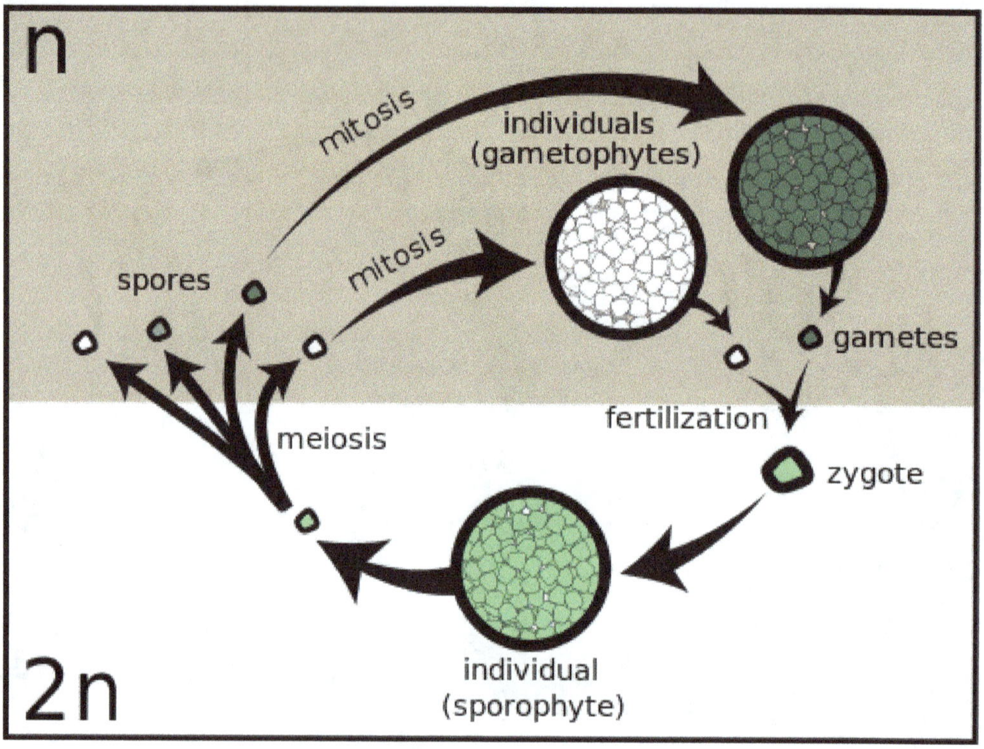

Mitosis/meiosis cycle. Menchi. CC BY-SA 3.0

Plant growth is divided into two generations that are diploid (2n) and haploid (1n). Higher plants (angiosperms) have a long-lived sporophytic generation that is the diploid sporophyte. The sporophyte is the growth you would easily recognize as a plant. Through the process of meiosis, the sporophyte produces haploid spores in the flower. The spores are the gametophytic generation. Meiosis occurs in the male flower parts to produce pollen (represented by the green circle) and the female floral organs produce egg cells (represented by the white circle). Spores grow by mitosis producing more haploid cells, this is the gametophytic generation. We get a brief glimpse of the gametophytic generation when pollen is released from the flower, the female gametes are hidden from direct view in the ovary. When the haploid gametes (male pollen and female egg cells) unite they reform the sporophytic generation producing a diploid (2n) zygote. The zygote grows into the embryo of the seed and eventually into the plant we see. Lower plants, mosses and ferns that are not flowering plants, also alternate generations, but the gametophytic generation is longer lived and separate from the sporophytic generation.

The plant doesn't magically transition to being haploid, but instead particular parts of the flower in the androecium and gynoecium develop and protect a limited number of haploid cells, called the male gametophyte and female gametophyte. A later chapter addresses how the male and female gametophyte include the haploid egg and sperm cells that must unite to form the diploid embryo in seeds. For now, know that meiosis is the gateway into the haploid phase. Meiosis is the type of cell division that starts with diploid cells and results in haploid cells. Without meiosis there is no egg and sperm, and thus no sexual reproduction.

Propagation and natural selection

From the natural selection perspective, how do asexual and sexual reproduction differ?

In asexual reproduction, the plants are genetic copies of the parent plant. Cell division is strictly mitosis. Except for rare mutations, the resulting progeny are identical to the parents. The fitness of the progeny will mirror the fitness of the parent. The downside of this type of propagation is that there is no genetic variance among progeny that might result in selection for plants that have greater fitness than the parent for characteristics such as increased cold hardiness, drought tolerance, or disease resistance. The upside is that if the parent has high fitness to begin with (and it must have had reasonable fitness to reach reproductive age), all progeny will also have that high level of fitness. If the environment remains the same as it was for the parent, the progeny will stand a good chance of reproductive success. But if the environment changes, the fitness of the progeny may no longer be optimum.

In sexual reproduction, since one gamete comes from the male parent and one from the female, and because in a population of cross-pollinating wild plants there are many potential parents, each with different genotypes, there are many potential genetic combinations of male and female gametes. Not only are the plants producing the gametes each genetically different, but each gamete from each plant is potentially unique. The many combinations of male and female gametes, and the uniqueness of gametes from the same plant, result in a substantial genetic variation among the progeny of plants that sexually reproduce. Some of these progeny will have greater fitness than others, and will be favored by natural selection — some will survive to reproductive age and have more progeny than other plants, while the rest will either not survive to reproduce or, if they do reproduce, it will be with low frequency. The DNA of the fittest plants will thus be represented more frequently in the next generation of plants than the DNA of the least fit plants, which may never survive to reproduce and pass on their DNA. This is the process of natural selection. The DNA of reproductively successful parents is passed on the next generation, while the DNA of reproductively unsuccessful parents is not.

Sex generates genetic variation. Genetic variation, generated by meiosis and sexual reproduction, is the fuel for the engine of natural selection.

Ploidy review

To review: if you count the number of chromosomes in a somatic cell, for instance a root tip cell, you will find that there is always an even number. These are diploid 2n cells that arose from mitotic cell divisions tracking all the way back to the zygote that formed the embryo of the seed. Listed below are the numbers of chromosomes found in somatic, diploid, 2n cells of a few commonly grown plants. Note that the number of chromosomes is even, never odd, and that it doesn't imply anything about the size or type of plant:

- Corn = 20
- Rice = 24
- Soybean = 40
- Green bean = 22
- Tomato = 24
- Potato = 48
- Apple = 34
- Rose = 14

Diploid cells always contain an even number of chromosomes because there are two copies of each chromosome, one contributed by the male sperm and one by the female egg. If you number each type of corn chromosome 1 through 10, there would be two 1s (a maternal and a paternal), two 2s, etc. Recall that the two (donated from the male and female) versions of the same chromosome in a diploid cell are called homologous chromosomes or homologs. In a diploid cell like corn where $2n$=20, there are 10 pairs of homologous chromosomes.

Also recall that the number of chromosomes in a gamete is half the number of chromosomes found in a somatic cell of the same plant. The gamete cells are haploid, abbreviated n. There may be an even or odd number of haploid chromosomes, depending on the diploid chromosome number. Beans have a diploid number of 22, so the gametes have an odd number of chromosomes (11). Tomato has a diploid number of 24, so the gametes have an even number of chromosomes (12).

When two gametes fuse and form a zygote, the zygote has the 2n chromosome number restored. From then on, the cell divisions that allow a plant to grow from zygote to full size are all mitosis, and all the cells are copies of the zygote formed by fusion of the two gametes.

Review questions

- In what sense is meiosis the gateway into the haploid or gametophytic stage of alternation of generations?
- Why does a diploid plant cell always have an even number of chromosomes?
- A diploid rose cell has 14 chromosomes. How many pairs of homologous chromosomes will you find in that diploid cell?

Meiosis mechanics

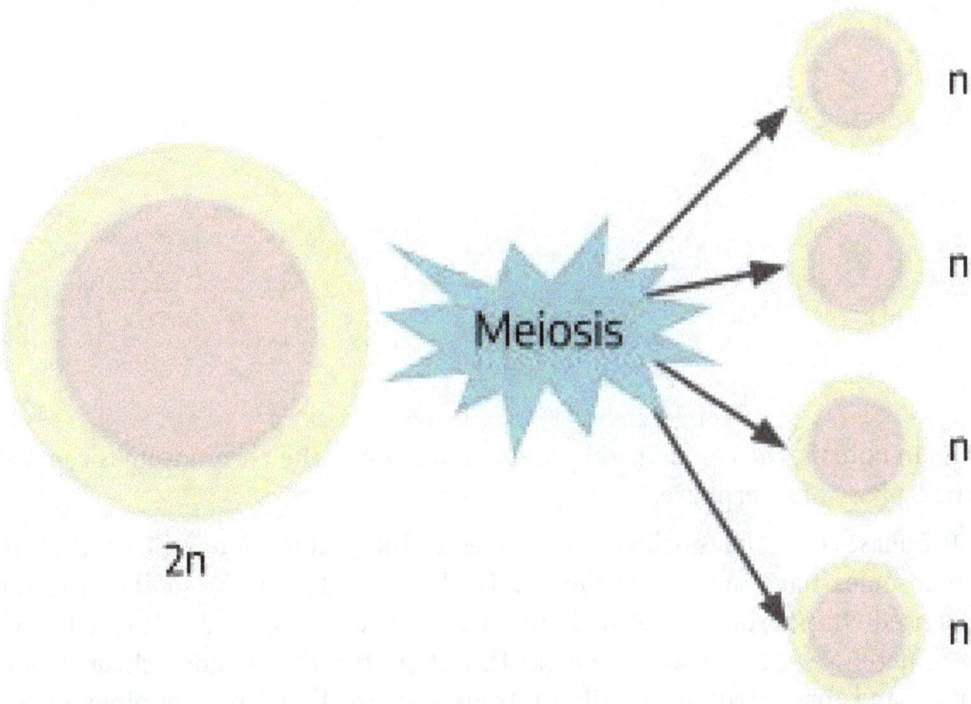

Meiosis. Emily Tepe

Meiosis starts with a diploid cell and results in haploid (n) cells that we could correctly call spores. In the illustration above, note that starting with one diploid cell and meiosis yields four haploid cells.

Below are the stages of meiosis. It won't be difficult to memorize them because you already know the stages of mitosis, and meiosis builds on mitosis. The illustration shows a hypothetical species with two pairs of chromosomes (2n=4) in the starting cell and n=2 in the resulting gamete cells. Meiosis has two chromosome divisions, so the stages are labeled I for those stages associated with the first division (e.g., Metaphase I) and II for those associated with the second division (e.g., Metaphase II).

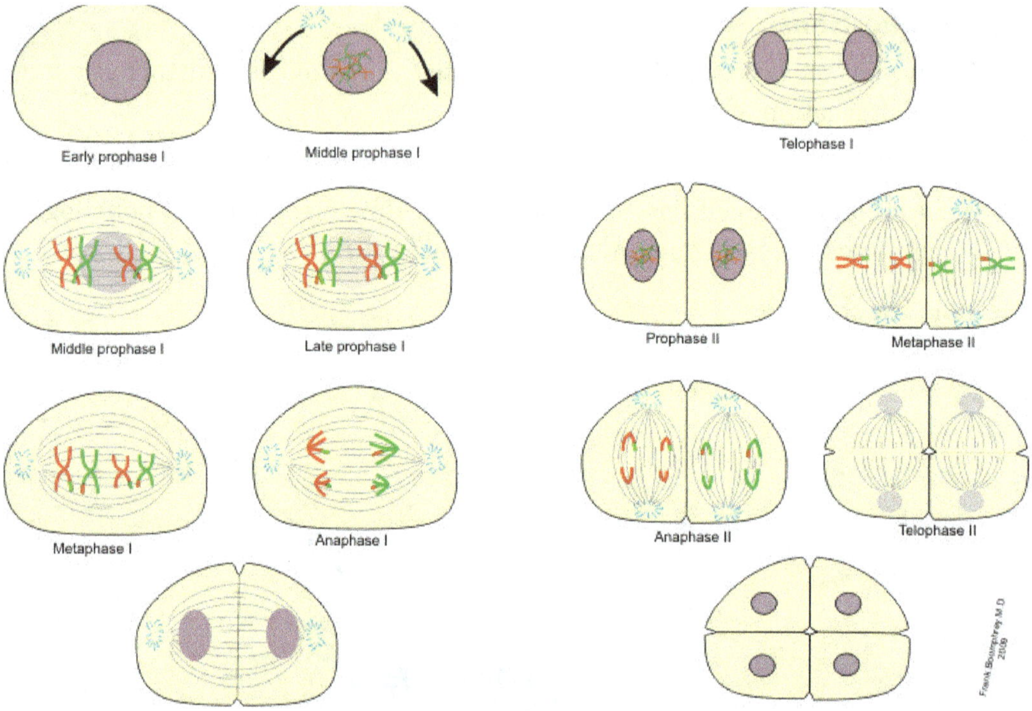

Stages of meiosis. Boumphreyfr. CC BY-SA 3.0

As in mitosis, the cell division process starts when the chromosomes replicate in the S phase of interphase.

Prophase I — the nuclear membrane disintegrates, and we see that the chromosomes have already replicated (in "S" of interphase) so that the now condensed chromosomes are made up of two sister chromatids attached at the centromere. New, however, is that in this stage the homologous chromosomes pair up and form structures called tetrads because they are groupings of four sister chromatids (two sister chromatids per homolog). This process of pairing and tetrad formation promotes chiasma (seen under the microscope as the point where sister chromatids of homologs lay over each other, forming an "X" shape) and crossing over between sister chromatids of homologous chromosomes. Crossing over results in an exchange of DNA between homologs and is another contributor to genetic variation in the gametes and resulting organisms.

Reread the preceding paragraph, making sure you understand how homologs pair, form chiasma, and cross over between sister chromatids. This type of pairing of homologs and subsequent chiasma formation doesn't happen in mitosis — a very important and essential difference between the two types of cell division.

Metaphase I — the tetrads are lining up on the metaphase plate, ready to divide. Recall that they are called tetrads because they are made up of four sister chromatids.

Anaphase I — the tetrads divide and homologs go to opposite poles. Note that sister chromatids stay intact and the centromeres do not divide yet. It is the homologous chromosomes that separate at Anaphase I. The sister chromatids may differ in some places along their arms due to crossing over that causes exchange of DNA between homologs.

Telophase I — the nuclear membrane reappears to separate the products of the first division.

The DNA relaxes in Interphase I, but no replication occurs. DNA condenses again in Prophase II, and we can see the chromosomes.

Metaphase II — the chromosomes (which are in the form of sister chromatids still connected at the centromere) line up at the metaphase plate, as in metaphase of mitosis.

Anaphase II — also like anaphase in mitosis, the centromeres split and sister chromatids are pulled to opposite poles.

Telophase II — nuclear membrane reforms, cytokinesis takes place, just like the telophase of mitosis. Note that the "II" stages of meiosis are just like the corresponding stages of mitosis, making them easy to remember.

Review questions

- Do the chromosomes replicate prior to meiosis?
- What happens during crossing over?
- The illustration above shows that some of the sister chromatids are combinations of red and blue rather than being all red or all blue. What does that represent?
- What separates in Anaphase I?
- What separates in Anaphase II?

Division

Memorizing this process helps you focus on clearly understanding the mechanics of the process, and recognize how it is that meiosis results in four n haploid cells instead of the two 2n cells that result from mitosis. You get four n haploid cells because the initial cell undergoes two divisions. The cell first divides into two nuclei, then those two divide again into four. The chromosome number drops from 2n in the original cell to n in each of the four haploid cells because the number of sets of chromosomes is reduced from 2 to 1 (that is, homologs separated to opposite poles) in the first meiotic division, and then the sister chromatids separated in the mitosis-like second meiotic division. There was chromosome replication before the first division, but no replication before the second division.

Here's a summary of what is dividing and when:

Two divisions:

Homologs separate in Anaphase I. Centromeres holding the chromatids do not split.

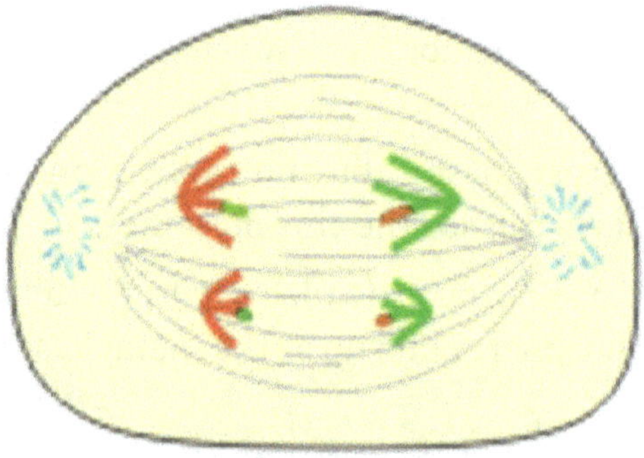

Anaphase I. Boumphreyfr. CC BY-SA 3.0

Chromatids separate in Anaphase II. Centromeres holding the chromatids do split.

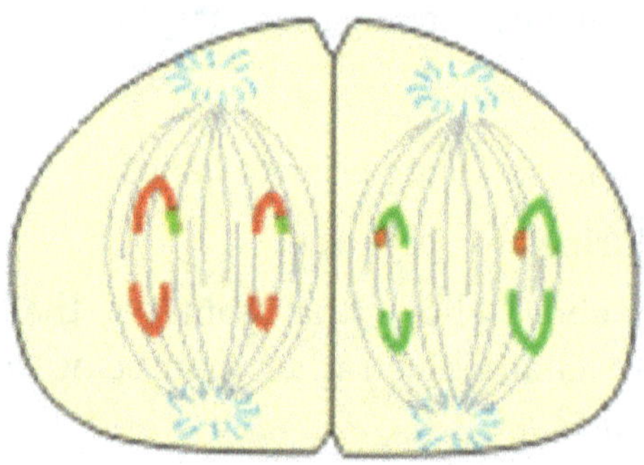

Anaphase II. Boumphreyfr. CC BY-SA 3.0

Genetic variation among gametes

Each gamete ends up with one of the homologs of the pair, not both.

- Imagine that there are two or three or even 30 pairs of homologous chromosomes. Each homolog pair making up the tetrad in Metaphase I separates in Anaphase I. One homolog from the pair heads to one pole and the other heads to the opposite pole. Each pair of homologs moves independently of all the other homolog pairs that are also separating. That is, all the paternal-source chromosomes making up the homologous pairs don't go to one pole and all the maternal-source go to the other pole (this is possible, but since it would be by chance the probability is very low).

Instead, paternal-source homologs of some chromosomes and maternal-source homologs of the other chromosomes are pulled to the poles, so that eventually there is a mix of maternal- and paternal-source chromosomes in the gametes, and that mix of maternal and paternal is generally thought to be random.

- This principle, where homologs move to poles independently, is called independent assortment, and leads to differences in the genotype of the gametes — one source of genetic variation among gametes.

Crossing over and exchange of DNA between homologous sister chromatids in Prophase I is another source of variation in gametes. If two cells in the sporangia are undergoing meiosis, the crossing over in each cell will probably happen in different places on the chromosome in each cell, resulting in exchanges taking place at different locations on the DNA backbone so that the gametes resulting from different cells going through meiosis will all be unique.

What is crossing over?

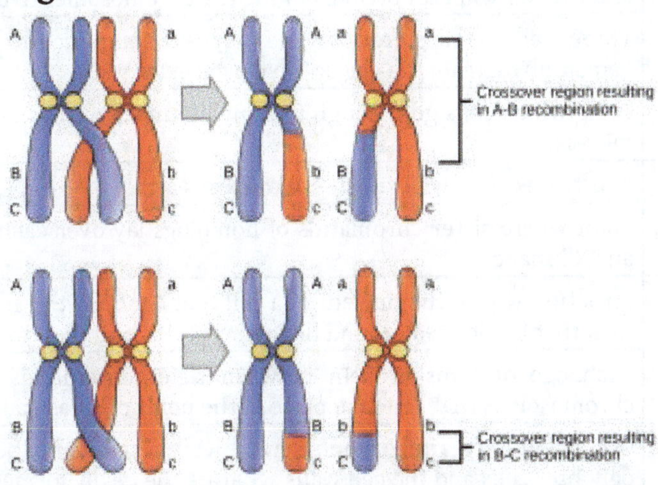

Crossing over. OpenStax. CC BY 4.0

The illustration above shows a tetrad where the red and blue homologous chromosomes of the same chromosome type are pairing. Each homolog here consists of two sister chromatids joined at the centromere. Note that the chromosomes have already replicated in interphase prior to the start of Prophase I.

- In Prophase I, homologous chromosome pairs come together (synapsis).
- Arms of sister chromatids from different homologs overlap (chiasma) and exchange DNA (crossing over).
- Note that all four of the resulting sister chromatids are now genetically different, with each potentially having a different DNA sequence.

Summary

- Meiosis is a type of cell division that starts with a diploid, 2*n* **cell.**
- The process includes two chromosome divisions and produces four haploid, *n* **cells.**
- The haploid cells are genetically different from each other due to crossing over in Prophase I and independent assortment in Anaphase I.
- Homologs separate in Anaphase I while sister chromatids separate (the centromeres divide) in Anaphase II.

TERMS

Flashcards

Alternation of generations	Cycle of diploid, asexual, vegetative generation alternating with the haploid, sexual generation.
Anaphase	Third phase of mitosis; the sister chromatids separate (now chromosomes) and the centromeres divide, pulling the chromosomes to opposite poles.
Antipodal cells	Three cells sequestered at the opposite end of the mature female gametophyte from the egg and synergid cells.
Cell cycle	Cycle which cells go through in their lifetime; consists of interphase and mitosis.
Centromere	Constricted spot where sister chromatids attach.
Chiasma	Point where sister chromatids of homologs lay over each other, forming an "X" shape.
Chromosome	Structure within the nucleus of a cell that contains the genes; made up of DNA that has looped around histone proteins and then coils and folds.
Crossing over	Exchange of arms of DNA between sister chromatids of homologous chromosomes that can take place at the point of chiasma formation.
Cytokinesis	Occurs directly after telophase; the cell plate forms between the two daughter cells and the cell walls separate the newly formed cells.
Diploid	Term used for zygote cells, where the cell has two sets of chromosomes; abbreviated 2n.
DNA	Basic biochemical compound that makes up the gene.
G1 stage of interphase	First stage of interphase; "G" stands for Gap/Growth.
G2 stage of interphase	Third and final stage of interphase; "G" stands for Gap/Growth.
Genetic code	Order of the four different combinations of the bases in DNA; AT, TA, GC, or CG.
Haploid	Term used for gamete cells that typically contain one set of each of the chromosomes; abbreviated n.
Histone protein	Protein around which the DNA surrounds.

Homologous chromosomes (homologs)	Matching chromosomes from the two different sets; carry the genetic information that affects the same characteristic or function at the same location on the chromosome; from sperm and egg cells.
Interphase	One of the two major parts of the cell cycle; consists of G1, S, and G2 stages.
Kinetochore	Point of attachment of the spindle and the centromere.
Metaphase	Second stage of mitosis; the spindle fibers grow and form attachments to the pairs of sister chromatids at the centromeres.
Metaphase plate	Equatorial plate formed along the midline of the cell between the poles.
Nucleosome	Made up of eight histone proteins and wrapped by a segment of DNA.
Ploidy	Number of sets of homologous chromosomes in a cell.
Polar nuclei	Two haploid nuclei contained within one cell membrane in the mature female gametophyte. One sperm cell will unite with these two polar nuclei to establish the triploid endosperm tissue.
Prophase	First stage of mitosis; chromatin begins to coil and condense to form chromosomes.
Purine	Consists of the base pairs Adenine and Guanine and contains two rings of carbon atoms.
Pyrimidine	Consists of the base pairs Cytosine and Thymine and contains one ring of carbon atoms.
Ribose-phosphate backbone	Chain of alternating ribose and phosphate that make up the sides of the DNA structure.
S stage of interphase	Second stage of interphase where the chromosomes replicate (DNA replicated).
Sister chromatids	Two chromosomes that are exact copies and are created during the S stage of interphase.
Somatic cells	Cells of flowering plants, other than the reproductive cells; always 2n.
Spindle apparatus	Microtubules associated with movement of the chromosomes during division.
Sporangia	Structures in the androecium and gynoecium where meiosis takes place and the gametophyte generation develops.
Spore	Haploid single cell produced by meiosis in the sporangium of a diploid sporophyte.
Telophase	Fourth and final stage of mitosis; the nuclear membrane forms around the chromosomes in each of the daughter cells.
Tetrads	Groupings of four sister chromatids.
Triploid	Term used for endosperm that has three sets of chromosomes; abbreviated 3n.

CHAPTER-14

VARIATION AND PLANT BREEDING

Plant propagation relies on gametogenesis for fertilization and the formation of fruit and seeds. Unlike asexual propagation, seeds from a single pollination can generation tremendous variation, depending on the diversity that exists within each parent. Geneticists take advantage of this diversity to understand how genes control the plant's phenotypes and to make improved varieties that are tastier and have improved disease resistance.

Learning objectives

- Identify the source and mechanisms that produce gametes.
- Explain double fertilization and the production of a zygote, embryo, and other seed structures.
- Contrast the difference between simple/qualitative and complex/quantitative inheritance and the basis of that difference.
- Predict the types of F_1 and F_2 offspring expected when crossing two parents with known genotypes and phenotypes.
- Understand how heritability is a measure of genetic influence, relative to other non-genetic influences.

14.1 GAMETOGENESIS

Recall, as shown below, that the flower is made up of a compressed, four-node stem called the receptacle that supports four whorls of modified leaves: calyx (sepals), corolla (petals), androecium (stamens), and gynoecium (carpels). The two whorls nearest the tip of the receptacle, the androecium and gynoecium, contain structures housing the (micro- and mega-) sporangia, where meiosis takes place and the gametophyte generations develop.

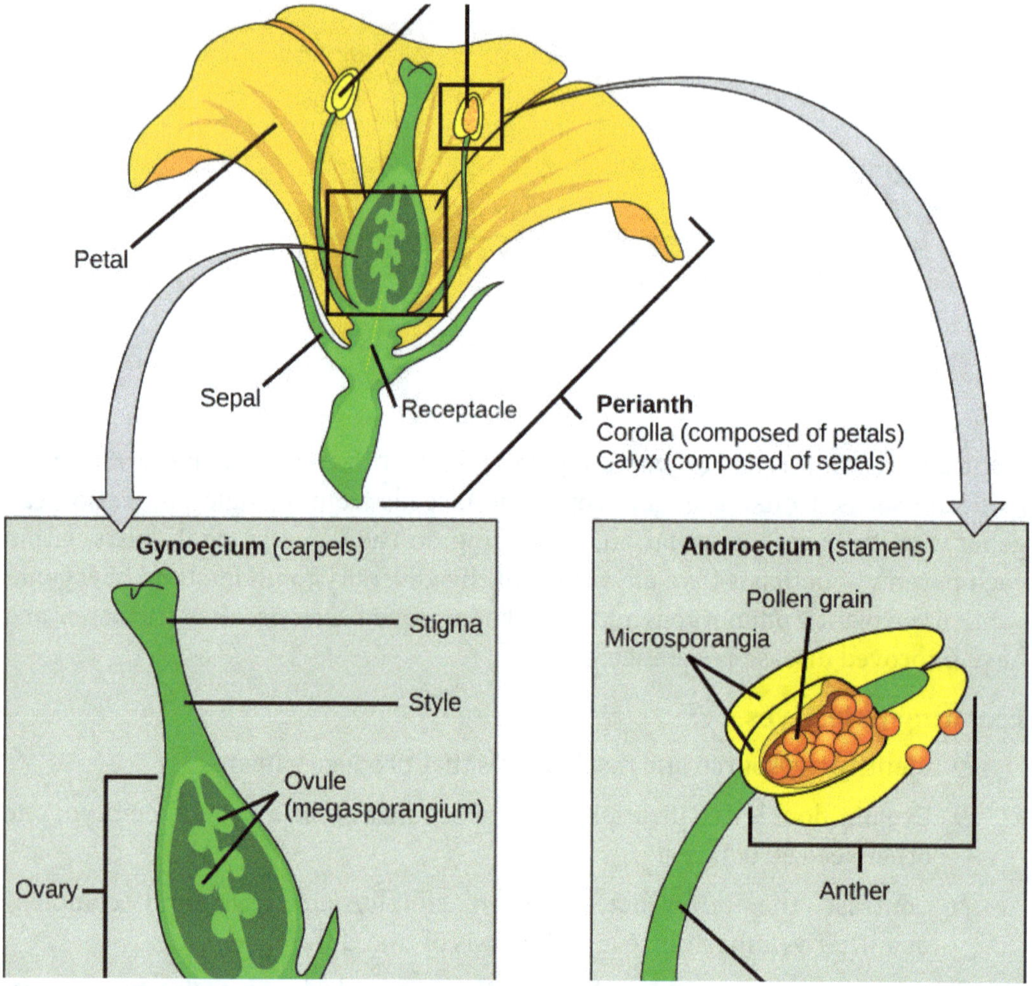

Parts of a flower. OpenStax Biology. CC BY 4.0

Sporangia

The sporangia of angiosperm plants are found in two places within the flower depending, on whether they lead to a female or male gametophyte. The sporangium in the anther (where the male gametophyte will be formed) is called the microsporangium. The sporangium in the ovary (where the female gametophyte will be formed) is called the megasporangium. Diploid microspore mother cells in the microsporangium and diploid megaspore mother cells in the megasporangium divide by meiosis to form four haploid micro- or mega- spores. These spores initiate the gametophyte phase. Two (male) to three (female) mitotic divisions later, the mature micro- and mega- gametophytes have been formed. Note how few divisions take place to form the gametophyte after the meiotic division of the microspore or megaspore mother cell, and that those two or three additional divisions are mitotic divisions. One round of meiosis is sufficient to generate haploid cells. From there on the divisions are mitotic, but in this case mitosis starts with one haploid cell and generates two haploid cells.

Helleborus foetidus cross-section. Simon Garbutt. Public domain

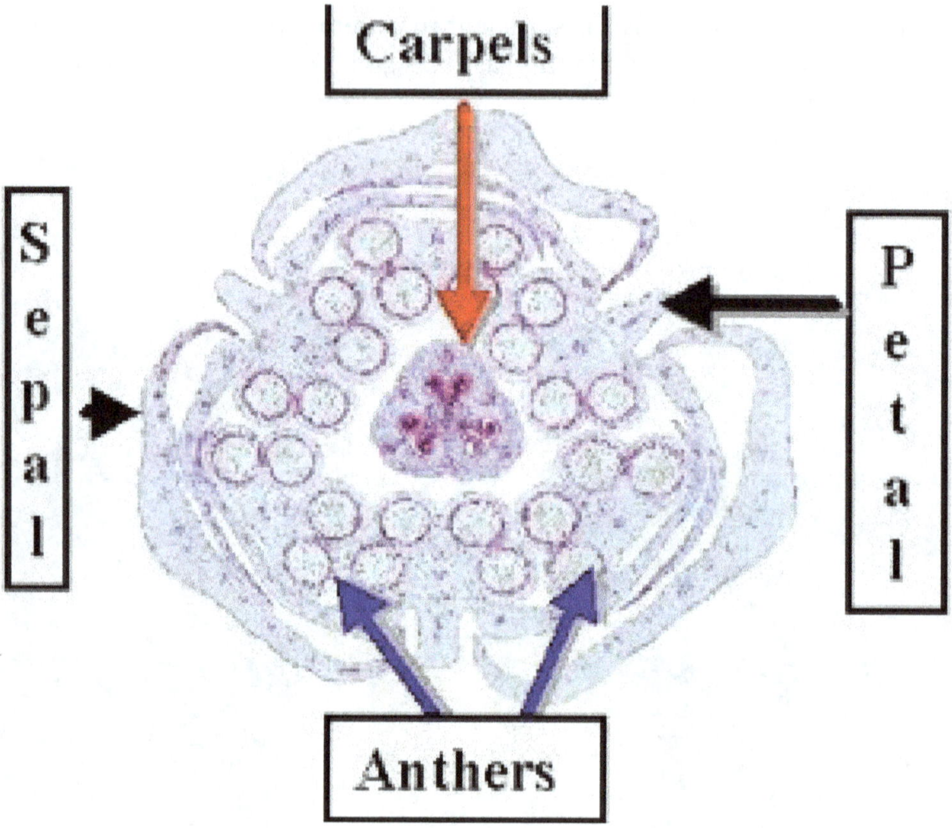

Lilium flower bud cross section. Photomicrograph by Michael W. Clayton, used with permission. © Board of Regents of the University of Wisconsin System.

The images above provide a visual orientation to the micro- and megasporangia. In the photo, the bud in the horizontal cross section shows the three chambers in the fused carpels as well as a number of anther cross sections. In the center of the vertical cross section of the flower bud you'll see a carpel with ovary, locule (interior chamber), and ovule within the locule. Also note the anther cross sections containing pollen.

Microsporogenesis

The illustration below shows the development of the male gametophyte within the anther. The process, called microsporogenesis, or male gametogenesis, starts in the top left hand corner. Study these steps and be able to draw them. Red "2*n*" or "*n*" notations indicate the ploidy of the nuclei and blue notations show where the mitotic cell divisions occur. To this point we've learned that mitosis starts with one diploid cell and results in two diploid cells. We'll modify that now to say that mitosis starts with a cell and results in two cells that are exact copies of the original cell. If the starting cell is diploid, there will be two diploid copies. If the starting cell is haploid, there will be two haploid copies.

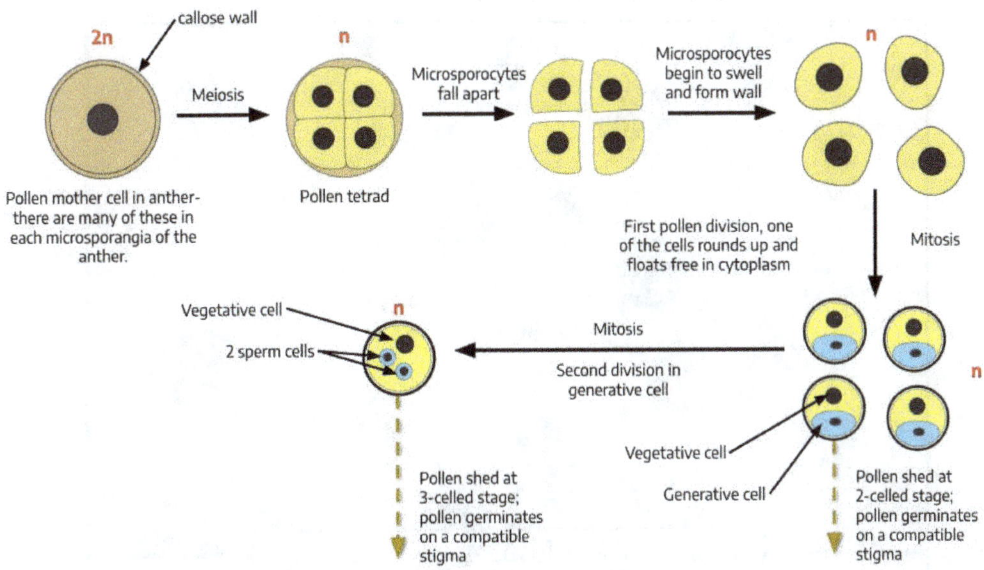

Pollen development and growth. Pollen can be shed at either the two-celled or three-celled stage; it varies from species to species. Image labeled and adapted by Emily Tepe. © Shutterstock. Used with permission.

The result of microsporogenesis, shown on the bottom row of the illustration above, is either a male gametophyte that is a two-celled pollen grain (with vegetative and generative cells), where the generative cell will later undergo another mitotic division to produce two sperm cells (shown to the right of the box), or, following mitosis of the generative cell and shown to the left of the box, a three-celled pollen grain with a vegetative (or tube) cell plus two sperm cells.

Below is a cross section of young anther showing the developing pollen grains in four chambers.

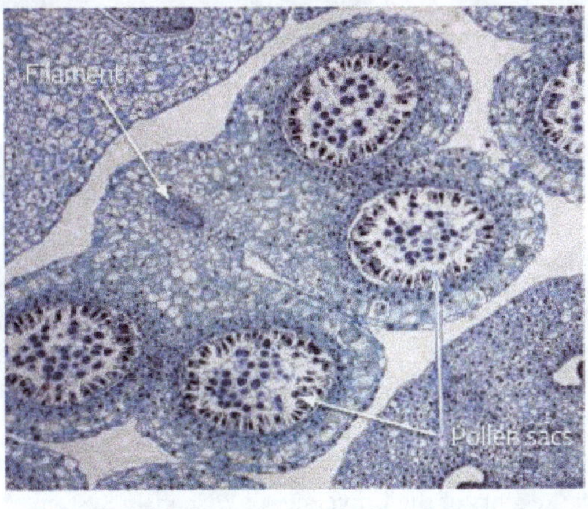

Lilium anther cross-section. Marc Perkins. CC BY-NC 2.0

Below are tetrads of cells following meiosis of the microspore mother cell. Note the sets of four cells still attached together.

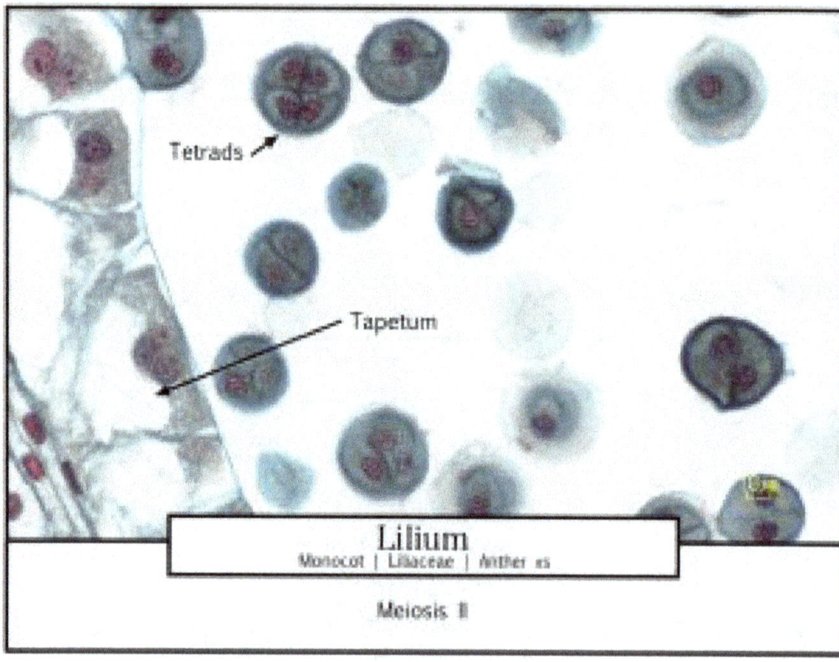

Lilium tetrads. Jen Dixon. CC BY-NC-SA 2.0

Below are nearly mature, two-celled pollen grains with the start of an exine coat. The exine coat is made up of protein and other compounds deposited by the inside wall of the anther, and it forms a protective coat around the three-celled microgametophyte.

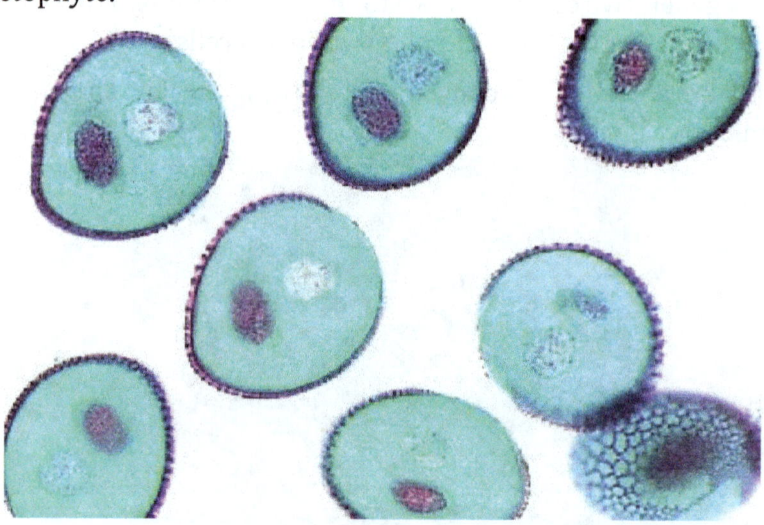

2-celled pollen grain of lily. Photomicrograph by Michael W. Clayton, used with permission. © Board of Regents of the University of Wisconsin System.

In this micrograph the exine coat looks like rough, geometrically shaped plates on the surface of the pollen. The exine coat patterns are characteristic of the species and can be used to identify the species of plant that produced the pollen. For more forensic botany, see this optional link from the Botanical Society of America.

Below is a mature, dehiscing (split and releasing pollen) anther.

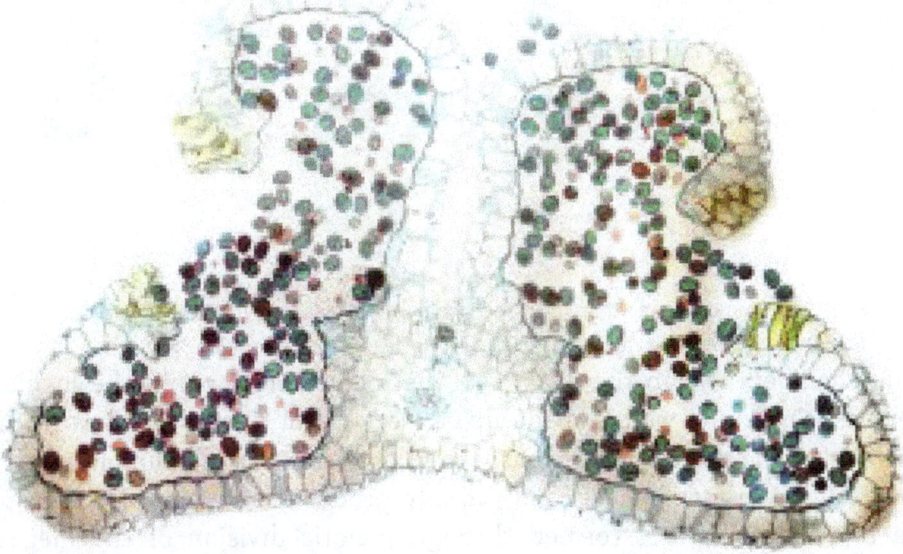

Lilium – cross section of dehiscing anther with two-nucleated pollen grains. Photomicrograph by Michael W. Clayton, used with permission. © Board of Regents of the University of Wisconsin System.

Review questions

- Where are the microsporangia located?
- Where is the mature male gametophyte in the Pollen Development and Growth illustration above?
- Starting with the microspore mother cell, how many meiotic and mitotic cell divisions are required to produce the mature male gametophyte?

Megasporogenesis

The illustration below shows the development of the female gametophyte, a process is called megasporogenesis **or** female gametogenesis. Study this diagram and be able to draw it.

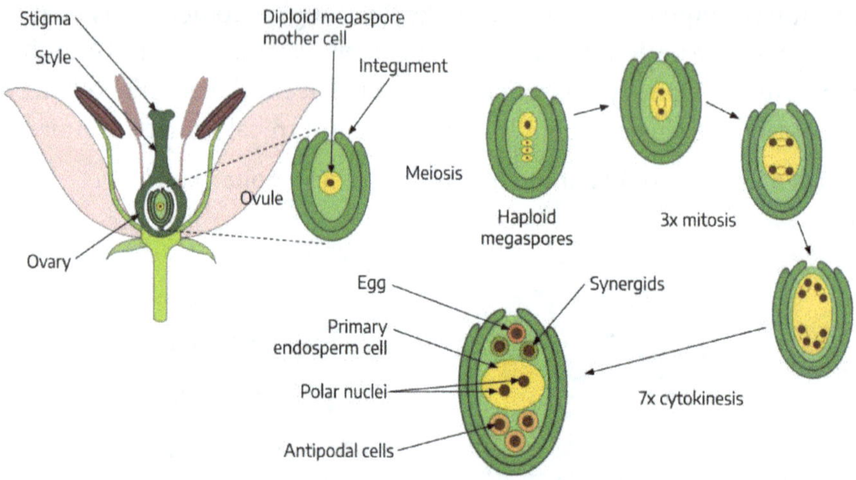

Development of the female gametophyte. Image labeled and adapted by Emily Tepe. Original image by Kazakova Maryia via Shutterstock. Used and adapted with permission.

The chambers within the ovary are the locules, and within the locules are the ovules. The ovule wall is made up of integument tissue, and it is from this integument tissue that diploid megaspore mother cells form. A diploid megaspore mother cell undergoes meiosis and initially produces four haploid megaspores. Three of the four spores formed through meiotic division of the megaspore mother cell disintegrate, as indicated by the "X" through three of the spores in the illustration. The lone surviving megaspore subsequently undergoes three mitotic divisions to form eight haploid cells that are held within the ovule in an embryo sac. These eight cells comprise the female gametophyte.

Only six of the nuclei are surrounded by their own cell membrane after the three rounds of mitosis. The two remaining nuclei are surrounded by one cell membrane. These two nuclei in one membrane become the polar nuclei and contribute two of the three nuclei to the endosperm tissue, with the other nucleus contributed by a sperm cell.

The megaspore mother cell is the swollen cell shown in the micrograph below. Note the linear tetrad of haploid megaspores resulting from meiosis of the megaspore mother cell. Three will disintegrate and one will undergo three rounds of mitosis.

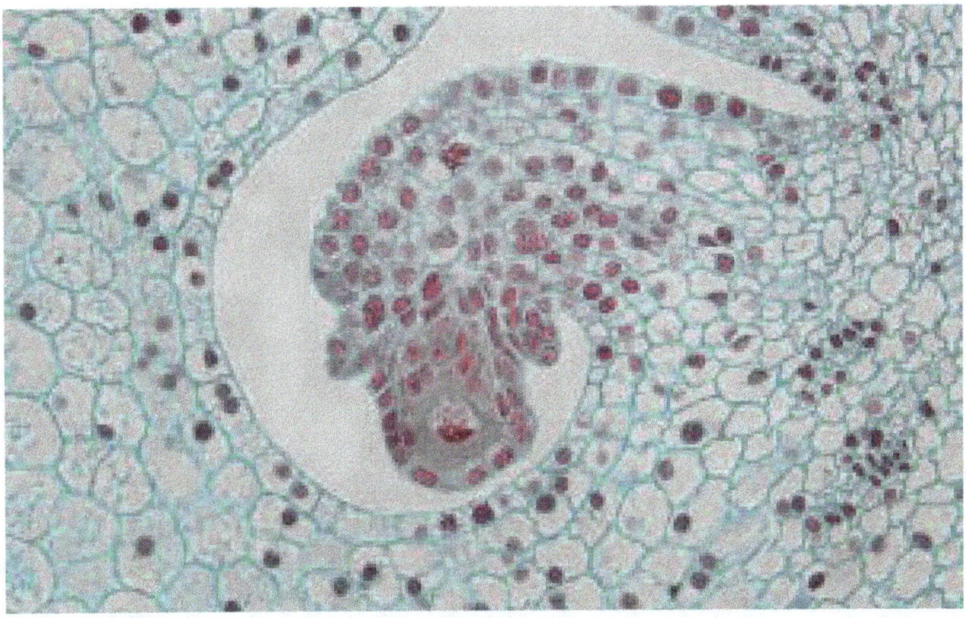

Lilium – ovule with megaspore mother cell. Photomicrograph by Micheal W. Clayton, used with permission. © Board of Regents of the University of Wisconsin System.

The four cells in the embryo sac in the left-hand image below are from two mitotic divisions of the one surviving megaspore. After one more mitotic division (center image below), the embryo sac, or female gametophyte, contains 7 cells (with 8 nuclei), and their arrangement is shown in the rightmost image below. The egg and two synergids are positioned on the bottom, two polar nuclei are shown in the center, and three antipodal cells are positioned at the top.

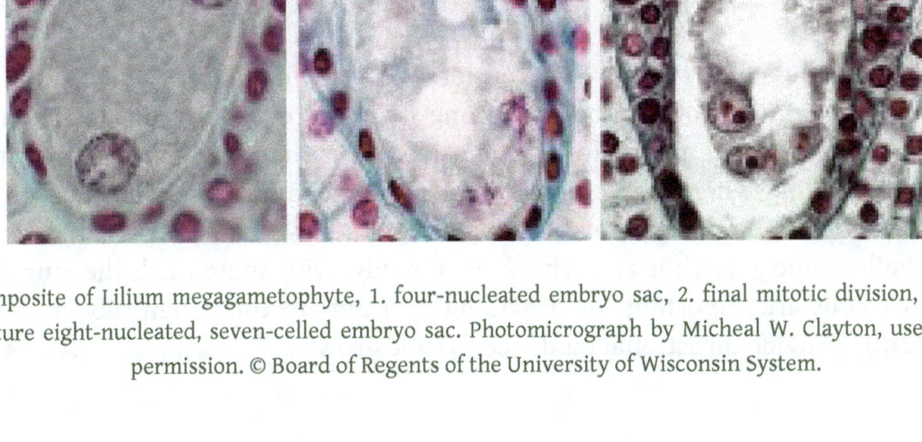

Composite of Lilium megagametophyte, 1. four-nucleated embryo sac, 2. final mitotic division, and 3. mature eight-nucleated, seven-celled embryo sac. Photomicrograph by Micheal W. Clayton, used with permission. © Board of Regents of the University of Wisconsin System.

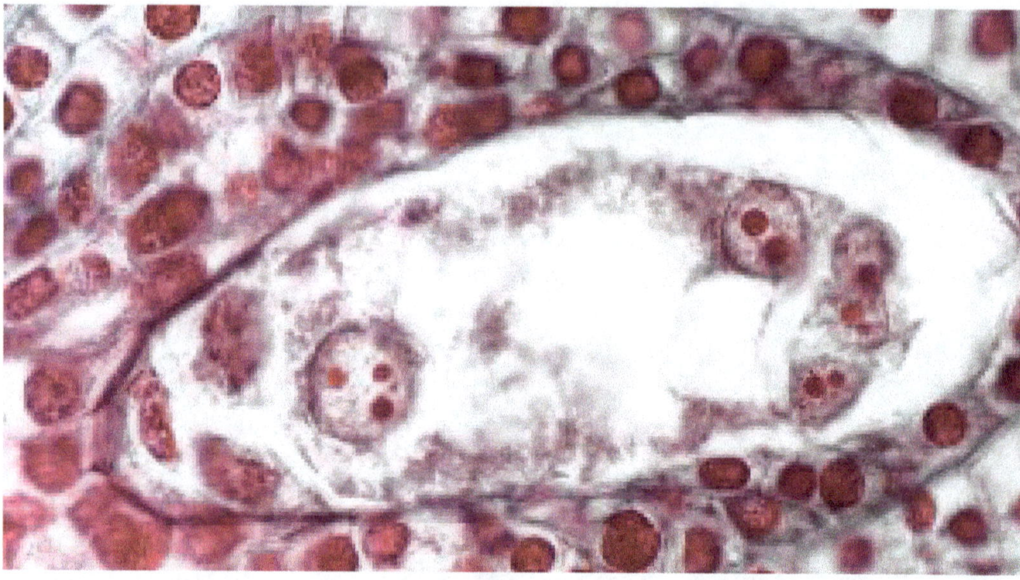

Mature embryo sac of Lilium. Photomicrograph by Michael W. Clayton, used with permission. © Board of Regents of the University of Wisconsin System.

Review questions

- Where are the megasporangia located?
- Where is the mature female gametophyte in the Development of the Female Gametophyte illustration above?
- Is the female gametophyte diploid or haploid? Is the integument tissue diploid or haploid?
- Starting with the megaspore mother cell, how many meiotic and mitotic cell divisions are required to produce the mature female gametophyte?

Fusion or fertilization

In the image below, pollen has fallen on the stigma and some grains are germinating. The next image shows the germination of pollen and tube growth in an artificial medium.

Think of the pollen tube as an extension of the cell membrane and cell wall of the pollen grain. The pollen tube isn't a long string of cells, but is one long skinny cell with two or three nuclei (one vegetative and either one generative or two sperm depending when the generative cell undergoes mitosis). The tube grows down the style through the intercellular (between cell) spaces rather than through cells. Pollen tube growth is an active area of study, with some work showing that the style cells are important for providing nutrients to the pollen tube and also apparently provide directional guidance to tube growth.

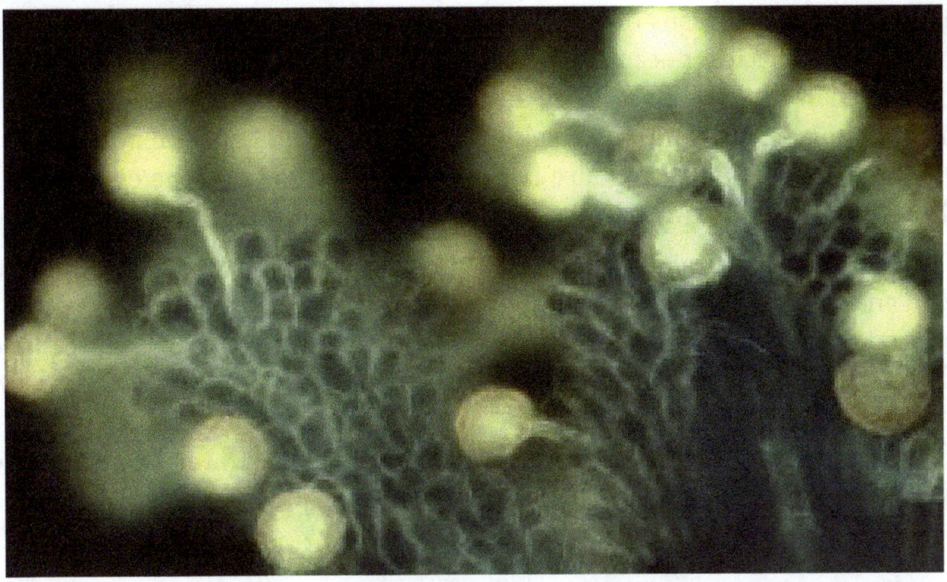

Stigmas and pollen tubes. George Shepherd. CC BY-NC-SA 2.0

Pollen grain and pollen tube (indicated by red arrow). Bruce Kirchoff. CC BY 2.0

In the illustration below we see a generative nucleus in a pollen grain. This subsequently undergoes mitosis to form two sperm cells. As noted earlier, in some species this happens during pollen development before it is shed from the plant; in other species it occurs in the pollen tube. The image on the far left below identifies all of the cells in both the female and male gametophyte: in the female gametophyte are three antipodal cells at the top of the ovule, the polar nuclei in the middle, and at the bottom are two synergid cells flanking the egg. There are eight cells or nuclei in total.

In the illustration, you can see the three cells within the pollen tube: two sperm and one tube or vegetative nucleus. The tube finds its way to the embryo sac through the micropyle and ruptures, and the sperm are delivered to the egg and polar nuclei. The rightmost section of the illustration is a closeup of the ovule and shows double fertilization. One sperm nucleus has fused with the egg nucleus to form a 2n zygote. The other sperm nucleus has fused with the two polar nuclei to form the 3n endosperm.

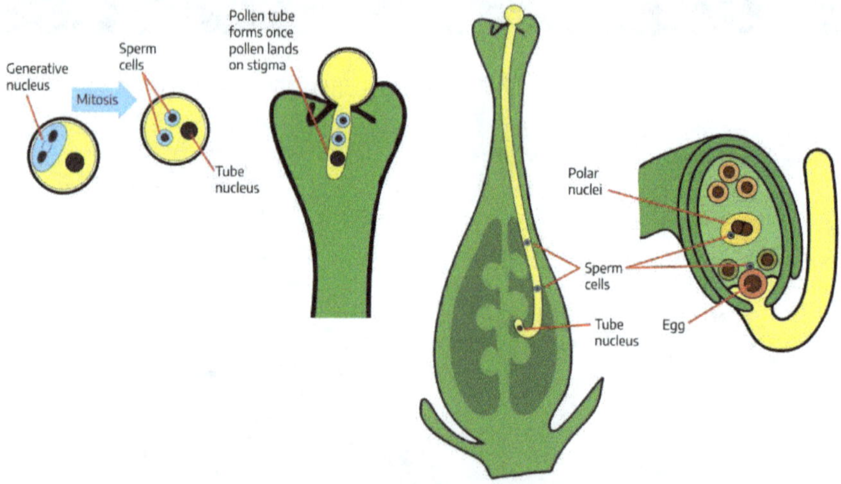

Pollen tube growth and double fertilization. Image labeled and adapted by Emily Tepe. Original image by Kazakova Maryia via Shutterstock. Used and adapted with permission.

Embryo growth

Embryo development is initiated when the zygote divides once mitotically to form an apical and a basal cell. The basal cell undergoes several additional mitotic divisions to form a suspensor that anchors the apical cell to the nucellus tissue on the inner surface of the maturing ovule wall (seed coat). Next, the apical cell begins to divide mitotically to form the embryo. The embryo goes through distinctive phases that look like particular shapes. This image shows the early embryo heart shape (second from right), and the torpedo shape (far right) stages. The lobes will become cotyledons. Between the lobes is the stem or shoot apical meristem, and at the base near the suspensor is the root apical meristem.

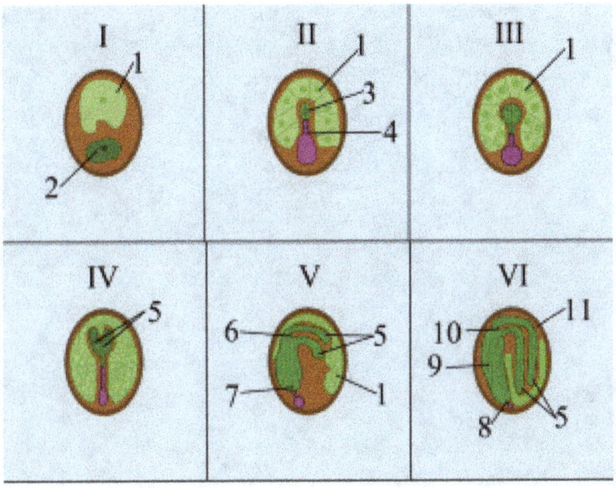

Seed development in plants consists of six stages. TheLAW14. CC BY-SA 3.0

Below is a heart stage embryo. Note the endosperm developing in another part of the embryo sac.

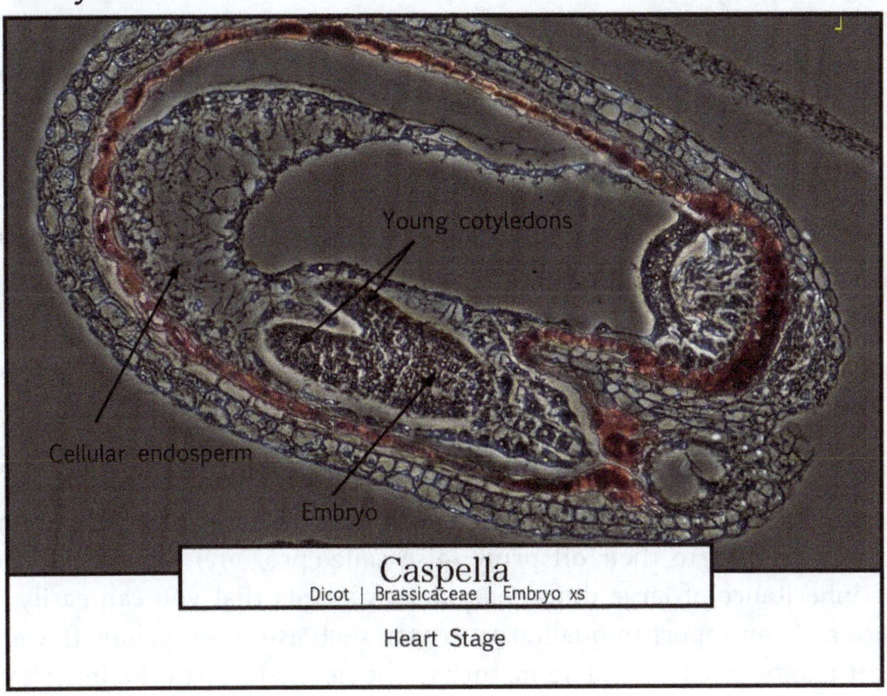

Heart stage embryo. Jen Dixon. CC BY-NC-SA 2.0

Finally, a later stage embryo with cotyledons readily apparent. This lesson has brought you from flower to seed.

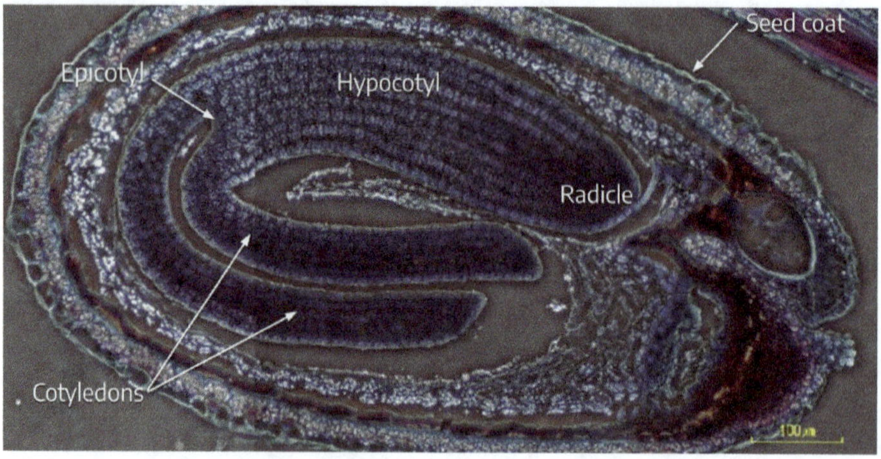

Micrograph of late stage Capsella embryo, mature with cotyledons. Botanical Voyeur, CC BY-NC-SA

14.2 INHERITANCE OF BIG DIFFERENCES

Learning objectives

- Compare the difference between simple and complex inheritance and explain how that difference is based on the number of genes involved and on the influence of the environment.
- Predict the types of offspring expected when crossing two parents that differ for their alleles at one locus, or when self-pollinating an F_1 **hybrid.**
- Predict the types of offspring expected when crossing two parents that differ for their alleles at two loci.
- Use a Punnett square to calculate the expected frequencies of different types of offspring.

Quantitative or qualitative differences?

This lesson and the next focus on the inheritance of characteristics that are passed down from parents to their offspring (also called progeny). This lesson focuses on the inheritance of large differences among plants that you can easily see or measure and can report in qualitative terms, such as red vs, yellow flowers, tall vs, short plants, or early vs, late maturity. The next addresses the inheritance of comparatively small differences that require more meticulous measurement and are reported in quantitative terms, such as seed yield in kg/ha, or milligrams of sucrose per gram of grape tissue. The expression of characteristics that are qualitatively inherited (larger, discrete differences) is influenced primarily by the plant's genes. The environment has little effect on the plant for these characteristics. In contrast to qualitative inheritance, plant characteristics that are quantitatively inherited are influenced by the environment, and sometimes the influence of the environment overshadows that of the genes themselves.

Recall that the genetic code lies in the order of the purine (Adenine and Guanine) and pyrimidine (Cytosine and Thymine) bases in the DNA double helix. Some of these base sequences don't appear to have a function, but others are translated into structural and enzymatic proteins that influence cell and plant growth. The sequences that are translated into proteins that influence cell and plant growth are called genes. In corn, for example, the sucrose manufactured in the leaves through photosynthesis is transported through the phloem to the ear and into the developing seed or kernel. There the sucrose is broken down to simpler sugars like glucose that are linked together into starch for storage in the corn kernel's endosperm. The starch synthesis process is catalyzed by a series of enzymes that are produced as a result of expression of a set of corn genes. If even one of those genes is defective and doesn't produce an enzyme used in the sequence of reactions leading to starch, starch production could be halted.

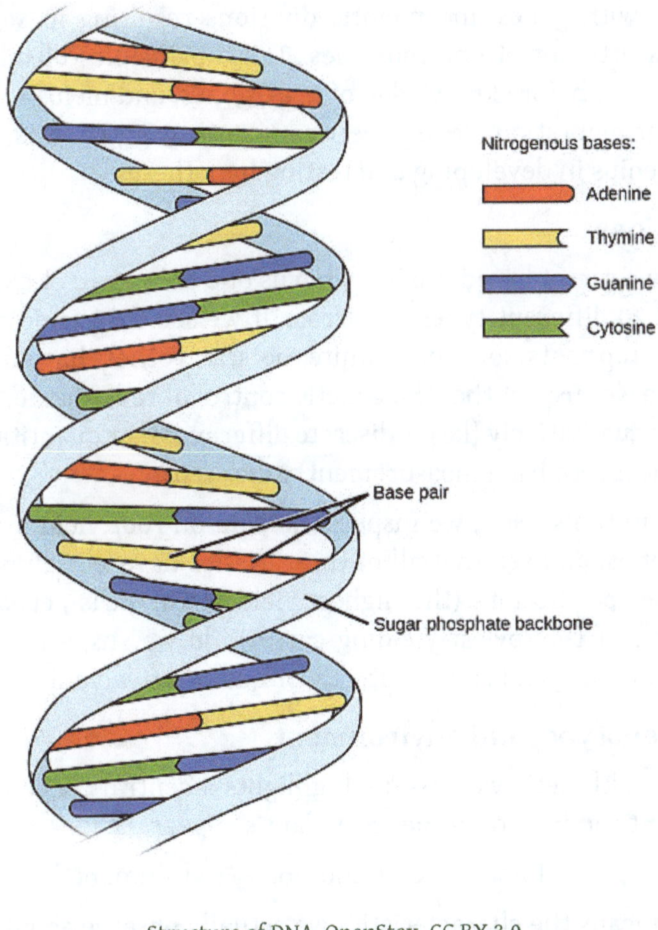

Structure of DNA. OpenStax. CC BY 3.0

Shrunken-2 (Sh_2) corn contains a DNA sequence (gene) mutation that renders nonfunctional an enzyme that is key to normal starch biosynthesis. Starch biosynthesis is interrupted at an early stage of kernel growth, so instead of accumulating starch, the kernel accumulates sucrose in the endosperm. Shrunken-2 (sh_2) corn has high kernel sucrose levels and is one of the common types of "supersweet" sweet corn. This is an example of a large difference (a qualitative trait), because you can easily distinguish the difference between a bite of starchy field corn from a bite of sweet corn, and because the difference in sweetness is the result of just one mutant gene. This mutation is not, however, good for the vigor of the plant in the early spring. A seed that has accumulated sucrose rather than starch is very susceptible to fungus infection at planting time in comparison to the starchy seed of field (commonly called dent) corn. Sweet corn is not nearly as vigorous during germination as regular dent corn, as the dent corn has stored high amounts of starch in its endosperm while the sweet corn has not.

The principles of qualitative inheritance are consistent with the concepts of DNA structure, with genes and meiotic divisions resulting in segregation, and independent assortment of chromosomes. These principles of inheritance were actually worked out before knowledge of DNA, genes, and meiosis, using carefully controlled experimentation, crosses between contrasting parents, observation of progeny, and genius in developing and testing hypotheses.

Review questions

- Let's say a carrot breeder is showing us one of her carrot experiments. She planted 20 different types of carrots. If we see that some of the varieties have red tap roots, some have white roots, and others have the usual orange roots, do we suspect that the genetic control of red, orange, and white root colors is qualitatively (large, discrete differences) or quantitatively (smaller differences requiring measurement) inherited?
- If, later in the season, we inspect the data on root yield for each of these 20 varieties, and see that all of the varieties have a fairly similar yield in kilograms per hectare (the highest yielding carrot is perhaps 20% higher than that of the lowest yielding carrot), do we suspect that the genetic control of root yield is qualitatively or quantitatively inherited?

Phenotype, genotype, and environment

This simple arithmetic expression highlights a central concept of how genes and environment combine to influence a plant's appearance:

$$\text{Phenotype} = \text{Genotype} + \text{Environment}$$

Phenotype means the characteristics we actually observe about the plant. Is it tall? Short? Green? Yellow? Starchy? Sweet? The starchiness or sweetness of corn from a garden is one example of a phenotype. Phenotype is the actual expression of a characteristic in the plant that can be measured or expressed in some way.

Genotype is the genetic composition of a plant, including chromosomes of the nucleus and the DNA in chloroplasts and mitochondria. The genotype of a plant is subdivided into genes, which are the hereditary units consisting of a sequence of DNA that occupies a specific location on a chromosome (locus) and determines a particular characteristic in an organism. Genes undergo mutation when their DNA sequence changes, which results in changes in a gene, for example a flower color gene that mutates from red to white in color. The alternative versions of the DNA sequences making up the gene are called alleles. Shrunken-2, introduced above, is a gene. That gene is found at a particular location of a chromosome, and that location is called the gene's locus. At that locus will be one of the gene's alleles. It will either be the allele for the normal enzyme that facilitates starch formation, or the mutant allele that blocks starch formation and leads to supersweet corn. The sequence of the purine (A, G) and pyrimidine (C, T) bases along the DNA strand and the differences the sequence imparts to the genetic code produce differences among gene alleles.

Environment is the total influence of known and unknown factors, other than genotype, that might affect this trait, like rainfall, soil type and fertility, temperature, insect predation, and other influences that we may not recognize, but might still affect the phenotype.

Phenotype is then the sum of the genotypic and environmental effects. If you are a sociologist, then you might be familiar with arguments about whether a certain type of human behavior is due to "nature" or "nurture." Roughly, "nature" corresponds to genotype while "nurture" corresponds to environment. A shorthand representation for the equation introduced above is $P = G + E$, which roughly corresponds to

$$\text{Behavior} = \text{Nature} + \text{Nurture}$$

For more on genotype and environment influences on a plant's phenotype, see this Wikipedia page (optional).

Genotype (G) has a much larger influence on phenotype (P) than does environment (E) when considering simply inherited characteristics that result in large, qualitative differences. Supersweet sweet corn is going to be sweet corn and not starchy field corn regardless of whether it is grown in Minnesota, California, or in a greenhouse on Mars. The exact amount of sucrose may differ a bit depending on environmental stresses, like how hot or cool the growing season is (hot seasons tend to result in slightly more sugar), but the genotype has a much larger impact on the sweet phenotype than the environment. It is quite common for large differences to be inherited through a single gene that has a large impact. This is called simple inheritance, major gene inheritance, or qualitative inheritance.

In contrast, environmental effects typically have a major impact on phenotype when considering inherited characteristics that are expressed as smaller, quantitative differences, such as those you might find when comparing common bean cultivated varieties (cultivars) for their seed yield, or spinach cultivars for the amount of chlorophyll in their leaves. The impact of the environment might even be larger than the effect of genotype. For example, the rank order of bean yield for three kidney bean cultivars grown in Minnesota might be Cultivar A > Cultivar B > Cultivar C, but if you grow them in Pennsylvania, it could be the reverse, due to a change in environment. It is common for small differences to be inherited through many genes each with a very small impact, where the cumulative effect of their acting together is noticeable. This is sometimes called complex inheritance, minor gene inheritance, or quantitative inheritance, because many genes each with small effect are involved.

Review question

- Thinking back to the carrot questions above, will environment have a greater impact on root yield or root color?

Inheritance of a qualitative trait

Gregor Mendel. Medical Heritage Library. CC BY-NC-SA 2.0

Gregor Johann Mendel, who lived from 1822 to 1884, was an Austrian monk and scientist who studied the simple inheritance of large, obvious differences among pea plants and developed the fundamental principles of modern genetics. He demonstrated that the large differences passed from parents to progeny are transmitted between generations through discrete units or packets of information that control the expression of specific characteristics, and that these discrete units are independently inherited. Take a look at this web page (optional) for background on Mendel, or browse here.

As noted above, Mendel chose pea, Pisum sativum L., as his model organism. Pea usually self-pollinates. It is one of those plants where the pollen is shed and the stigma is receptive even before the flower opens (recall that this is called cleistogamy). Egg and sperm come from the same plant and, following fertilization, form the embryo. The flower is conveniently large enough though that if you very carefully open up a flower bud a few days before the pollen sheds, you can remove the anthers and pollinate the stigma with pollen from another plant, making it fairly easy to cross if you emasculate the flower and supply another source of pollen. If left to its own devices, however, the pea flower self-pollinates.

Mendel looked at a number of qualitative traits controlled by single genes, including:

Gene	Phenotype
Seed shape	smooth vs wrinkled
Seed color	green vs yellow
Flower color	purple vs white
Several other obvious pod and plant characteristics that can be seen by eye	

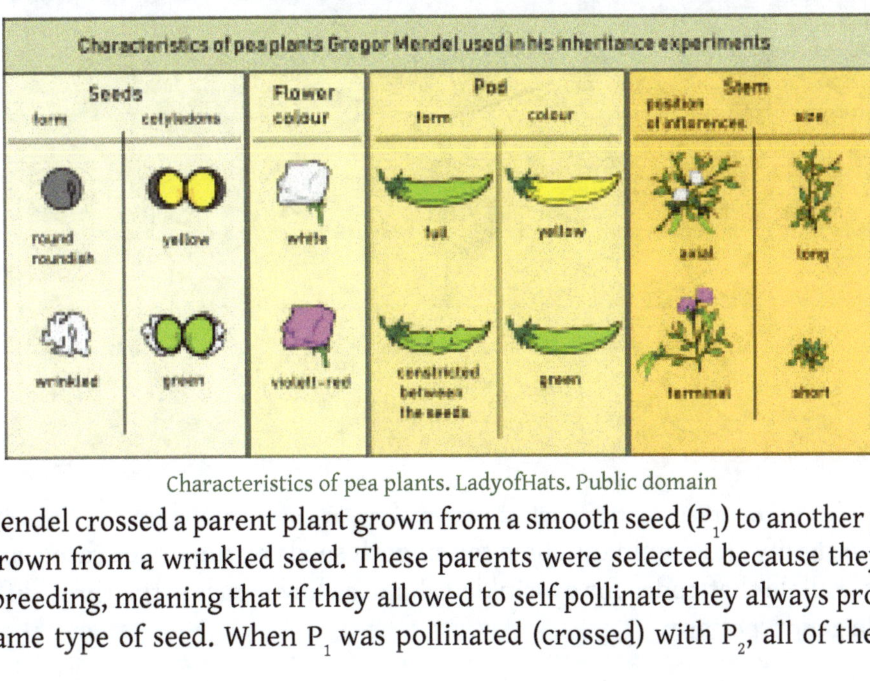

Characteristics of pea plants. LadyofHats. Public domain

Mendel crossed a parent plant grown from a smooth seed (P_1) to another parent (P_2) grown from a wrinkled seed. These parents were selected because they were true breeding, meaning that if they allowed to self pollinate they always produced the same type of seed. When P_1 was pollinated (crossed) with P_2, all of the seeds resulting from the cross (called F_1, short for first filial generation) were smooth-seeded.

Here's a summary of the experiment so far:

P_1 (smooth) X P_2 (wrinkled) F_1 (all smooth)

Mendel then planted the F1 seeds, let them naturally self-pollinate, and collected the seeds produced by these plants. He counted a total of 7324 F2 (second filial generation) seeds and found that 5474 were smooth and 1850 wrinkled. This is roughly a ratio of 3 smooth : 1 wrinkled. This, and other similar results, led him to the development of a model where seed plumpness is controlled by one gene with two different versions (or alleles) of the genetic code for that gene. Furthermore, based on the results, when one of each allele was present in the F1 progeny, the allele donated from the smooth seeded P1 parent was dominant to the recessive allele donated from the wrinkled-seeded P2 parent. Dominant here means that if two different alleles are present in an organism, but only one is expressed (like the smooth seed being expressed even though alleles for both smooth and wrinkled are present), the allele that is expressed is called the dominant allele, and the allele that is not expressed is called recessive. Recessive alleles are only expressed if no dominant alleles are present.

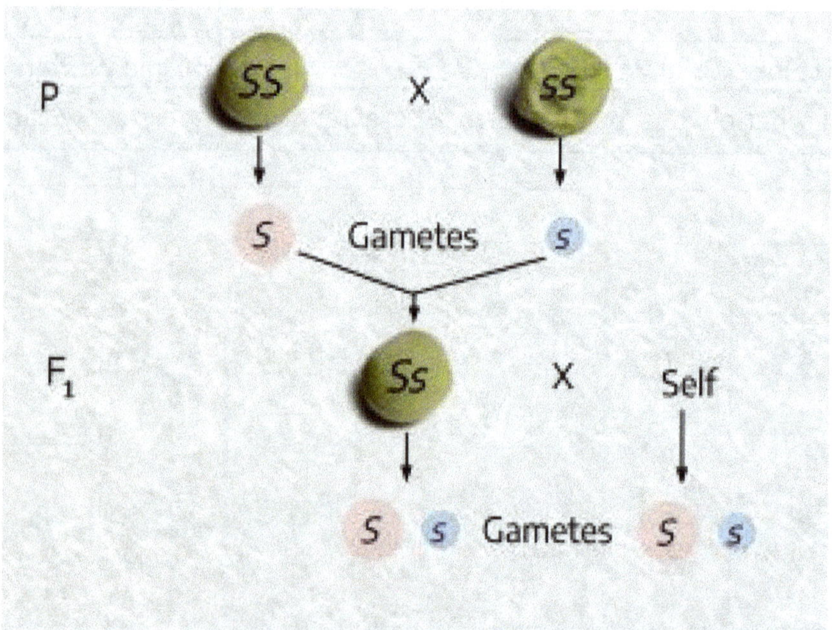

F1 generation – Round/wrinkled pea crossing. Alejandro Porto. CC BY-SA 3.0

If we signify the dominant allele for smooth seeds as the upper case S and the recessive allele that codes for a wrinkled shape as lower case s, the three different possible genotypes are:

- SS diploids that are smooth (homozygous dominant).
- Ss diploids that are also smooth. Here, in the heterozygous genotype, you can tell which allele is dominant to the other. Both alleles are present (S and s), but only one is expressed — the S — so S must be dominant to s.
- ss diploids that are wrinkled (homozygous recessive).

Note that, by convention, the designation for the allele (like S **or** s) and the genotype (like Ss) is underlined.

Above is a diagram showing the smooth (SS) and wrinkled (ss) parents crossing to form Ss (smooth) F1 progeny. Recall that if an SS parent goes through meiosis, only an S type of gamete is formed. There cannot be any s gametes, because the parent only has the S allele. Likewise, only s gametes are possible when an ss parent undergoes meiosis and forms gametes. So Ss progeny is the only possible result from crossing an SS parent to an ss parent. The phenotype of all of these progeny will be smooth because S is dominant to s. Notice that, in this example, which is one of big, qualitative differences, there is no influence of the environment, so Phenotype = Genotype.

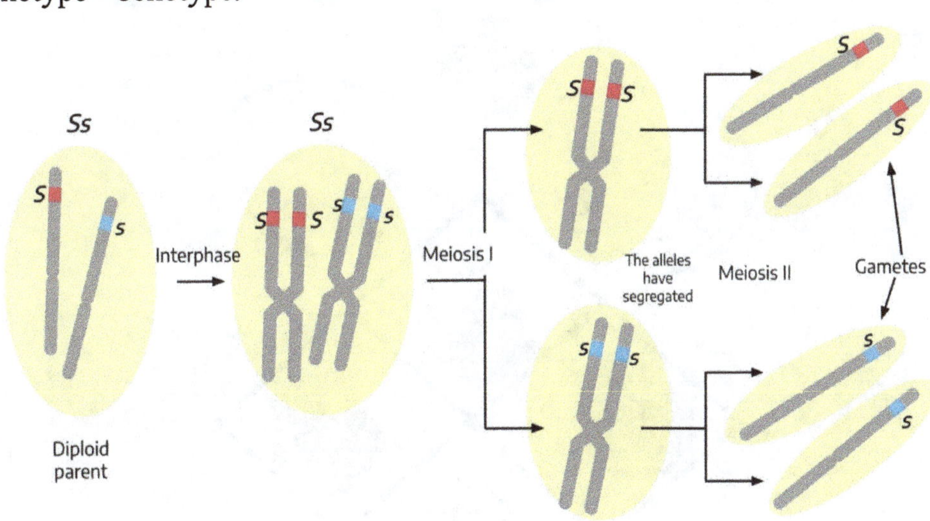

Image by Emily Tepe.

The next step in the experiment is to allow the smooth, Ss F1 progeny to self-pollinate. The lower part of the above diagram shows that the F1 plant will produce two types of gametes, S and s, in equal numbers. Think back to meiosis and gametogenesis. Just before a diploid cell with genotype Ss heads into meiosis, the homologous chromosomes (one carrying the S allele and the other carrying the s allele) replicate in interphase so that there are four sister chromatids, two with the S allele and two with the s allele. The illustration to the left shows how the S and s alleles segregate. The process of meiotic cell division will result in each of these sister chromatids winding up in a separate spore, so of the four male gametes formed during microgametogenesis, two are carrying S and two are carrying s. The one surviving gamete from megagametogenesis has an equal probability of being either S or s (recall that of the four spores, only one survives), so if you consider many egg cells, roughly half will be S and half will be s. Mendel's results, and his explanation, are thus consistent with what we know about meiosis.

You can use a simple tool called a Punnett Square to visualize the types of zygotes, and their expected frequency, formed from the male and female gametes resulting from self-pollination of an Ss individual, as shown below:

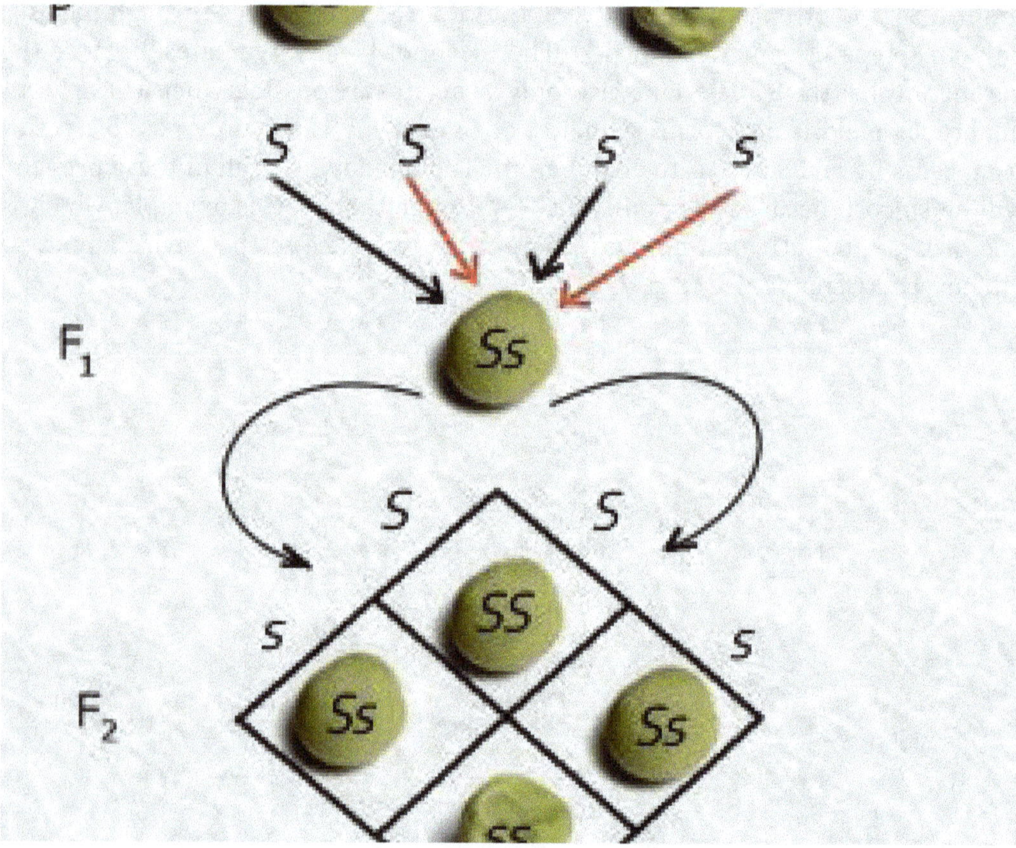

F2 Generation. Alejandro Porto. CC BY-SA 3.0

The male gametes, S and s are shown in the two columns across the top of the square, and the female gametes, also S and s, are listed down the left margin. Each cell shows the contribution of one male gamete and one female gamete.

In this example, self-pollinating (also called selfing) an Ss F1 plant means that we list S and s as possible male gametes and also as possible female gametes, since both sperm and egg are from the same F1 plant. The potential zygotes are shown in the four cells inside the square.

If you count up the results of from the Punnett square, you find that the genotypic ratios of the F_2 progeny are:

1 SS : 2 Ss : 1 ss

and the phenotypic ratios are:

3 S- (Smooth) : 1 ss (Wrinkled)

which is precisely the ratio that Mendel found between smooth and wrinkled seeds in his F2 pea generation. Recall that the difference between the genotypic and phenotypic ratios occurs because S is dominant to s so you can't tell the phenotypic difference between SS and Ss genotypes.

Mendel's First Law — the law of segregation — states that during gamete formation each member of the allelic pair separates from the other member to form the genetic constitution of the gamete. That is, in the F1, the Ss diploid produces S and s gametes, not Ss gametes. Mendel's work paved the way for figuring out this process.

Review question

- Using the Punnett square diagram above, explain why the expected frequencies for the three possible genotypes should indeed be 1 SS : 2 Ss : 1 ss.

Simultaneous inheritance of two qualitative traits

Here is a slightly more complex situation, but one that can still be understood through your knowledge of meiosis. We'll simultaneously consider seed shape (or plumpness) and another characteristic, seed color, that Mendel studied. Like seed shape, seed color is controlled by one gene with two alleles where:

- YY diploids are yellow
- Yy diploids are also yellow
- yy diploids are green

You can tell from the heterozygote Yy that the Y allele for yellow color is dominant to the recessive y allele for green.

Note that the gene for seed shape is on a different chromosome than the gene for seed color. Mendel didn't know this, as the concept of chromosomes had not yet been proposed. Because they are on separate chromosomes, seed color and seed shape will assort independently.

Let's start by crossing a plant that grew from a smooth, yellow seed known to have the genotype SSYY to a plant that grew from a wrinkled green seed with genotype ssyy.

SSYY X ssyy → ?

The only type of gamete that can be produced by the SSYY parent is SY, and the only type that can be produced from the ssyy parent is sy. The F_1 progeny can only be the double heterozygote SsYy, which has smooth, yellow seeds.

SSYY X ssyy → SsYy

Next, allow the SsYy F_1 progeny to self pollinate. Below is a diagram of key stages of meiosis for a cell with genotype SsYy. Note that the seed shape gene is on one type of chromosome (straight) and the seed color on another (squiggled).

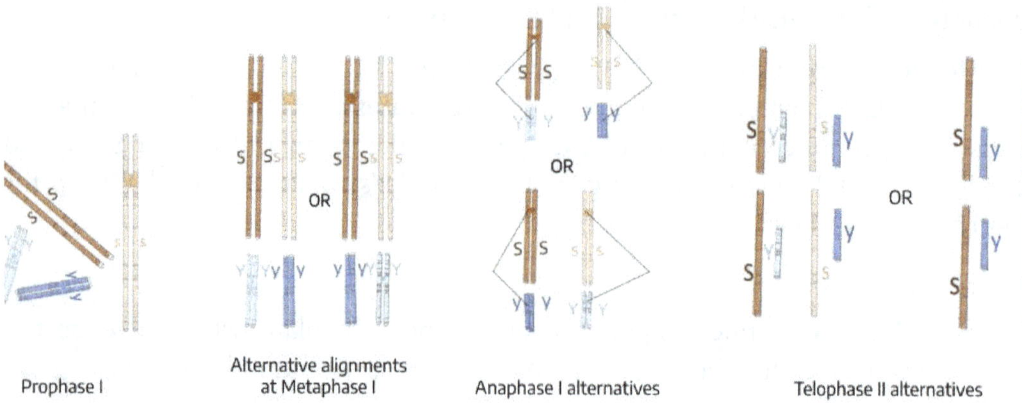

Image by Matthew Clark.

- At Prophase I the chromosomes are replicated and show sister chromatids.
- At Metaphase I we see that the homologs have synapsed and lined up on the metaphase plate. There are two ways that the recessive and dominant alleles can line up on a metaphase plate. The left side of Metaphase I shows that both dominant gene homologs are on the same side; the alternative alignment on the right side shows that the dominant homolog for one gene could just as likely line up with the recessive homolog for the other gene.
- The way they line up at Metaphase I determines how they separate at Anaphase I, so the diagram of Anaphase I shows two alternatives for how the homologs migrate to the poles. This has an impact on the types of spores produced in Telophase II.
- In the Telophase II frame above, the left side shows that. based on one of the Anaphase I alternatives, the result can be for the do♠minant S and the dominant Y to be together in one type of spore and both recessive genes together in other spores. Equally likely, as shown on the right, the result can b eone dominant gene with one recessive gene.

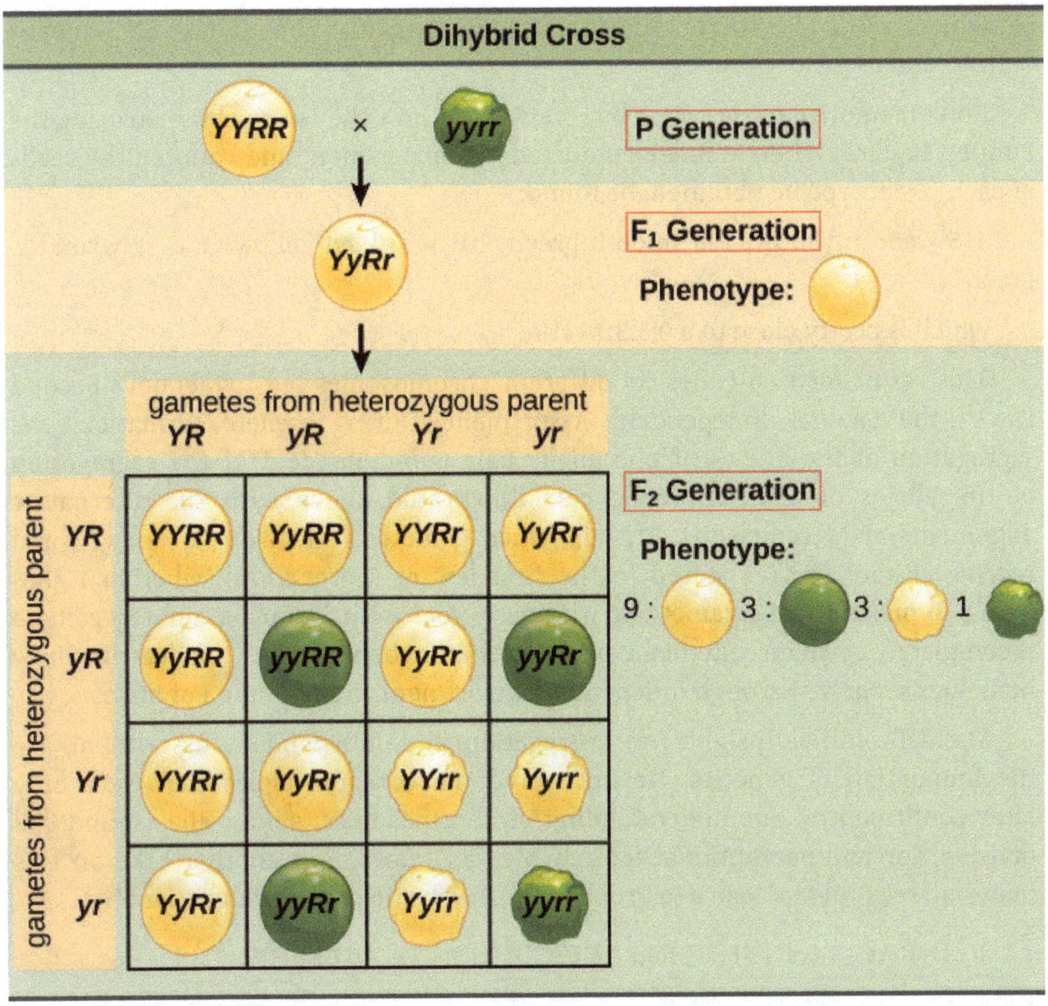

Yellow/green/wrinkled/round pea Punnett Square. OpenStax. CC BY 4.0

Meiosis and gamete formation in the SsYy F1 thus results in an equal likelihood of four types of gametes: SY, sY, Sy and sy. If we create another Punnett Square to model self pollination of the SsYy F1 and put these four gametes both across the top and down the margin (since we are allowing the plant to self pollinate, and the frequency and genotype of the gametes will be the same whether egg or sperm), we get the results shown.

Looking at the diagrams of the seeds in each cell, you can count the number of times each of the four possible phenotypes appears. The phenotypic ratios are:

9 S-Y- (smooth yellow) : 3 S-yy (smooth green) : 3 ssY- (wrinkled yellow) : 1 ssyy (wrinkled green)

The dash "-" in the genotype, like S-, means that the allele represented by the "-" could be either the dominant or the recessive allele because, regardless of which allele is there, the phenotype remains the same. So S- means both SS and Ss genotypes.

Don't memorize these ratios, but understand how they were obtained using the Punnett Square. When Mendel conducted this experiment and counted the seeds of each phenotype he obtained, he found:

315 smooth yellow, 108 smooth green, 101 wrinkled yellow, and 32 wrinkled green

which is pretty close to a 9:3:3:1 ratio.

This result for two genes on different chromosomes led to Mendel's Second Law— the Law of Independent Assortment. During gamete formation, the segregation of the alleles of one allelic pair is independent of the segregation of the alleles of another allelic pair (illustrated above with the alternative alignments at Metaphase I). That is, when gametes form from an SsYy F_1, and receive either S or s (as stated in the first law of segregation), whether Y goes with S to make up a SY gamete or whether y goes with S to make up a Sy gamete is completely by chance. Remember the mechanics behind this, from learning how homologs separate through independent assortment in Anaphase I of meiosis.

Mendel's initial insights on segregation and independent assortment are the foundation of genetics. He uncovered the core principles that were only later confirmed by our understanding of chromosomes, genes, alleles, and cell division. For another optional look at Mendel's laws, and another take on this material, see Phil McClean's (a Prof at NDSU) web site on Mendelian Genetics.

14.3 LINKAGE AND INHERITANCE OF SMALL DIFFERENCES

Learning objectives

- Know the meaning of linkage and the impact of crossing over on linked loci.
- Recognize the difference between qualitative and quantitative inheritance.
- Understand how heritability is a measure of the genetic influence over a quantitative trait, relative to other non-genetic influences.

Review

The previous section included examples of simple inheritance, where large, obvious, discrete differences among plants are controlled by one gene and where one of the alleles is completely dominant over the other. Based on the inheritance patterns of round vs. wrinkled pea seeds described in that lesson, Mendel developed the Law of Segregation.

Often called Mendel's First Law, the Law of Segregation in modern terms states that, during gamete formation, the two alleles for the gene of interest (recall that in diploid cells there are two homologs for each type of chromosome, so there are potentially two versions or alleles of each gene) separate from each other during meiosis and are passed individually to the next generation through the egg and sperm. This is apparent in the mechanics of meiosis, where the homologs separate and migrate to opposite poles in Anaphase I and, following the separation of sister chromatids in Anaphase II, are packaged into separate gametes. In the self-pollination of a Smooth x Wrinkled F1 pea example, the Ss diploid produces S and s gametes, not Ss gametes. The S and s are segregated into separate gametes.

The previous lesson also tracked the joint inheritance of two genes on separate chromosomes that affect different traits (Smooth vs Wrinkled seeds and Yellow vs Green seeds), which leads to Mendel's Second Law — the Law of Independent Assortment. In modern terms, this law roughly states that, during gamete formation, the allocation of the alleles of a gene on one chromosome is independent of the allocation of the alleles of a different gene on a different (non-homologous) chromosome. So when gametes form from an SsYy F_1, and the gametes receive either S or s (as stated in the First Law of Segregation), then whether Y goes with S to make up a SY gamete, or whether y goes with S to make up a Sy gamete, is completely by chance. This is also seen in the diagram of meiosis and how the chromosomes divide at Anaphase I. Homologs separate at Anaphase I, and in the case of two different pairs of homologs, the way in which the first pair of homologs splits doesn't influence how the second pair of homologs splits. What Mendel discovered by examining segregation ratios (the ratios of each genotype resulting from a particular cross) can thus also be explained through the mechanics of meiosis. Mendel's thinking was revolutionary at the time, and a great example of thinking carefully about the outcomes of an experiment and developing a model of the process that explains the observed data.

Linkage

What if those two genes are positioned closely on the same chromosome rather than on separate chromosomes, as in the pea shape/pea color example? This situation is called linkage.

Imagine a case in which the hypothetical gene "A" (with alleles "A" and "a") and gene "B" (with alleles "B" and "b") occur at loci near each other on the same chromosome. Imagine further that one of the chromosomes has the dominant "A" and the dominant "B" allele while its homolog has both recessive "a" and "b" alleles. This situation is illustrated below.

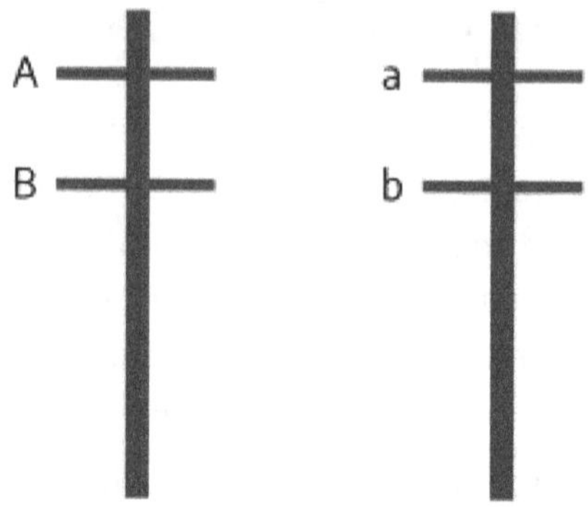

Alleles that cannot independently assort. Image by Emily Tepe.

When these chromosomes go through meiosis the alleles at the two loci cannot independently assort because they are on the same chromosome — they are physically connected. As a result, the gametes will be either AB or ab. There will be no Ab or aB gametes, as you would normally expect from independent assortment, because the A allele is physically attached to the B allele, and the a allele attached to the b allele — they are on the same chromosome. The only way to get Ab or aB gametes together is through a physical breakage in the chromosomes between the two genes and the exchange of arms between homologs; recall that this is called crossing over.

The illustration shows homologs in Prophase I following synapsis (pairing) and where sister chromatids crossed over, with the cross-over event occurring between the loci for the two genes. The chromatids cross over and will exchange arms. When they separate at Anaphase I, the homolog that migrates to the left side will have one sister chromatid with the alleles AB and one with aB. The homolog that migrates to the right will have one chromatid that is Ab and another that is ab.

The frequency of crossing over is one way to describe the distance between the two genes on the chromosome. If crossing over is very infrequent — in this example, if there are very few Ab and aB gametes — the two genes must be very close to each other. This situation is called a tight linkage. If crossing over is quite frequent, approaching 50%, the two genes must be very far apart. If it appears that AB, ab, Ab, and aB gametes have equal frequency, this indicates that the genes are located on different chromosomes or are on the same chromosome, but separated so much that the frequency of crossing over is 50%, which has the same result as if they were on separate chromosomes.

The Wikipedia site on linkage provides a bit of additional information.

Review questions

Why can two linked genes NOT independently assort?

In the example above, will tight linkage result in many or few Ab gametes?

Would you expect the frequency of aB gametes to be roughly the same or quite different from the frequency of Ab gametes?

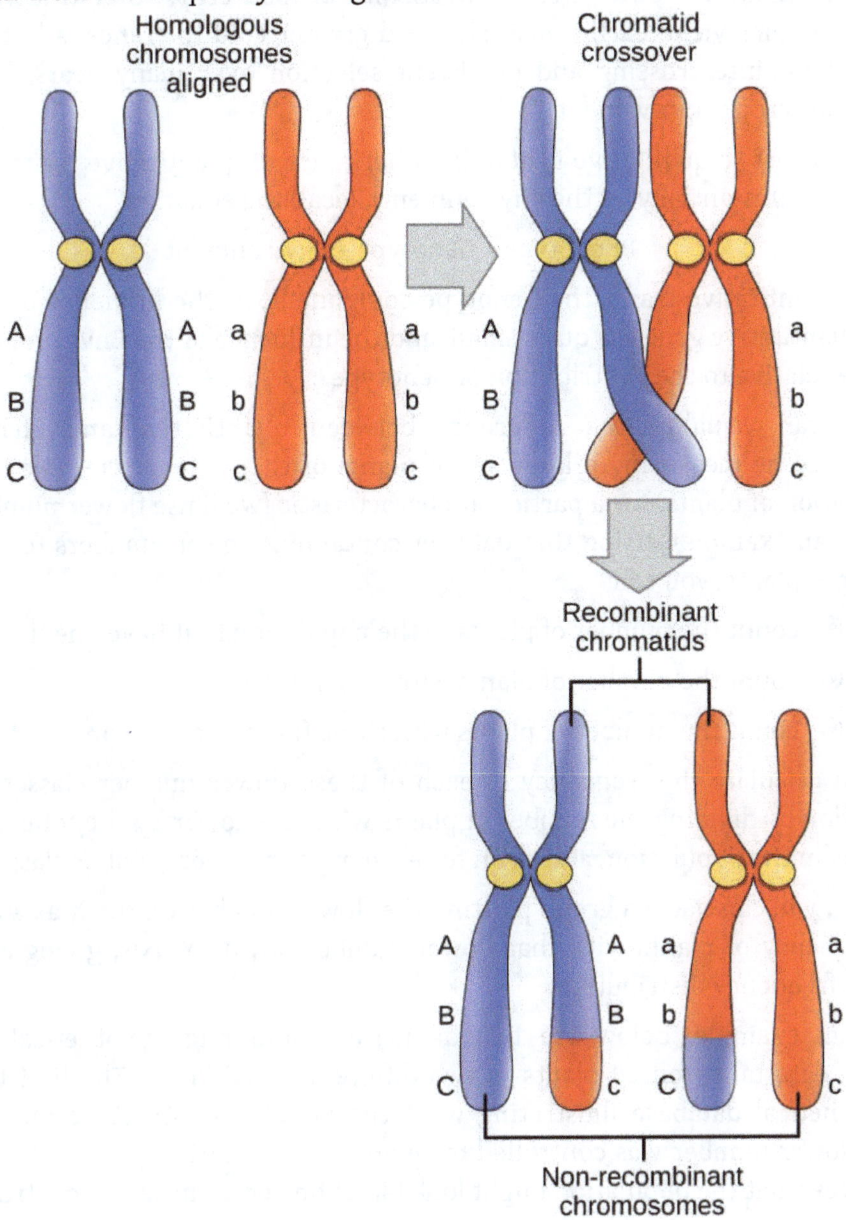

Homolgs in prophase 1 crossing over. OpenStax. CC BY 4.0

Quantitative traits — inheritance of small differences

Recall that inheritance does not always involve large, qualitative differences. Inheritance of small differences that require more meticulous measurement and are reported in quantitative terms such as seed yield in kg/ha, number of chloroplasts per mesophyll cell, and grape sucrose content are very important in horticulture. Indeed, these small differences, when accumulated, can be more important than the large qualitative differences in horticultural food crops. Quantitative traits result in higher yield, earlier maturity, and greater cold tolerance, all of which, with patient intercrossing and persistent selection over many years, leads to continuous crop improvement.

In contrast to qualitative traits, the inheritance of quantitative traits may be influenced substantially by the environment. Recall the equation

Phenotype = Genotype + Environment

For quantitative traits, the Genotype component, or the influence of each of these quantitative genes, is quite small, and the influence of the Environment can be large relative to the contribution of Genotype.

You can visualize the difference between quantitative and qualitative inheritance by measuring a large set of plants of the same species (we'll call it a population of plants) for a particular characteristic (we'll use flower number per plant as an example). Using that data set containing flower numbers for a large number of plants, you can

- count the number of plants in the population that have one flower,
- count the number of plants with two flowers,
- count the number of plants with three flowers, and so on,

then calculate the frequency of each of these flower number classes in the population by dividing the number of plants with one flower by the total number of plants in the population, and so on for each of your flower number classes.

Next, you can make a graph plotting the flower number on the X axis against the frequency of plants with that flower number on the Y axis, giving a figure called a frequency distribution.

As an example, below are two databases containing hypothetical flower number data taken on 60 plants in two different populations. The first table is a hypothetical database illustrating what the population structure might look like if flower number was controlled by quantitative inheritance, and the second illustrates what the population might look like if flower number was controlled by qualitative inheritance. Both tables have three columns; the first shows the number of flowers, ranging from 1 to 15 per plant; the second shows the number of plants in the sample of 60 with the corresponding number of flowers; and the third shows the frequency of plants in the sample with the corresponding number of flowers per plant. In the quantitative database on the left, in the row corresponding to plants with four flowers, read across to see that there were two of these plants, and that their frequency was 2/60 = 0.0333, rounded to 0.03. This is called a frequency distribution table because it shows the frequency of plants with a particular class of a characteristic, in this case flower number per plant.

Notice that the frequencies in the quantitative table are spread out over a wide range of flower numbers, with a peak around seven flowers per plant. In the qualitative table, in contrast, the frequencies are clustered around two discrete flower numbers, with one large peak at about five flowers and a smaller peak at 13.

Quantitative example			Qualitative example		
# Flowers	# Plants	Frequency	# Flowers	# Plants	Frequency
1	0	0	1	0	0
2	0	0	2	0	0
3	1	.02	3	0	0
4	2	.03	4	3	.05
5	5	.08	5	30	.50
6	7	.12	6	9	.15
7	16	.27	7	3	.05
8	15	.25	8	0	0
9	6	.10	9	0	0
10	4	.07	10	0	0
11	2	.03	11	1	.02
12	2	.03	12	1	.02
13	0	0	13	12	.20
14	0	0	14	1	.02
15	0	0	15	0	0
Total	60	1.0	Total	60	1.0

These data are easier to understand in a graph, with the flower number on the X axis and the frequencies on the Y axis:

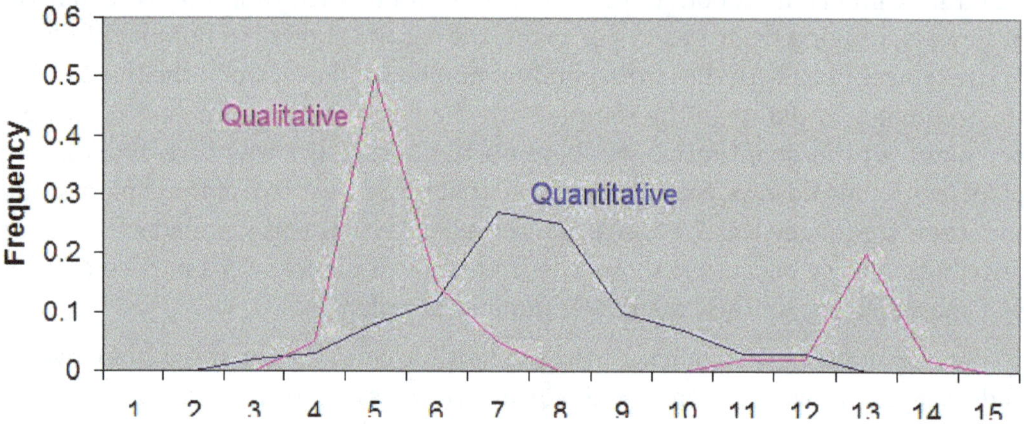

Image credit: Tom Michaels

Notice that the magenta line illustrating the qualitative inheritance distribution has two peaks at 5 and 13 flowers per plant, and that the dispersion around those peaks is quite narrow. Most values are either 5 or 13. In contrast, the blue line illustrating quantitative inheritance has a peak in the 7–8 flower per plant region, and there is a great deal of dispersion. There are many different values of flowers per plant, although the most frequent is in the 7–8 range.

The dispersion around the peaks is in part due to the influence of the environment. For qualitative traits (magenta line), this influence is small. For quantitative traits (blue line), it is larger. Also notice that if you added up the frequencies under the two magenta curves, you would find that the curve with the peak at five flowers has three times the frequency of plants as the curve peaking at 13 plants. This 3:1 ratio is indicative of a single qualitative gene where the dominant allele is associated with, on average, five flowers per plant.

Review questions

- Would you expect that a frequency distribution with two sharp peaks represents a qualitative or quantitative trait?
- Would you expect that a frequency distribution which looks like a mound with one high point represents a qualitative or quantitative trait?
- Would you expect that a frequency distribution with four sharp peaks represents a qualitative or quantitative trait? (hint: 9:3:3:1)

Heritability

Inheritance of quantitative traits is often associated with the term heritability. If a parent and its offspring have very similar values for a quantitative trait, the trait is considered to be highly heritable. Highly heritable traits tend to be under strong genetic control with little influence from the environment. In contrast, if the offspring values don't seem to have any relationship to that of the parent, the trait is considered to have low heritability. Low heritability usually results when the influence of the environment is quite high relative to the impact of the genotype on the trait being studied. If you are trying to improve a plant characteristic through breeding, you'll have much greater gain from selection if the trait has high heritability rather than low heritability. If a plant characteristic has high heritability, dramatic changes to the population from natural selection will also be more rapid (require fewer generations) than for a characteristic that has low heritability. See the graph below.

The X axis represents the number of flowers on the parent plant, while the Y axis represents the average number of flowers on the offspring of those parents. A dot represents each parent-offspring data pair. For example, the maroon dot with the black arrow pointing to it represents a data point for a parent with 10 flowers whose offspring had, on average, two flowers (read the 2 off the Y axis).

The gold dots represent cases where the trait has high heritability. Note that you can draw a line with a positive slope through the gold dots, with the dots being relatively close to the line. This indicates high heritability, and means that the offspring's performance can be predicted based on the performance of the parent. If the parent has a high flower number, the offspring will as well. Plant breeders can make selections with the expectation that if they save seeds from superior plants, the offspring growing from those plants will exhibit the same superiority.

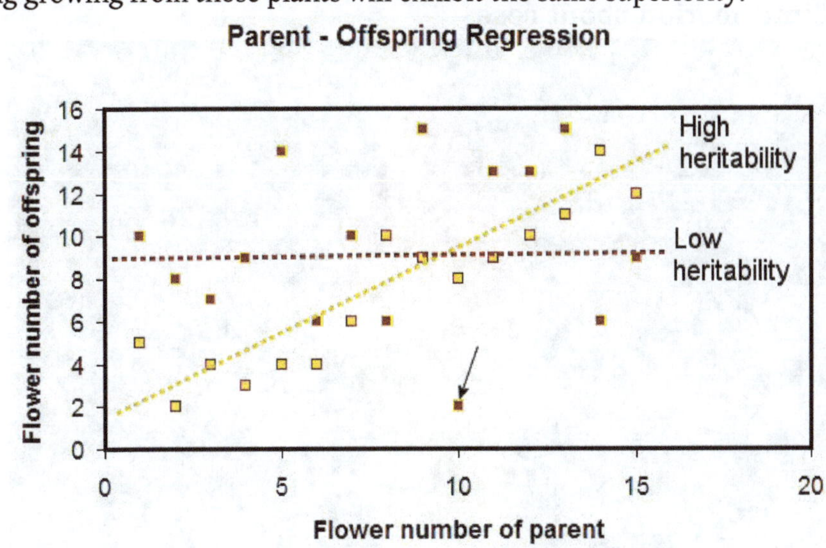

Image credit: Tom Michaels

The maroon dots, in contrast, represent the situation of low or zero heritability, and have no line of good fit. Some parents with high flower numbers had offspring with low flower numbers and others with high flower numbers had offspring with high numbers. You can't predict the offspring flower numbers based on the parent flower numbers. This represents a nightmare for breeders. Some offspring of superior plants will be good as the parents, others will be better, and others will be wildly inferior.

14.4 PLANT BREEDING

Learning objectives

- Understand how to get started breeding two common ornamental and food species.
- Know the purpose of emasculation when making hybridizations (crosses) between parents.
- Recognize the difference that propagation method makes in a plant breeding program.

Overview

This lesson presents examples of plant breeding of two common garden plants, rose and tomato. The strategies for the two plants differ because of how the final plant will be propagated. Roses are propagated asexually, while tomatoes are propagated by seed. For both, breeding starts with a cross between two different plants (called a bi-parental cross).

General information about rose

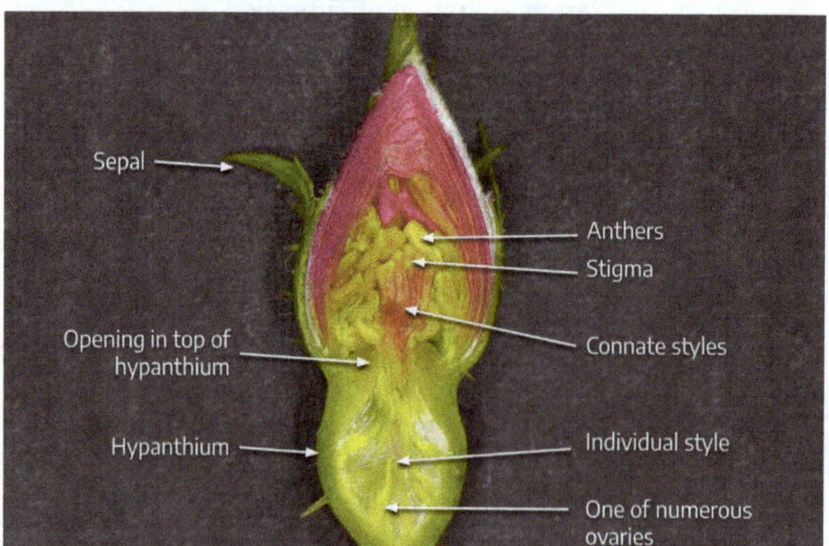

Photo by Eastfield College Microscopy Lab, Mesquite, TX, CC BY-NC.

Rose (genus Rosa with many species) is a perennial, mostly deciduous (they annually lose their leaves), mostly temperate-climate shrub in the Rosaceae family. The rose flower is perfect. The calyx, corolla, and androecium whorls are fused at their bases to form a small cup-shaped structure called a hypanthium that surrounds the ovary. (Refer to Chapter 8 if these are not familiar terms.) The base of the hypanthium is attached to the receptacle. The hypanthium is the structure that ripens into a bright red or red-orange fruit called a "hip," recall that this is also the tissue that forms the fleshy part of the apple. Within the hypanthium are hard achenes containing the rose seeds. In summary, the rose hip is an accessory fruit (parts other than ovary wall constitute the fleshy ripe portion) and an aggregate fruit (one flower, many carpels forming separate fruits) where the subsidiary fruits are achenes (optional).

Nevada rose. T. Kiya. CC BY-SA 2.0

The species of rose differ in their chromosome number, or ploidy. Roses have seven different types of chromosomes, so the total number of chromosomes in a rose is normally a multiple of seven. Some species, particularly wild species, are diploids with two sets of chromosomes, and so have a total of 14 (2 x 7 = 14) chromosomes. The large-flowered hybrid tea roses have four sets of chromosomes, so they are tetraploids with 28 (4 x 7 = 28) chromosomes. There are even triploid roses with 21 chromosomes. This optional journal article by Cédric Grossi & Maurice

Jay on Chromosomes studies of rose cultivars provides more information on chromosome numbers. Some roses are sterile because of triploidy or an imbalance in chromosome numbers, and never form rose hips. Sterile roses cannot be used to breed new roses. From a practical standpoint, if the rose is fertile you can use it in backyard breeding without much concern about whether it is diploid or tetraploid.

Roses can self-pollinate or cross-pollinate. A breeding project should be started with a cross. Because the flowers are perfect, you need to emasculate (remove the anthers from) the flowers you intend to use as the female. Emasculation happens at the bud stage before the pollen has been released and before the stigma is receptive. If you wait too long, the pollen will be shed and the stigma may have received pollen from its own anthers. The commercial roses that you grow in your garden are normally the result of cross-pollination and are genetically highly heterozygous (the opposite of inbred), which in this case refers to most of the loci of plant being in a heterozygous condition or having two different alleles. High heterozygosity results in a more vigorous plant because the genome has increased diversity. Roses are asexually propagated by rooting or grafting cuttings of desirable plants so that superior genotypes are maintained and not broken up by meiotic cell division, as would happen through seed propagation.

General information about tomato

Tomato, Solanum lycopersicon, is a member of the Solenaceae family, the same family that contains potato, pepper, tobacco, and nightshade. The crop has a fascinating history, starting with its likely origin in Peru, importation to Europe with returning explorers, gradual introduction into European cuisine (some cultures associated tomato with nightshade and considered it poisonous), and introduction to North America. This optional tomato history page describes one version of this history.

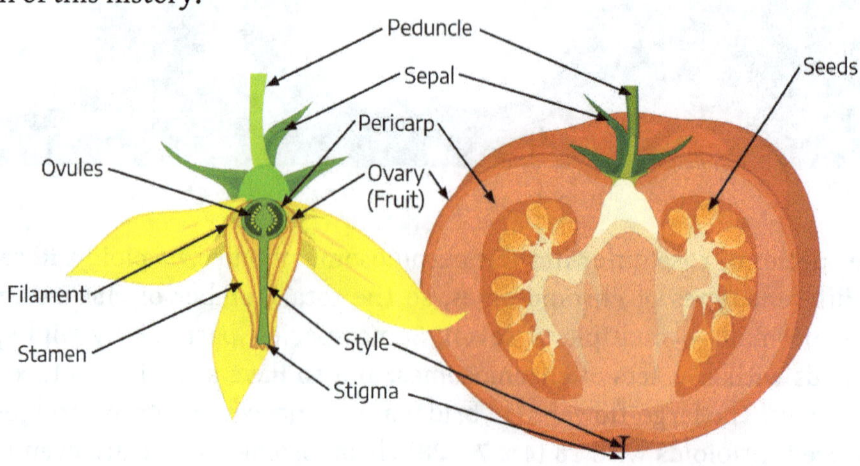

Tomato flower and fruit. © Shutterstock. Adapted and labeled by Emily Tepe.

The tomato flower is perfect (in the botanical sense). The anthers form a cone that completely surrounds the gynoecium. In wild tomatoes, the stigma and style protrude up above the cone of anthers, so these wild types tend to cross-pollinate. The stigma of domesticated tomatoes is either just slightly above the cone, or buried within the cone. Domesticated tomatoes self-pollinate, so we will also need to emasculate the female plants in our crosses. The pistil is made up of several fused carpels, and the mature fruit is a berry.

In contrast to rose, all tomatoes, whether wild or domesticated, are diploids where 2n = 24. They will all cross with each other. Domesticated tomatoes that you grow in your garden are either highly homozygous (inbred) or are F1 hybrids and highly heterozygous. Heirloom and older style tomatoes are typically inbreds (sometimes called pure lines or [incorrectly] open pollinated in seed catalogs), while modern and commercial tomatoes are F1 hybrids. You can use either heirloom or commercial hybrid types in your breeding experiments. Tomatoes are propagated by seeds.

Review questions

- Compare and contrast rose and tomato for their ploidy and method of propagation.
- How does the variation in ploidy affect rose breeding? Does it have the same impact on tomato breeding?
- What is emasculation and why is it done? Which parent do you emasculate, or do you emasculate both parents?

Rose breeding

The overall strategy:

- Identify your objective.
- Choose your parents.
- Make controlled crosses among parents.
- Plant the F_1 **seeds.**
- Evaluate the offspring over a few years.
- Keep seeds from the best plants and compost the rest.
- Asexually propagate the very best one or two by stem cuttings.
- Sell the rights to your award-winning rose to a nursery who distributes it worldwide.

You should start a breeding project with an objective in mind. At first, your objective will probably be just to see if the process works. Later, you will have specific characteristics in mind that you would like to emphasize in the progeny from the cross. Step two is to choose your parents; at least one of the parents should have the trait you are interested in — like flower color, disease resistance, or cold hardiness for surviving Minnesota winters. The other parent should have complementary traits, like size, branching, floral scent, or long vase life.

In a rose, you might choose as the female a nice shrub rose with an attractive red flower and that has made it through several winters without special care. If it has nice full rose hips at the end of the year, you know it's fertile. Perhaps you'd like the new rose to have the same flower type and color, but in a dwarf plant. And perhaps your neighbor has a dwarf rose that sets hips and must be fertile, and you can use some pollen from that rose even though you don't like the color of its flower. Hopefully, you'll get some offspring that show the same red as your female, but in the dwarf growth habit. (You'll probably get some sprawling, gangly rose plants with ugly flowers as well, but you can toss them in the compost.)

Try the cross in the opposite direction as well (switch which plant is male and which is female). This is called a reciprocal cross. Sometimes the cross is easier to make in one "direction" (A x B vs B x A) than the other. When writing a pedigree for the cross, which shows the names of the two parents, remember to write the female first — cross A x B, for instance, means plant A was the female. Breeders sometimes write the pedigree with a slash ("/") rather than an "X:" A/B.

By making several crosses that meet your objectives, you increase your chance of obtaining the specific combination of alleles for the specific traits. If a plant meets your objectives it can be asexually propagated, not by seed, so that you keep the genotype intact in your propagules.

Rose hybridization

The main steps in crossing are:

- Choose buds at the right stage.
- Emasculate the female.
- Collect pollen from the male.
- Transfer pollen from anther of male to stigma of female.
- Identify the cross with a tag.
- Protect the carpel until the ovary begins to swell.

Rose hybridization is a sufficiently popular hobby among gardeners that there are websites with reasonably easy-to-follow instructions. This is a good example from a rose breeding project **(12:18 min).**

The Santa Clarita Rose Society in California has an informative hybridization page, with photos and descriptions of the rose flower before and after emasculation, pollen transfer with an artist's brush, and the resulting hips and achenes.

While you may get your first bloom when your new hybrid is just a seedling, you won't know whether you have a really good rose until you have grown it outdoors for two or three years. Once you think you have something you want to propagate, use stem cuttings **to propagate the selected plant.**

Roses are a good example of a common garden plant you can hybridize. You germinate the resulting seeds, evaluate the progeny, and then asexually propagate the plants you like. It will take at least three years or more to identify plants that maintain the traits you desire to continue through multiple years and environments.

Review questions

- When making a cross, why is staging the flower development important for emasculation and pollination?
- In the cross MN5125 / ND163 which of the two parents do you emasculate?
- What is a reciprocal cross?

Tomato breeding

The overall strategy:

- Identify your objective.
- Choose your parents.
- Make controlled crosses among parents.
- Plant the F_1 **seeds.**
- Evaluate the offspring for the characteristics you are most interested in.
- Keep the seed from the best plants and begin inbreeding.
- Grow the seed from the best plants next spring or in a greenhouse.
- Select the best plants, allow their flowers to self, and keep the seed.
- Repeat the last two steps for about 5–7 generations. Notice that the progeny are now very similar to each other.
- Somewhere around year 5 or 6, identify the best plant that will be the founder of your new, pure line (true breeding) tomato cultivar.
- Sell the rights to your award-winning tomato to a seed company who distributes it worldwide.

Notice that the rose and tomato strategy lists are not identical. With roses, we planted out the seeds from the hybridization, took a few years identifying the best plants, and propagated them asexually. With tomatoes, we want to develop pure lines by inbreeding. Inbreeding is producing seed by selfing over 5–7 generations to develop plants that are highly homozygous at most loci in their genomes. Homozygosity results in identical plants propagated by seed. This is done so the resulting progeny look more and more alike until, after 4–6 years, plants grown from seeds harvested off the same plant are indistinguishable from each other. At that point you have an inbred pure line that will be true breeding if self-pollinated, and you can be pretty sure that you know what you're going to get when you plant seed from year to year. The seed from inbred pure lines are the same because all loci are homozygous, so all gametes are the same and fertilization produces clonal seeds.

Again, identify your objectives and then choose the parents. You might like the cherry tomato that you've been growing on the patio because the fruits are perfect for salads. But being a loyal Gopher, you'd like a gold fruit rather than one that's Badger red. So you reciprocally cross the red patio tomato with the gold beefsteak type tomato and begin assessing progeny.

Tomato hybridization

- Choose buds at the right stage.
- Emasculate the female.
- Collect pollen from the male.
- Transfer pollen from anther of male to stigma of female.
- Identify the cross with a tag.
- Collect seed at maturity.

Tomato is a diploid. This series of web pages (optional) walks you through the outcomes of meiosis in a diploid (click on the links at the bottom of the page to get the whole story), and has graphics illustrating how self-pollination leads to inbred pure lines. There is also a page that runs you through calculations to estimate the number of plants you will need to have a reasonable probability of finding the combination of characteristics you seek. Don't sweat the probabilities of success. Despite the statistics, some of breeding is a chance event at combining the correct set of alleles. Choose your parents wisely, make the cross, evaluate your progeny, and plant out seed from crossing the best progeny the next year. Keep it simple.

Be sure to keep accurate notes, and to tag and identify your crosses, seeds, and plants. Recall the rose breeding website, and how the breeder clipped the sepals of the female flower he was pollinating so he could tell which rosehip contained hybridized seeds. That's an efficient way of avoiding self-pollinated seed.

Dehybridizing

Most new commercial tomatoes, including new garden tomatoes, are F1 hybrids. The seeds you plant in the field are the result of crossing two parents, as described above. By planting the F1 hybrid, the grower capitalizes on the extra vigor associated with a highly heterozygous hybrid genotype. This website (optional) discusses hybrid tomatoes vs. saving the seed from the hybrid to plant. It's an interesting discussion, but remember that science is a systematic study and requires replications, something you don't get from planting seeds from a hybrid in a single year. If you collect seed from a hybrid tomato fruit, whether from a hybrid plant in your garden or a fruit from the grocery store, those seeds are F2 seeds. By planting those seeds out, selecting the most vigorous seedlings, and later in the year keeping seeds from the plants with the best fruits, you are following the same steps laid out in the tomato breeding section, except that you fast-forwarded past the crossing and F1 grow-out stage and went directly to the F2. If you continue to grow out the seed and select the best plants for a few more years, you will end up with a stable variety of your own. This process is called dehybridizing because you are starting with a hybrid, but after several generations of production, selection, and seed saving you are creating new pure lines from that hybrid.

This approach will work well with hybrid tomatoes and hybrid peppers because they are naturally self-pollinating. Garden catalogs will tell you whether the seed you are buying is hybrid. If you are getting your fruits from the store, you can count on them being hybrids unless they are marked as heirloom. The upside of dehybridizing is that it's the easiest way to breed your own crop because you don't have to emasculate and cross. The downside is that you are breeding "blind." You didn't choose the parents for the F_1 cross to achieve your objective, and you don't know the parents' characteristics, so the phenotypes that you get from dehybridizing are anyone's guess.

TERMS

Flashcards

Dominant allele	When one allele is expressed over the other alleles present.
Generative nucleus	Nucleus in the immature male gametophyte that will later divide by mitosis to produce two sperm cells.
Genes	Hereditary units consisting of a sequence of DNA that occupies a specific location on a chromosome (locus) and determines a particular characteristic in an organism. Genes undergo mutation when their DNA sequence changes.
Genotype	Genetic composition of an organism, including chromosomes of the nucleus and the DNA in chloroplasts and mitochondria.

Heritability	Measurement of a quantitative trait that passes from parent to offspring and is measured in high and low; high being similar between parent and offspring and low being dissimilar between parent and offspring.
Heterozygote	Plant with two different alleles of a particular gene and giving rise to varying offspring; offspring are generally more vigorous than offspring from homozygote.
Homozygote	Plant with identical alleles of a particular gene; gives rise to identical, or nearly identical, offspring.
Integument	Cells that form the ovary wall. Nucellus cells on the interior of the ovule wall develop into megaspore mother cells.
Linkage	When two genes are on the same chromosome.
Locus	Location on a chromosome where a particular gene is found.
Megaspornatium	Place in the ovary where the female gametophyte will be formed.
Mendel's First Law – the law of segregation	Principle that during gamete formation each member of the allelic pair separates from the other member to form the genetic constitution of the gamete; e.g., Ss diploid produces S and s gametes.
Mendel's Second Law – the law of independent assortment	Principle that during gamete formation the segregation of the alleles of one allelic pair is independent of the segregation of the alleles of another allelic pair.
Microsporangium	Place in the anther where the male gametophyte will be formed.
Phenotype	Physical appearance of an organism; expressed as Phenotype = Genotype + Environment
Punnett square	Simple database used to visualize the types of zygotes and their expected frequency that form from male and female gametes.
Qualitative differences	Large differences that can easily be seen or measured in qualitative terms; e.g. fruit color.
Quantitative differences	Small differences that are measured numerically; e.g. yield in kg/ha. Can be influenced by the environment.
Recessive allele	Allele(s) that are not expressed if a dominant allele is present; will be expressed if there is no dominant allele.
Suspensor	Produced by multiple mitotic cell divisions of the embryo's basal cell; the suspensor anchors the apical cell of the embryo to the ovule wall.
Synergid cells	Cells flanking the egg cell in the mature female gametophyte.
Tube nucleus, or vegetative nucleus	Nucleus in the male gametophyte that is associated with pollen tube growth.

CHAPTER- 15
INVASIVE PLANTS AND GMOS

Although the discovery of DNA as the genetic material seems like a long time ago (~1928), plant genetic modification has been going on for thousands of years as farmers make selections of individuals to tailor crops to their needs. We continue manipulating plants in the discipline of plant breeding. Genetically Modified Organisms (GMOs) are very recent, but offer a new way to improve plants. Although GMO technology is controversial, it has been rapidly adopted by farmers.

A short lesson on invasive plants introduces this issue and discusses how plant propagators can help. As consumers we must be aware of potential risks some plants pose to the environment.

Learning objectives

- Define what GMO technology is and how it is applied to plant improvement.
- Discuss why GMO technology is controversial and develop an informed opinion about it.
- Know several examples of non-native invasive plants and their impact on the environment.

15.1 INVASIVE PLANTS

Learning objectives

- Be able to explain the consequences of introducing non-native plants that become invasive.
- Be able to map how some plants become invasive.
- Know several examples of non-native invasive plants in Minnesota.

Overview

This lesson introduces some of the features and impacts of non-native invasive plants, using a Minnesota perspective. Be aware, however, that just as a plant native to an exotic continent may become a weedy invasive plant in Minnesota, some plants native to Minnesota have been taken to other continents and become invasive in their new homes. Invasive plants are defined by the U.S. Department of Agriculture as those non-native to an ecosystem whose introduction causes or is likely to cause economic or environmental harm or harm to human health. One example is the oriental bittersweet (Celasturs orbiculatus); the photo below shows, from near Red Wing, Minnesota, shows it smothering ground and tree vegetation.

Bittersweet. Ross. CC BY-NC-ND 2.0

Invasive plant species have contributed to the decline of about half of the endangered and threatened plant species in the U.S. Their impacts include, but are not limited to, the following:

- Competition with native plant species for moisture, sunlight, nutrients, and space.
- Decrease in biodiversity.
- Degradation of wildlife habitat, agriculture lands, and water quality.
- Increase in soil erosion.
- Decrease in recreational opportunities.

- If sexually compatible native plants are present in the invaded regions, hybrids can form resulting in genetic contamination or pollution to native environments.

This lesson reviews some of the common features of invasive plants in Minnesota, and includes specific information about several species.

Where do invasive plants come from?

People and plants have a close relationship, evolving a co-dependence over thousands of years. As people move around the globe they bring with them the plants that are useful for fuel, food, feed for their animals, enhancing the aesthetics of their new home, and providing fiber for clothing and construction. While most of these plants have been positive forces in the lives of people and do relatively no harm, some escape from cultivation and become invasive. In Minnesota, for example, the most reported invasive plant is common tansy (*Tanacetum vulgare*). It was likely first brought to the **U.S.** by John Winthrop, Governor of the Massachusetts Bay colony, about 400 years ago. Although humans are a primary dispersal means for invasive plants and often move them intentionally, such plants may also arrive unintentionally as contaminants in food plants or as seed contaminants in fodder that travels with people and their animals.

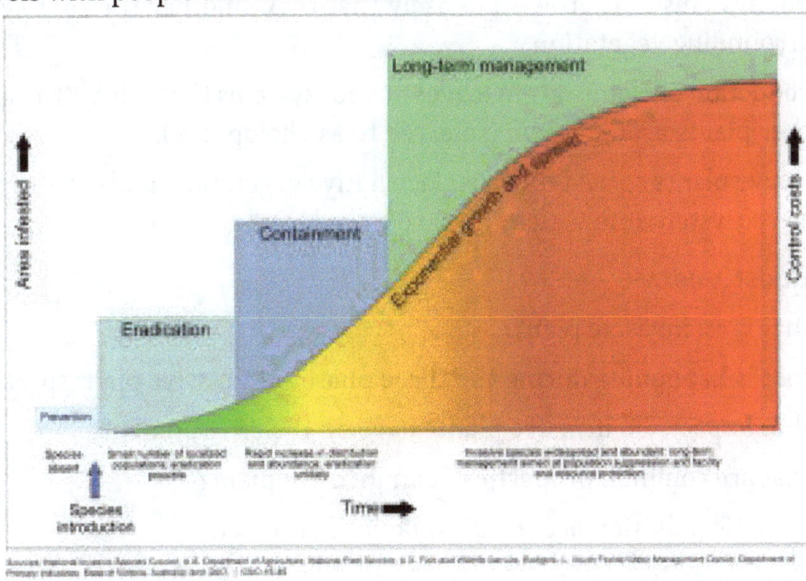

Invasion curve. US Government Accountability Office. Public domain.

The graph above shows the relationship between an invasive plant's spread and time. Initially, the spread is slow and is in a "lag phase" due to the small number of plants in the initial introduction. This lag phase (left third of the graph) is a period in which, if the plant is detected and addressed, control could result in its eradication. Most invasive plants, however, have a high rate of reproduction, either asexual or sexual. In many cases they spread by prolific fleshy fruit production (as with the bittersweet shown above) that is eaten by birds and other animals. As numbers of the invasive plant increase, it enters the exponential rate of spread (middle third of the graph). Think of exponential spread as a rate similar to 2, 4, 16, 256, 65536. If left unchecked, as some invasive plants are, spread will reach a maximum (shown in the last third of the graph). The horticultural industry is a common entry point for many invasive plants. The properties that make a plant attractive to growers and consumers for their yard or garden – including novelty, robust growth, abundant flowering over a long period of time, and easy sexual or asexual propagation – are the very qualities that make many ornamental introductions invasive.

In addition to large quantities of seed, invasive characteristics listed by the U.S. Department of Agriculture's Invasive Plant website include:

- Thriving on disturbed soil.
- Distribution by birds, wind, or humans over great distances.
- Aggressive root systems that spread long distances from a single plant.
- Root systems that grow so densely that they smother the root systems of surrounding vegetation.
- Production of chemicals in leaves or root systems that inhibit the growth of other plants around them (referred to as allelopathy).

As invasive plants spread they displace native vegetation, and can dramatically impact the ecosystem into which they are introduced.

Review questions

- What is an invasive plant?
- What is happening during the three phases of invasive plant spread?
- What impacts do invasive plants have on the environment?
- What are common properties of an invasive plant?
- If genetic pollution occurred, could it be reversed?

Invasive plant examples

Common tansy (*Tanacetum vulgare*)

Tansy, an herbaceous perennial in the family Asteraceae, is the most reported invasive plant in Minnesota. It was introduced to North America from Eurasia about 400 years ago, and has been used as a medicinal plant and a funerary herb because of its aroma. Rumor has it that the first president of Harvard University was buried with tansy, and when his grave was moved almost 200 years later, the tansy was still fragrant. Tansy's invasion of pasture lands is problematic due to its toxicity to mammals. For more information on common tansy see the Minnesota Department of Agriculture website.

Tansy. Robert Flogaus-Faust. CC BY 4.0

Oriental bittersweet (*Celastrus orbiculatus*)

Oriental bittersweet was intentionally introduced as an ornamental for dried floral arrangements and fall decorations (left-hand photo, below). It is a woody vine that climbs trees and structures and is capable of girdling the trees as the vines tighten around the truck (right-hand photo). Girdling plus the accumulated growth of heavy vines can bring down large trees. Oriental bittersweet thrives in a wide range of habitats, light levels, and soil types, and can grow to over 20 meters (65 ft) in length. Although it is on the Minnesota Department of Agriculture's list of plants that are prohibited and must be eradicated, it is still grown and propagated in Wisconsin for use in floral arrangements.

American Bittersweet.
Peggy. CC BY-ND 2.0

Bittersweet girdling a tree.2.0
mwms1916. CC BY-NC-ND

Knotweeds (*Fallopia sp.*)

The knotweed complex consists of two species and their hybrid, includes Japanese knotweed (Fallopia japonica), giant knotweed (F. sachalinense), and their hybrid Bohemian knotweed (F. × bohemica). (Remember that an "×" between the genus and specific epithet indicates an interspecific hybrid.) The knotweed was originally imported as a novel ornamental. With its fast growth and late fall flowering it can make a very showy specimen planting. Each taxa of knotweed has very rapid growth. As an herbaceous perennial it resumes growth from large rhizomes (an underground stem) that produce prolific shoots and adventitious roots, making it difficult to control with herbicides.

Japanese knotweed. David Short. CC BY 2.0.

Knotweeds can also cause damage to structures. With their long-lived rhizomes capable of adventitious rooting, and dioecious flowers capable of prolific seed set, knotweeds are a double threat to native plants and the environment.

Knotweed can grow through cracks in pavement, walls, and foundations, causing structural damage. Gordon Joly.

Japanese barberry (*Berberis thunbergii*)

As a popular landscape plant, Japanese barberry (*Berberis thunbergii*) has been extensively bred, and many crosses have been made with related and sexually compatible barberries. It is planted in most areas of the University of Minnesota Twin Cities campuses. Barberry produces a fleshy drupe that is consumed and dispersed by birds. It has many sharp spines along the shoots and forms dense thickets. These thickets can prevent students from cutting across landscapes and provide excellent protection to small rodents (mice). As barberry populations increase, so do those of mice and other small rodents. Increased mouse populations are associated with increased tick populations (ticks are small, blood-sucking parasitic bugs) that may increase tick-borne diseases.

Japanese Barberry. James Gaither. CC BY-NC-SD 2.0.

Winged burning bush or winged euonymus (*Euonymus alatus*)

Winged burning bush or winged euonymus (Euonymus alatus) is small tree or shrub from eastern Russia, central China, Korea, and Japan. It is very popular for the interesting wings on its stems and its brilliant red fall foliage. Its popularity for shade or sun landscape planting and ability to escape from cultivation means that escaped populations often occur in nature. This video shows a paddling trip to the Tellico River in Tennessee and the discovery of a large winged euonymus escaped from cultivation. Winged euonymus is adaptable to many growing conditions and forms dense canopies, reduces native native plant diversity. It is a popular landscape plant and still being propagated and sold in Minnesota. Fortunately, several researchers are breeding Euonymus cultivars with reduced or no seed set. How might these cultivars be propagated if they were sterile?

Euonymus alatus 'Compactus'.
Leonora (Ellie) Enking. CC BY-SA 2.0

Winged burning bush young stem.
Wisconsin First Detector Network. CC BY-NC 2.0

Euonymus alatus.
Katja Schulz. CC BY 2.0.

Common buckthorn (*Rhamnus cathartica*)

Common buckthorn is native to Europe and Asia and is a highly invasive perennial understory shrub or tree. Although propagation and sale of common buckthorn (Rhamnus cathartica) are prohibited in Minnesota, it is common in both urban and natural habitats. There is, for example, a large and very old male buckthorn tree (a dioecious plant) at the corner of Cleveland Avenue North and Buford Avenue on "The Lawn" of the St. Paul Campus of the University of Minnesota.

Subopposite leaves on common buckthorn. Eli Sagor. CC BY-NC 2.0

Buckthorn fruit. Peter O'Connor. CC BY-SA 2.0

Goats grazing on buckthorn. USFWS Midwest Region. Public domain.

Buckthorn was introduced to North America as an ornamental shrub and used for living fence rows and wildlife habitat. It has since spread aggressively across most of the northeast and upper Midwest through production and distribution of prolific fleshy fruit (middle photo, above). Common buckthorn has become a serious threat to native forest understory habitats, where it outcompetes native plant species. With such a large distribution, novel control methods are being tested such as goat grazing (above), which is showing success and has a reduced environmental impact compared to herbicide control.

These examples are a small sample of the plants that are known to be invasive in Minnesota. Fortunately for the horticulture industry, consumers, many scientists, and communities are engaged in eliminating invasive plants that have escaped into their environments.

15.2 GMOS

Learning objectives

- Understand why and how GMO technology is applied to plant improvement.
- List examples of GMO crops and explain why they were adopted.
- Explain why GMO technology is controversial.

Overview

A GMO is a genetically modified organism — a plant carrying one or more transgenes as part of its genome. GMOs are produced through genetic engineering, where a transgene from any organism is manipulated to produce a trait in a plant after it has been introduced. Transgenes are DNA that are manipulated to function in a plant to produce a specific trait. They can originate from any other organism, and the new DNA need not be introduced through sexual reproduction as in making crosses for plant breeding. (In non-GMO crops, all genes originate from sexual reproduction.) As discussed in the chapter on DNA, the genetic code is the sequence of bases along a strand of DNA, and is universal among all living organisms, enabling the transfer of a gene from a bacteria, virus, or any organism to a plant.)A transfer of DNA across kingdoms would never occur by sexual reproduction, because interkingdom crosses cannot be made.) GMO technology offers, for example, the possibility of engineering resistance to the corn rootworm by introducing a bacterial transgene into corn or the ability to test a gene from any source in a plant.

This lesson offers insight into how GMOs offer a novel mechanism for plant improvement that can add significant traits to forestry, agronomy, and horticulture crops. Their contributions include herbicide tolerance for improved weed control, resistance to insect and viral pests, and improved health benefits for crops.

Unlike rose and apple breeding, in which hobbyists can breed new crops in their gardens, GMO technology integrates knowledge from genetics, molecular biology, and tissue culture to produce a new GMO crop. These skills would be difficult to combine in the garden, but the plants that are produced are propagated identically to their conventional counterparts. It would be difficult to distinguish the GMO "Innate" potato from a non-GMO potato, unless they were tested for browning, acrylamide production, or disease resistance. GMO potatoes look and behave like other potatoes. Adoption of GMO technology by a farmer requires no new equipment or cultivation techniques.

What is a GMO?

Papaya tree infected with Papaya ringspot virus (PRSV).. Scot Nelson. Public domain.

GMOs begin where plant breeding ends. An excellent example of GMO technology occurred after the introduction of Papaya Ringspot Virus (PRSV) to Hawaii. PRSV devastated papaya production to the where point growers could not produce a crop. Attempts to use plant breeding to produce a virus-resistant plant were unsuccessful. No sexually compatible plant with resistance to PRSV could be located, so no amount of traditional plant breeding by crossing would produce resistance. This viral disease prevented papaya production in Hawaii until Dennis Gonsalves, of the US Department of Agriculture, adopted a GMO strategy that had previously been used to produce potatoes resistant to several viruses, and made papayas resistant to PRSV. Today, any papaya produced in Hawaii is likely to be either "SunUp" or "Rainbow," the GMO cultivars. Although the virus-resistant papaya seems like a win for growers and consumers, it is very controversial. In the 1980s, the early development of GMO technology was tied to large agricultural companies that patented and used the technology in corn, soybeans, and cotton. Others see the "unnatural" movement of a gene from a bacteria to a plant as a process that we should not use.

The GMO papaya is a good example for presenting the process involved in producing a GMO. A first step in developing GMO papaya was to understand the virus, its genome, and its replication. Virologists, those devoting their lives to the study of virus biology, knew that the virus' genome had a gene encoding a coat protein that surrounded the virus' genetic material and that was essential to its infection of plant cells. The GMO strategy added a copy of the viral coat protein gene into the plant's chromosomes. The coat protein protects the virus as it is transmitted from plant to plant, and is essential for replication. The viral gene is engineered to be in the opposite orientation from its orientation in the virus. This opposite orientation of the viral gene, now acting as plant gene, effectively shuts down viral replication before it can cause disease, making the plant resistant.

John Sanford.

FMSResearcher. CC BY-SA 4.0.

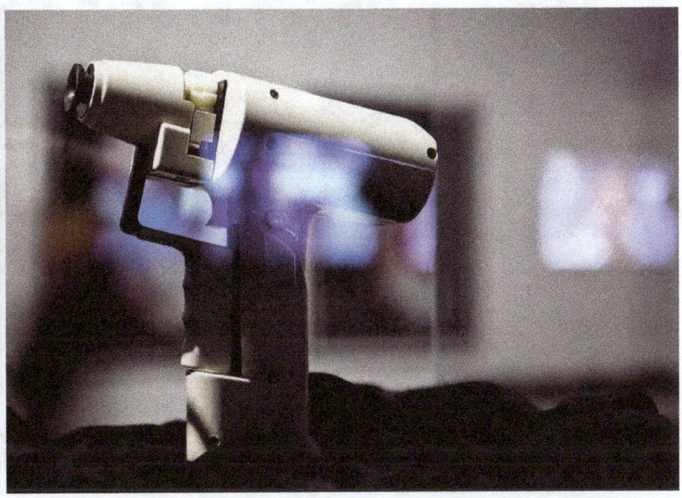

Gene gun.

Z33 House for Contemporary Art, Design & Architecture. CC BY-NC-ND 2.0.

To introduce the engineered viral gene into the plant's chromosome without crossing, horticulturist John Sanford invented a means of introducing genes into plant cells by literally shooting the gene into the nucleus of a cell. (John graduated in 1976 from the University of Minnesota with a BS in horticulture, and continued in horticulture to become a professor and researcher at Cornell University.) In this process, DNA of the gene of interest is coated onto particles much smaller than the plant cell's nucleus. The gun works off of compressed air, similar to a BB gun. The particle coated with the DNA (gene of interest), now in the nucleus, diffuses away from the particle and integrates into the plant's chromosome. The gene is then expressed and is hertible in the same qualitative manner as other single gene traits. The process of introducing a transgene into a plant is called transformation. The plant cell or the whole plant carrying the engineered gene is said to be transformed or transgenic (synonyms for GMO).

Alan pointing to gall on a redwood in Muir Woods that may have been produced from *A. tumefaciens.*

Could this process ever happen in nature? Agrobacterium tumefaciens is a bacterial plant pathogen that produces crown gall disease. This bacterium transfers a small amount of its DNA (several genes) to the plant as part of its pathogen attack. The photo to the left shows a large gall on a redwood that may have been caused by Agrobacterium. The pathogen's genes have been engineered through evolution to take over the plant cell's normal regulation and produce a gall — a callus of rapidly dividing undifferentiated cells similar to a callus. The introduced bacterial genes, now functioning in the gall or callus, produce compounds that feed the Agrobacterium, which produces more pathogenic cells to infect other plants. This is crown gall disease of plants. Biologists studying Agrobacterium's life cycle realized they could use the natural DNA transfer process to introduce genes into plants by piggy-backing a gene of interest from any organism onto the Agrobacterium's transfer process. This has been very successful, and is the most used method of gene introduction into many crops. Although the genes to produce a GMO virus-resistant papaya were introduced with the gene gun, most of those in GMO crops were introduced using Agrobacterium.

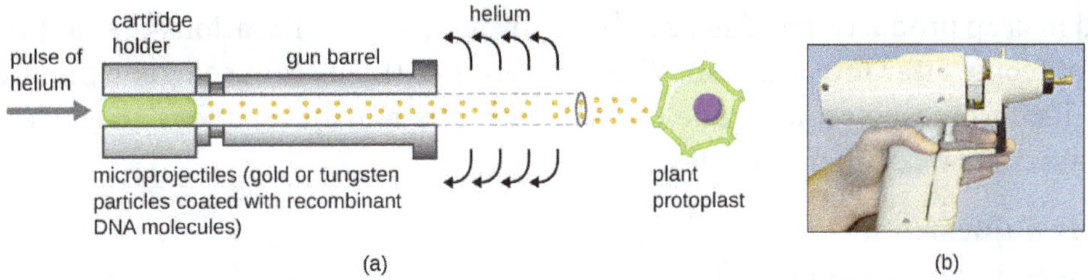

Diagram of gene gun process. OpenStax. CC BY 4.0.

The gene gun (above) and Agrobacterium-mediated gene introductions are inefficient processes that introduce the gene into only a very small number of plant cells. Since only a single cell receives the gene, plant tissue culture is used to regenerate an entire plant from that single cell. Plant tissue culture uses synthetic growth media to provide the environment for mitotic cell divisions and organization of those cells into shoot and root meristems. When the meristems are formed, the tiny plants can be removed from tissue culture and transferred to soil. This is possible because plant cells are totipotent — they can regenerate a whole plant from a single cell. (Think about about adventitious rooting, where parenchyma cells can divide and differentiate into a root meristem when buried in potting mix.) This capability has been identified in only a few plant species using very specific components in the tissue culture medium to coax the cells to divide and regenerate a whole plant.

As in plant breeding, GMO plants must be assessed and selected for the traits of interest. Introducing a single gene into a plant can produce a large phenotypic change — as in the naturally occuring Shrunken-2 allele that transforms field corn to sweet corn, or the introduction of the Papaya Ringspot Virus coat protein gene to provide virus resistance. Introducing a new gene using GMO technology does not, however, change other genes or traits of the plant's genome. Being able to maintain all of a cultivar's characteristics and add a gene for resistance from any organism is a major advantage of GMO technology.

The art of GMO technology is discovering how non-plant organisms (virus, bacteria, fungi, insects, and so on) might contribute a gene to improve a plant, be it through weed control, increased nutrition in foods, or resistance to pests or viruses. GMO crops are produced by interdisciplinary teams with members expert in gene identification, gene engineering, plant breeding, gene introduction, and tissue culture. These skills mimic the steps required for GMO technology: identifying a gene or genes and their source to solve a problem, engineering the gene for plant expression, using the gene gun or A. tumefaciens to introduce into the crop of interest, and assessing the new plant. It is possible that the predicted gene action may not occur and that other genes or gene modifications are necessary. All genes used in crop production today have been through several iterations that improved the outcome after introduction. When a plant has the introduced gene integrated into its chromosomes, it can be crossed with other plants to move the new gene and its traits into other varieties.

Review questions

- What is a GMO?
- What are the two significant advantages of GMO technologies over traditional plant breeding?
- What are the differences between the two processes used to introduce a novel gene into a plant?
- Can a GMO seed, seedling, or mature plant be distinguished from a non-GMO one just by looking at it? What distinguishes the two, either visibly or genetically?
- Where do transgenes come from and how are they introduced into a GMO crop? Why do farmers pay extra for GMO seeds?

GMO technology has been rapidly adopted by farmers in the US and worldwide (the graph below shows adoption rates for the US in 1996–2018). In the US, corn, soybean, and sugar beet crops are GMO for herbicide tolerance (which allows for improved weed control), insect-resistant GMO cotton is extensively adopted to prevent boll weevil damage. For a more comprehensive list of crops and GMO traits, see the GM Approval Database from the ISAAA.

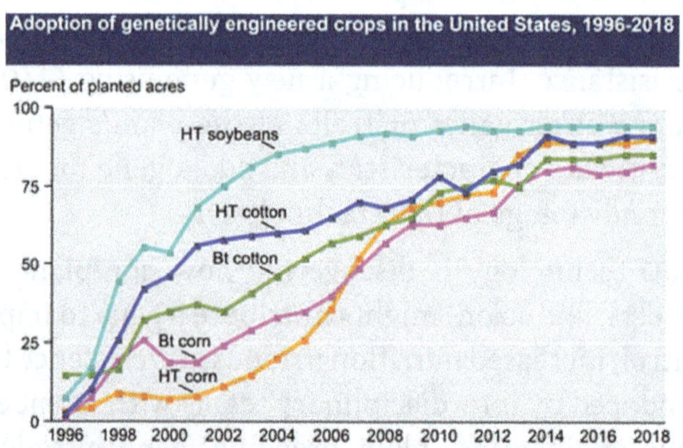

Note: HT indicates herbicide-tolerant varieties; Bt indicates insect-resistant varieties (containing genes from the soil bacterium Bacillus thuringiensis). Data for each crop category include varieties with both HT and Bt (stacked) traits.

From:https://www.canr.msu.edu/news/why-many-growers-are-quick-to-adopt-genetic-modification-technology

As it implies, the "approval" in the name of this database indicates that GMO crops must go through extensive testing and a review process before being allowed into production and into our food stream. The rapid adoption of GMO technology triggered questions of risks that resulted in a review process for all GMO crops by the US Department of Agriculture, the Food and Drug Administration, and the Environmental Protection Agency. The overwhelming scientific evidence indicates GMO crops are safe for the environment and for humans. To read more (optional) on the safety of GMO technology, see the Academics Review's site on Genetic Roulette for science-based information and critical review of GMO technology risks.

GMO crops have economic, environmental, and convenience advantages for farmers. Several economic analyses have shown that increased yields from GMO crops are a major impetus for adoption by farmers, even though GMO seeds cost more than conventional seeds. A good example is the rapid adoption of GMO sugar beets that were modified for resistance to the non-selective herbicide Roundup. This allowed the beets to be sprayed with Roundup, killing weeds in the field and leaving the beets unharmed. A 2008 survey of GMO sugar beet growers showed the highest weed control ratings in the history of the survey and the near elimination of mechanical and manual weeding. Weed control had previously been inefficient, requiring herbicide and mechanical controls. Making beets herbicide tolerant increased yield and made weed control convenient. Adoption was greater than 90% one year after the GMO herbicide tolerant seed was made available.

The most significant environmental benefits of GMO technology come from GMO insect resistance. Most insecticides are non-specific, killing beneficial insects as well as the pest, and are toxic to humans, birds, fish, and other organisms. The basis of GMO resistance is the introduction of the Bacillus thuringiensis gene, Bt, into the plant, so the plant produces Bt. You may have heard of Bt as an insect control used in organic food production or mosquito control. Bt has a very low environmental impact because it is highly specific for the target insect, unlike conventional insecticides that can have wide ranging collateral damage when applied. Bt is also very labile, breaking down rapidly in the soil. The Bt gene from B. thuringiensis has been introduced into several crops. Currently, Bt crops include corn (field and sweet), cotton, potato, eggplant, tobacco, and soybean for control of several insect pests. Each crop has a specific Bt that targets the insect pest in that crop. Where the GMO crop is grown in place of conventional crops, insecticide applications have been greatly reduced. The battle continues, however, as insects develop resistance to Bt, leaving the crop susceptible to insect damage.

Although using GMO technology requires a significant investment to develop a crop, propagation and use by a farmer is identical to that of conventional counterparts. The simplicity and effectiveness of GMO crops has accelerated their adoption by farmers. GMO herbicide tolerance, for example, allows for a single post-germination application that provides improved weed control, eliminating the need to apply several different herbicides at different times and to use mechanical weed removal. This single application reduces fuel cost for farmers, reduces carbon emission, and provides excellent weed control. It is called herbicide tolerance because genes are introduced into the crop to make it tolerant (resistant) to a specific herbicide that kills weeds. The convenience to the farmer comes from using a single herbicide with flexible timing of application to the crop.

Review questions

- What advantages does a GMO crop present to a farmer?
- Do any of the advantages of a GMO crop for a farmer translate to advantages to the consumer?
- Why were GMO crops adopted so rapidly by farmers?
- What are two examples of GMO crops that were rapidly adopted?
- What does a farmer need to do to adopt a GMO crop with insect resistance?

GMO crops benefiting human health

Early in the technology's development, most GMO work focused on the farmer and/or on production problems. Herbicide tolerance, insecticide resistance, and virus resistance each protected the crop or made farming easier. There are now, however, several GMO crops developed primarily for the consumer. The iconic, increased-nutrition GMO crop is Golden Rice. It contains introduced genes that synthesize carotene in the endosperm. You can easily distinguish Golden Rice from conventional rice by its yellow color from the accumulated carotene, the same pigment found in carrots (see photo below). When humans consume the carotene in Golden Rice, it is converted to vitamin A, an essential vitamin for human health. Conventional milled-rice contains no vitamin A, which results in vitamin A deficiencies in millions of adults and children who consume most of their calories by eating rice. The Golden Rice Project estimates the number of child deaths caused by vitamin A deficiency at 1.15 million/year. Vitamin A deficiency also causes loss of sight, increased susceptibility to a number of diseases, and reduced intellectual development. While Golden Rice would greatly mitigate the problem and has been crossed into many regional cultivars, adoption has been slowed or blocked by anti-GMO organizations. Research continues with the development of golden bananas and cassava.

Golden Rice grain compared to white rice grain in screenhouse of Golden Rice plants. International Rice Research Institute.

A recently approved GMO crop for production, the "Innate" potato has introduced genes that reduce the production of the neurotoxin acrylamide. Acrylamide forms in many cooked foods from the reaction of amino acids, sugars, and heat. The risk of consuming acrylamide is somewhat controversial, but the fact that it is a neurotoxin makes this an important improvement to the potato. The "Innate" potato has two other introduced genes that make it resistant to the potato blight fungus and prevents browning of the tuber after being cut or bruised. It remains to be seen whether this potato is accepted by consumers and the fast food industry, where many potatoes are consumed. One of the first GMO crops marketed was the "New Leaf" potato, which had both insect and virus resistance. Consumers and the fast food industry, led by anti-GMO organizations, mounted a campaign of GMO fear that effectively ended cultivation of this potato. The "Innate" potato reopens the discussion of GMO foods.

Review questions

- What are the differences in traits that benefit the farmer vs. benefiting consumers?
- What human health issues can be addressed using GMO technology?

Why are GMOs controversial?

March Against Monsanto in Lethbridge Alberta. John Novotny. CC BY 2.0.

The overwhelming evidence from peer-reviewed science is that GMO crops are safe for growers, consumers, and the environment. There are several anti-GMO organizations that have been effective in sensationalizing concerns and controversy concerning GMO technology. The first concern about GMOs is that the technology is not natural. A gene from a bacteria would have slim chances of integrating into a plant's chromosome without technology intervening — or would it? Concerns have been raised about genetic pollution, where genes from GMO crops would escape to conventional crops or weedy relatives, producing superweeds. The second leading cause of controversy is that several large chemical and agricultural companies invested heavily in GMO technology, and then produced the first GMO crops. GMO technology was patented by these large companies, increasing concerns due to the environmental records of these companies. The lack of a federal requirement to label foods as GMO also fanned fears of potential risks.

Other concerns have originated in the technical nature and rapid development of GMOs, which many argue prevented a full assessment of risk. Arguments centered around a new technology's potential to have unintended or unknown consequences. Some speculated on the possibility that the introduced genes could lead to the production of toxins, allergens, or carcinogens, despite thorough testing for these compounds in the review process. Adding to these concerns was the US Department of Agriculture's decision to not require labeling of food from or containing GMO plants. Additionally, growers' concerns led to the rejection of GMO crops in certified organic production. Each of these issues has been amplified by anti-GMO organizations such as the Non-GMO Project, the Center for Food Safety, and Greenpeace. The message from these organizations was and continues to be that GMO foods are unsafe for people and the environment, despite the overwhelming scientific evidence to the contrary. This is the partial story of the development and adoption and controversy of GMO crops. You can look forward to hearing more about the GMOs, especially the "Pinkglow" that Jimmy Kimmel says "tastes exactly like a pineapple."

TERMS

Flashcards

Bi-parental cross	Cross between two different plants.
Conventional crop	Crop that is not GMO, with all of its genes originating from sexual reproduction.
Emasculate	Act of removing the anthers before pollen has been shed from a flower that is used as the female in breeding; used to reduce self-pollination when wanting to cross.
Exponential spread	Very rapid spread of an invasive plant as numbers and reproductive rates accelerate.
Genetic engineering	Manipulation and introduction of a transgene into a plant for a specific trait.
Genetic pollution	Where genes from GMO crops may escape to conventional crops or weedy relatives.
GMO	Genetically engineered organism or a plant containing a transgene.
Inbreeding	Producing seed by selfing over 5–7 generations to develop pure lines.
Invasive plant	Plant that is non-native to an ecosystem, and whose introduction causes or is likely to cause economic or environmental harm or harm to human health.
Lag phase	Phase during invasive plant spread that is slow due to a low number of plants being introduced.
Plant tissue culture	Method that uses synthetic growth media to provide the environment for mitotic cell divisions of plants and is used to regenerate a single cell into a whole plant.
Ploidy	Number of sets of homologous chromosomes in a cell.
Pure line	True breeding plant produced by inbreeding so it is homozygous at most loci and produces identical plants by seed.
Reciprocal cross	Matching cross where the pollinator becomes the female and female of the former cross becomes the pollen donor.
Rhizome	Horizontal stem growing just below the soil surface.

Super weed	Weed produced by crossing with a GMO crop; inherits the GMO trait, like herbicide tolerance.
Totipotent	Ability of a single plant cell to grow into a whole plant.
Transformation	Process of using the gene gun or Agrobacterium tumefaciens to introduce a transgene into a plant.
Transformed	Synonym for GMO, or plant carrying a transgene.
Transgene	Gene introduced into a plant from another organism, not through sexual reproduction.
Transgenic	Synonym for GMO, or plant carrying a transgene.
Viral coat protein	Protein that surrounds the viral genome, protecting it; essential to virus replication.

GLOSSARY OF TERMS

Accessory pigments: Light-absorbing pigments, other than chlorophyll, that are found in chloroplasts.

Accessory tissues: Tissue of the fruit that is from non-carpel origin, usually in epigynous and perigynous flowers. E.g. the flesh of an apple is hypanthium tissue and the ovary is the papery core that encloses the seed.

Adhesion: A force where dissimilar molecules stick together; in plants this force of adhesion between water and the walls of the xylem helps hold the water in the xylem against the downward force of gravity.

Adventitious: Tissue arising from an organ other than expected.

Adventitious roots: Roots that emerge from the stem rather than roots.

Aggregate fruit: Fruit formed from the ripened ovaries present in one flower with numerous simple carpels.

Aggregates: "Clumps" in the soil; see soil structure definition.

Agriculture: The science or practice of farming, including cultivation of the soil for the growing of crops and the rearing of animals to provide food, fiber, and other products.

Agronomy: The science and technology of producing and using plants for food, fuel, fiber, and land restoration on an extensive scale. The value per acre is lower than for a typical horticultural crop.

Alternate leaves: Leaves are attached on alternate sides as they go up the stem.

Alternation of generations: Cycle of diploid, asexual, vegetative generation alternat ing with the haploid, sexual generation.

Anaphase: Third phase of mitosis; the sister chromatids separate (now chromosomes) and the centromeres divide, pulling the chromosomes to opposite poles.

Androecium: One of the whorls of a flower and is all of the male reproductive parts; stamens.

Angiosperms: Group of flowering plants whose seeds develop inside an ovary.

Annual plant: Plant that is produced from seed in the spring and dies at the end of the growing season.

Annual rings: The demarcation between small-celled later summer and large-celled spring secondary xylem.

Antenna complex: Structure of chlorophyll and accessory pigments that are embedded in the thylakoid membranes; it captures and routes energy from sunlight to a collector called a reaction center.

Anther: The pollen-bearing component of the stamen.

Anthocyanin pigments: Red pigments that are produced primarily in the autumn in response to bright light and excess plant sugars in leaf cells.

Anticlinal division: The type of cell division where the new cells have divided so that the wall of the cells is perpendicular to the outside of the stem.

Antipodal cells: Three cells sequestered at the opposite end of the mature female gametophyte from the egg and synergid cells.

Apex: Tip of the stem.

Apical bud: Bud located on the tip of the stem.

Apical meristem: Group of more or less continually dividing cells located at the tip of a shoot or root.

Apomixis: A form of clonal reproduction where vegetative cells in the flower develop into zygotes to form seeds.

Apoplast: Space outside the cell membrane where water and minerals can move freely. It is interrupted by the casparian strip in roots.

Approach graft: A type of grafting where two independent plants are grafted together and severed only once the graft has "taken."

Asexual propagation: A form of propagation that results in plants with identical genetics to the parent plant.

ATP: A principle molecule for storing and transferring energy in cells; it is created in the LR.

Axil: Upper angle between a lateral structure and the stem to which it is attached.

Axillary bud: Bud borne in the axil of a stem.

Axillary meristem: Group of more or less continually dividing cells located at the axils of a stem.

Bark: All of the tissues exterior of the vascular cambium, which includes the primary and secondary phloem, phelloderm (if present), cork cambium, and cork.

Basal root: Root that emerges from the region just above where the main stem stops and the root begins.

Bi-parental cross: Cross between two different plants.

Binomial nomenclature: System of naming in which two terms are used to denote a species of living organism, the first one indicating the Genus and the second the Specific Epithet.

Bisexual: The term used for a flower that has both the androecium and gynoecium; also called hermaphroditic or a perfect flower.

Bract: A modified leaf or scale, usually small, with a flower or flower cluster in its axil.

Branch: Vegetative growth coming from a node on the main stem.

Bridge graft: A type of repair graft used when a plant has been girdled; scion pieces are inserted above and below the girdled site and act to repair the disruption of the cambium.

Bud: Immature vegetative or floral shoot or both, often covered by scales; also called a meristem.

Budding: It is a form of grafting where a single scion is used rather than an entire stem.

Bulb: A specialized, underground organ with a short, fleshy basal stem enclosed by thick, fleshy scales modified for storage.

Callus: A growing mass of unorganized parenchyma cells produced in response to wounding.

Callus bridge: Parenchyma cells that lie between the cambium of the rootstock and the scion and differentiate into cambium cells.

Calyx: One of the whorls of a flower and is located at the base of the receptacle and contains all the sepals.

Cambium: Lateral meristem in vascular plants, including the vascular cambium and cork cambium, that forms parallel rows of cells resulting in secondary tissues.

Carbohydrates: One of the three major types of nutrients found in seeds; they provide energy in the form of starch and sugar.

Carotenoid pigments: Yellow and orange pigments that are present in the leaf all growing season, but during the warm part of the season these colors are hidden by the high concentration of green-colored chlorophyll. They take longer to break down than chlorophyll.

Carpel: Composed of three parts: stigma, style, and ovary.

Casparian strip: A band-like deposit of waterproof suberin that wraps around each cell in the endodermis and forces water to move through the cells rather than the intercellular spaces.

Cell cycle: Cycle in which cells go through in their lifetime; it consists of interphase and mitosis.

Cell division: The process in mitosis where one plant cell divides into two identical cells.

Cell membrane: Made up of layers of protein and lipid (fats and oils are examples of lipids) and is semi-permeable, meaning that it allows select compounds in and out, but blocks other types of compounds.

Cell wall: A rigid membrane that contains cellulose (a carbohydrate that is indigestible for humans) and is the outer covering of the cell.

Cellulose: Main constituent of the cell wall and is key to a plant's structural integrity; it is a long chain of glucose. It sequesters atmospheric CO2 for long-term storage.

Centromere: Constricted spot where the sister chromatids attach.

Chasmogamy: When the anther matures after the flower opens and pollen is shed before the stigma becomes receptive; it's common in non self-pollinating crops.

Chiasma: Point where sister chromatids of homologs lay over each other forming an "X" shape.

Chimera: When two different genotypes are growing on a single plant.

Chlorophyll: Green photosynthetic pigment found in plants, algae, and cyanobacteria that captures light for photosynthesis.

Chlorophyll a: Type of chlorophyll; it mainly absorbs violet and red light while reflecting green light.

Chlorophyll b: Type of chlorophyll; it mainly absorbs blue and orange light while reflecting green light.

Chloroplast: An organelle that contains chlorophyll where light energy is captured and where the first steps are taken in the chemical pathway that converts the energy in light into forms of energy that the plant can transport and store, like sugar and starch.

Chromoplasts: Cellular organelles that contain types and colors of pigments other than the chlorophyll found in chloroplasts.

Chromosome: Structure within the nucleus of a cell that contains the genes; it is made-up of DNA that has looped around histone proteins then coils and folds.

Class: Taxonomic rank below Division and above Order.

Clay: Smallest particle in soil and has high nutrient holding capacity.

Cleft grafting: A form of grafting where the rootstock is much larger than the scion; both are dormant.

Cleistogamy: When the anther matures, pollen is shed, and the stigma is receptive before the flower opens; it's common in self-pollinating crops.

Cohesion: A force where similar molecules stick together; in plants this occurs with water molecules bonding together.

Coleoptile: Protective sheath that covers in the plumule and epicotyl in the Poaceae family.

Coleorhiza: Protective sheath that covers the radicle in the Poaceae family.

Collenchyma: An elongated cell type with thicker walls and usually arranged in strands; provide support.

Companion cells: Associated with sieve tube members (direct the metabolism) and contain a nucleus (alive).

Complete flower: Where all four whorls are present: calyx, corolla, androecium, & gynoecium.

Compost: A type of organic matter that builds soil structure and assists in retaining moisture and nutrients.

Compound inflorescence: Inflorescence with a group of flowers and includes a rachis.

Compound leaf: Leaf with a blade margin that is completely interrupted and segmented into separate leaflets.

Control: Used to verify or regulate a scientific experiment by conducting a parallel experiment or by comparing with another standard

Conventional crop: Crop that is not GMO and all of its genes originated from sexual reproduction.

Cork: The outer protective tissue of bark; also called phellem.

Cork cambium: A lateral meristem that is responsible for secondary growth that replaces the epidermis in roots and stems; also called phellogen.

Cork cells: The cells located in the cork that are lined with suberin and are dead at maturity.

Corm: A condensed stem, storage organ, typically grown underground and covered in scale leaves.

Corolla: One of the whorls of a flower consisting of all the petals.

Cortex: Also known as the ground meristem, is found just inside the epidermis and extends toward the interior of the stem and root, and is made up of three types of cells: parenchyma, collenchyma, and sclerenchyma.

Cotyledon: An embryonic leaf in seed-bearing plants, one or more of which are the first leaves to appear from a germinating seed.

Cotyledonary node: Food storage structure used in germination.

Cover cropping: Crop used to benefit the soil rather than the main crop species.

Crossing over: Exchange of arms of DNA between sister chromatids of homologous chromosomes that can take place at the point of chiasma formation.

Crown: Compact stem tissue at or near the soil surface.

Cultivar: A plant variety that has been produced in cultivation by selective breeding. The term comes from combining the words 'cultivated' and 'variety'.

Cuticle: Protective waxy coating of cutin on epidermis cells that restricts water loss.

Cutin: Water-resistant substance that coats the wall of the cell exposed to the environment and helps limit the loss of water that is inside of the plant to the atmosphere.

Cytokinesis: Occurs directly after telophase; the cell plate forms between the two daughter cells and the cell walls separate the newly formed cells.

Cytoplasm: The fluid inside the cell membrane in which the organelles and other plant cell parts are suspended.

Dehiscent: Used to categorize fruits with seeds that separate from a dried pericarp.

Demonstration experiments: A very valuable method for actively learning the body of scientific knowledge that has been previously discovered and communicated by others; and it is specifically orchestrated for teaching and learning, not for the discovery of new information about the world around us.

Derivative (cells): Other sister cells that, after the initial meristematic initial cells are created, divide once or twice more and then differentiate.

Dicotyledon (dicot): Seed plant that produces an embryo with paired cotyledons, floral organs arranged in cycles of four or five, and leaves with net-like veins.

Differentiation: Process by which cells or tissues undergo a change toward a more specialized form or function.

Dioecious: When an entire plant has only male or only female flowers; means two houses.

Diploid: Term used for zygote cells, where the cell has two sets of chromosomes; abbreviated 2n.

Discovery experiments: Focus on uncovering new relationships and solving problems, follow scientific method, test hypotheses and their predicted outcomes, and utilize a careful design in order to maintain meaningfulness and credibility.

Division: Highest taxonomic category, consisting of one or more related classes, and corresponding approximately to a Phylum in zoological classification.

DNA: Basic biochemical compound that makes up the gene.

Dominant allele: When one allele is expressed over the other alleles present.

Dormant/Dormancy: Term used when seeds are alive and don't germinate when provided with favorable conditions for germination.

Double fertilization: Where one haploid male sperm cell fuses with the female haploid egg cell to form the diploid zygote, and the second haploid male sperm cell fuses with two egg cells to form a triploid endosperm.

Ecodormancy: When external factors, usually environmental, prevent a seed from germinating.

Emasculate: Act of removing the anthers before pollen has been shed from a flower that is used as the female in breeding; used to reduce self-pollination when wanting to cross.

Embryo: Nascent (new, young) plant resulting from the combination of genes from the male sperm transmitted by the pollen to the female egg held in an ovule in the ovary.

Embryo axis: Embryonic root and shoot.

Emergence: Germination, when the embryo becomes active and the radicle grows through the seed coat.

Endocarp: Inner layer of the pericarp.

Endodermis: The innermost cells of the cortex.

Endodormancy: Internal factors within the seed prevent germination.

Endosperm: Tissue that results from the second haploid male sperm cell fusing with two egg cells during fertilization.

Epicotyl: Portion of the stem of a seedling or embryo located between the cotyledons and the first true leaves.

Epidermis: The outermost layer of cells in the plant.

Epigeal: Type of seedling emergence where cell division in the hypocotyl is initially more active and rapid than cell division in the epicotyl. Cotyledons are brought above the soil surface as the hypocotyl expands.

Epigynous: When the perianth and androecium is positioned above the ovary; also called an inferior ovary.

Evaluation experiment: Typically used during the development of new technologies to identify the best products for the desired purpose (eg. which pesticides are effective against the target insect, but not harmful to non-target insects), but are not used to discover new knowledge about how the world works so they typically don't advance our understanding of the natural world. Used to pick a winner from among a number of options.

Evapotranspiration: Movement of water in the plant from the root through the stem to the leaf and out the stomata to the atmosphere; also called transpiration.

Exocarp: Outer layer of the pericarp.

Experimental design: Process of planning an experiment to test a hypothesis.

Experimental unit: The entity to which a specific treatment combination is applied.

Exploration experiments: Focus on detailed observation of organisms and habitats, increase our information about the natural world, identify potential relationships that need to be tested, and are essential to sound and testable hypothesis-building.

Exponential spread: Very rapid spread of an invasive plant as numbers and reproductive rates accelerate.

Family: Taxonomic rank below Order and above Genus.

Fascicular cambium: The cambium within the vascular bundle.

Fertilizer analysis: N-P-K content of a bag of fertilizer; it is shown in percentages by weight.

Fibrous root: Root system where the radicle grows and then rapidly slows or completely halts in growth. Once this happens roots will emerge above the radicle and from the stem tissue located below the soil.

Filament: The stalk that holds up the anther so that pollen grains can be effectively released.

Floret: Single flower in a compound inflorescence.

Floriculture: Discipline of horticulture concerned with the production and marketing of plants valued for their flowers.

Flower: Reproductive structure in a flowering plant.

Forestry: The science or practice of propagating, planting, managing, and caring for forests, which includes harvesting.

Fructose: Simple sugar; it can be produced via photosynthesis.

Fruit: Ripened ovary together with the seeds within the ovary.

Funiculus: Stalk that connects either an ovule or a seed to the placenta.

G1 stage of interphase: First stage of interphase; "G" stands for Gap/Growth.

G2 stage of interphase: Third and final stage of interphase; "G" stands for Gap/Growth.

Gas exchange: Movement of oxygen and carbon dioxide through stomata in the plant.

Generative nucleus: Nucleus in the immature male gametophyte that will later divide by mitosis to produce two sperm cells.

Genes: Hereditary units consisting of a sequence of DNA that occupies a specific location on a chromosome (locus) and determines a particular characteristic in an organism. Genes undergo mutation when their DNA sequence changes.

Genetic code: Order of the four different combinations of the bases in DNA; AT, TA, GC, or CG.

Genetic engineering: Manipulation and introduction of a transgene into a plant for a specific trait.

Genetic pollution: Where genes from GMO crops may escape to conventional crops or weedy relatives.

Genotype: Genetic composition of an organism.

Genus: Group of species possessing fundamental traits in common but differing in other lesser characteristics. The taxonomic rank below Family and above Specific Epithet.

Geophyte: New growth begins underground and the function of the underground growth is storage of food, nutrients, and water during adverse environmental conditions.

Glucose: A simple sugar; it can be produced via photosynthesis.

GMO: Genetically engineered organism or a plant containing a transgene.

Graft union: Location where the rootstock and scion meet.

Grafting: Art and science of connecting two pieces of living plant tissue together in such a manner that they will unite and subsequently grow and develop into one composite plant.

Grana: Stacks of thylakoids.

Granular aggregation: Interaction of small soil aggregates; it is important to have a mixture of large and small holes between the aggregates to allow for water and gas exchange.

Gravitropism: Growth in response to gravity.

Green manure: Crop grown to purposefully be tilled back into the soil to increase the organic matter (and thus change the soil structure). They can also smother weeds.

Ground meristem: The new, primarily parenchyma, cells lying between the protoderm and procambium that will mature to become the cortex tissue.

Guard cells: Located on the epidermis and regulate the size of the stomata.

Guttation: Dew-like drops of water that are forced out of the leaves of some plants due to root pressure.

Gymnosperms: Group of plants whose seeds are produced without the protection of an ovary.

Gynoecium: One of the whorls of the flower and is all of the female reproductive parts; carpels.

Haploid: Term used for gamete cells that typically contain one set of each of the chromosomes; abbreviated n.

Heartwood: The older, darker xylem in the stem that is clogged with resins that limit the transport of water.

Herbaceous: Plants whose above-ground parts die back to the soil surface at the end of the growing season.

Herbaceous annual: Plants that completely die over winter. These plants complete their life cycle from seed to flower to seed in one year.

Herbaceous perennial: Plants where only the above-ground growth dies over winter. The underground portion lives for more than two growing seasons (two years).

Heritability: Measurement of a quantitative trait that passes from parent to offspring and is measured in high and low; high being very similar between parent and offspring and low being dissimilar between parent and offspring.

Hermaphroditic: The term used for a flower that has both the andreocium and gynoecium; also called a perfect flower or bisexual.

Heterozygote: Plant with two different alleles of a particular gene and gives rise to varying offspring; offspring are generally more vigorous than offspring from homozygote.

Hierarchy: System of grouping where each classification is a subset of a superior grouping, and may contain subordinate categories.

Histone protein: Protein around which the DNA surrounds.Homologous chromosomes (homologs) Matching chromosomes from the two different sets; they carry the genetic information that affects the same characteristic or function at the same location on the chromosome; from the sperm and egg cells.

Homozygote: Plant with identical alleles of a particular gene and give rise to identical, or nearly identical, offspring.

Horticulture: The art and science of the development, sustainable production, marketing, and use of high-value, intensively cultivated food and ornamental plants.

Humus: Sticky material made from organic matter that helps bind soil particles together into aggregates; it can absorb and hold up to 6x its weight in water, it releases nitrogen, and holds positively charged cations for plant growth.

Hydrogen bond: When water molecules are near each other and the negative region of one molecule is attracted to the positive region of another; it is a weaker bond than covalent bonds.

Hypocotyl: Embryonic shoot below the cotyledons.

Hypocotyl roots: Roots that emerge above the basal roots.

Hypogeal: Type of seedling emergence where the cotyledons remain below the surface of the ground.

Hypogynous: When the perianth and androecium are attached below the ovary; also called a superior ovary.

Hypothesis: Scientific means of forming a question or proposed explanation made on the basis of limited evidence as a starting point for experimentation. In science, it is a testable statement.

Imbricate bulb: Underground storage organ formed primarily of modified leaves (scales) without a papery covering. Individual scales do not encircle the entire bulb.

Imperfect flower: The term used for flowers that have only the androecium OR only the gynoecium present.

Incomplete flower: The term used for flowers missing one or more of the four whorls.

Indehiscent: Used to categorize fruits with seeds that are retained within the dried pericarp.

Indeterminate: When the apical meristem remains a vegetative meristem that is capable of forming new nodes and internodes throughout the season. Once the hormonal signals are right, reproductive axillary meristems at the nodes below the apical meristem produce inflorescences.

Inferior ovary: When the perianth and androecium is positioned above the ovary; also called an epigynous flower.

Inflorescence: Complete flower structure of a plant and includes the flower, pedicle, rachis, and peduncle.

Initials (cells): Meristem cells that remain meristematic because they continue to initiate new cells.

Integument: Cells that form the ovary wall. Nucellus cells on the interior of the ovule wall develop into megaspore mother cells.

Interfascicular cambium: The cambium between the vascular bundles.

Internode: Stem region between nodes in plants.

Interphase: One of the two major parts of the cell cycle and consists of G1, S, & G2 stages.

Invasive plant: Plant that is non-native to an ecosystem, and whose introduction causes or is likely to cause economic or environmental harm or harm to human health.

Kinetochore: Point of attachment of the spindle and the centromere.

Lag phase: Phase during invasive plant spread that is slow due to a low number of plants being introduced.

Lamina: Another name for a leaf blade.

Lateral meristem: Specialized meristems that are made-up of cells that undergo mitotic cell division.

Lateral or secondary roots: Roots that extend horizontally from the primary root and serve to anchor the plant securely into the soil. This branching of roots also contributes to water uptake, and facilitates the extraction of nutrients required for the growth and development of the plant.

Leaf: A usually green, flattened, lateral structure attached to a stem and functioning as a principal organ of photosynthesis and transpiration in most plants.

Leaf axil: Upper angle between a leaf petiole and the stem to which it is attached.

Leaf blades: Broad portion of a leaf and does not include the petiole.

Leaf margin: Edge of the leaf blade.

Leaf primordia: Young leaves, recently formed by the shoot apical meristem, located at the tip of a shoot.

Leaf scar: Mark indicating the former place of attachment of petiole or leaf base.

Leaf sheath: Structure where the blade attaches to an envelope of leaf tissue that wraps around the shoot of the plant and attaches to a lower node on the stem.

Leaflet: Small leaf-like structure that is found on compound leaves. Multiple leaflets make-up a single compound leaf.

Lenticel: Small opening in the cork of woody stems that allows for gas exchange.

Light absorption: Process in which light is absorbed and converted to energy.

Light Independent Reaction: Second half-reaction in photosynthesis and occurs without the presence of light and uses the energy produced in the Light Reaction to grab the carbon from carbon dioxide and use the carbon to build simple sugars; abbreviated LIR.

Light reaction: First half-reaction in photosynthesis and occurs with the presence of light and uses light energy to split water, which transforms the energy from the sun into hydrogen ions and electrons; abbreviated LR.

Light reflectance: Light wavelengths that are not absorbed, but are reflected back.

Light wavelength: Length of the wave from one peak to the next; it is measured in nanometers.

Linkage: When two genes are on the same chromosome.

Lipids: Compact plant oils that store energy; also called triglycerides.

Locule: A chamber in the ovary.

Locus: Location on a chromosome where a particular gene is found.

Megaspornatium: Place in the ovary where the female gametophyte will be formed.

Mendel's First Law - the law of segregation: Principle that during gamete formation each member of the allelic pair separates from the other member to form the genetic constitution of the gamete e.g. Ss diploid produces S and s gametes.

Mendel's Second Law - the law of independent assortment: Principle that during gamete formation the segregation of the alleles of one allelic pair is independent of the segregation of the alleles of another allelic pair.

Meristem: Group of continuously dividing cells; also called a bud.

Mesocarp: Middle layer of the pericarp.

Mesophyll: The site of most photosynthesis reactions in the leaf and is located in the middle layer of the leaf.

Metaphase: Second stage of mitosis; the spindle fibers grow and form attachments to the pairs of sister chromatids at the centromeres.

Metaphase plate: Equatorial plate which is formed along the midline of the cell between the poles.

Microsporangium: Place in the anther where the male gametophyte will be formed.

Middle lamella: A material containing pectin that forms between cells and that cements the cell wall of one cell to the cell wall of an adjacent cell.

Midrib: Main vein, generally in the center of the leaf from which secondary veins emerge.

Mitochondria: An organelle where the stored sugars are metabolized to produce forms of energy that the plant can use for growth. It is the cell's power plant.

Mitosis: Cell division where a cell divides into two identical daughter cells.

Monocotyledon (monocot): Seed plant that produces an embryo with a single cotyledon and parallel-veined leaves; includes grasses, lilies, palms, and orchids.

Monoecious: When an entire plant has both male and female parts (can be perfect or imperfect); means one house

Multiple fruit: Fruit formed from the ripened ovaries from a cluster of flowers that are in close proximity in an inflorescence that coalesce into one unit.

NADPH: Energy created in the LR and is used to drive the LIR.

Nitrogen: One of the most important elements for plant growth (by quantity); it is a key element found in protein; abbreviated N.

Node: Stem region of a plant where one or more leaves attach and is the location of lateral buds.

Nomenclature: Formal system of names attached to the taxonomic groupings.

Nucleosome: Made up of eight histone proteins and is wrapped by a segment of DNA.

Nucleus: An organelle that contains the chromosomes. Chromosomes contain the genetic code that is carried within each cell and that directs which chemical reactions are turned on and off in the cell.

Olericulture: Discipline of horticulture concerned with the production and marketing of plants or plant parts valued for culinary use as vegetables.

Opposite leaves: Where the leaves grow directly opposite each other on the stem.

Order: Taxonomic rank below Class and above Family.

Organelle: The generic term for a plant organ.

Organic material/matter: Material that has come from a recently living organism (such as plants) that may be partially or fully decomposed.

Organic molecule: Chemical compound associated with living organisms that contain carbon atoms.

Ovary: Part of the carpel and contains ovules which develop into seeds.

Ovary wall: Provides protection to the ovules; also called the pericarp.

Ovule: The part of the ovary that contains an embryo sac and is surrounded by the nucellus, which develops into a seed after fertilization.

Palisade mesophyll: The densely packed, columnar-shaped, elongated cells full of chloroplasts. It is analogous to cortex parenchyma cells in the stem, but in the leaf are specialized for light energy capture.

Palmate venation: Where several veins radiate from the point where the petiole attaches to the blade. The veins fork, then travel a bit, then fork again, travel, fork and so on until the veins reach the margin of the leaf.

Palmately compound leaf: Compound leaf where the petiolules of the leaflets connect directly to the petiole (no rachis).

Parallel venation: Distribution or arrangement of a system of veins in a leaf blade in a non-intersecting network. The veins are parallel to each other and the long edge of the leaf.

Parenchyma: A cell type with thin cell walls, is unspecialized but carries on photosynthesis and cellular respiration and can store food, and form the bulk of the plant body.

Pedicel: Short stalk that holds up the flower.

Peduncle: Large, central stalk that attaches the rachi to the stem of the plant.

Perennial: Plant that lives for more than two growing seasons (more than two years); perennials may be woody or herbaceous (the latter with underground perenniating structures).

Perfect flower: Term used for a flower that has both the andreocium and gynoecium; also called hermaphroditic or bisexual.

Perianth: Both the calyx and corolla.

Pericarp: Ripened ovary wall and it is made-up of three parts: exocarp, mesocarp, and endocarp.

Periclinal division: Type of cell division where the new cells are formed either to the outside or inside and the cell wall that separates the two new cells is parallel to the outside of the stem.

Pericycle: A single layer of tightly packed cells located in the vascular cylinder that retain the ability to divide and produce new cells. This layer of cells is the source of lateral roots.

Periderm: Consists of the cork cambium, phelloderm, and cork.

Perigynous: When the ovary is surrounded by the fused bases of the perianth and androecium.

Petals: Modified leaves that make-up the corolla; they are showy and attract pollinators.

Petiole: Stalk by which most leaves are attached to a stem; it is part of the leaf structure, not the stem.

Petiolule: Stalk that connects the leaflet to the top of the petiole.

Phellem: Another name for cork.

Phelloderm: New cells that are laid down toward the inside of the stem or root by the cork cambium.

Phellogen: Another name for cork cambium.

Phenotype: Physical appearance of an organism.

Phloem: Tissue consisting of sieve tube and companion cells in the vascular system of plants that moves dissolved sugars and other products of photosynthesis from the leaves to other regions of the plant.

Phosphorus: One of the most important elements for plant growth (by quantity); it is a key component in energy transfer molecules like ATP and as part of the DNA backbone; abbreviated P.

Photoautotrophs: Name given to living things, namely plants, that use energy from light to produce organic molecules with which they build their cells and store energy; they are self-nourishing.

Photon: Particle representing a quantum of light; they provide the energy that drives photosynthesis.

Photosynthesis: Process of capturing light energy and producing carbon-based organic molecules.

Phylum: Taxonomic rank below Kingdom and above Class, used in zoological classification.

Pinnate venation: Type of webbed venation where there is a strong midrib and the secondary veins fan out opposite of each other.

Pinnately compound leaf: Compound leaf where the leaflets are arranged opposite of one another on the rachis.

Pistil: Term used when several carpels are fused together.

Pistillate flower: When an imperfect flower only contains the gynoecium.

Pith: Occupies the central part of the stem and is composed of thin-walled parenchyma cells often with larger intercellular spaces than you would find in the cortex.

Placenta: Part of an ovary where the funiculus attach.

Plant tissue culture: Method that uses synthetic growth media to provide the environment for mitotic cell divisions of plants and is used to regenerate a single cell into a whole plant.

Ploidy: Number of sets of homologous chromosomes in a cell.

Plumule: First true leaves of the plant and emerge from the seed, rise above the soil surface, and start to collect energy from the sun.

Polar nuclei: Two haploid nuclei contained within one cell membrane in the mature female gametophyte. One sperm cell will unite with these two polar nuclei to establish the triploid endosperm tissue.

Pomology: Production and marketing of plants or plant parts valued for their culinary use as fruits including nuts); propagated by cuttings, grafting (asexual propagation).

Potassium: One of the most important elements for plant growth (by quantity); it is a key part of the mechanism for moving nutrients into and out of cells; abbreviated K.

Primary (cells): Term used for the cells that originate from cell divisions of the apical meristem

Primary growth: Growth that results from activity by an apical meristem; causes the elongation of the cells in the apical meristem region, which leads to increasing plant length.

Primary meristem: Apical meristems on the shoot and root apices in plants that produce plant primary tissues.

Primary phloem: Phloem tissue that results from differentiation of derivative cells (procambium).

Primary root: Root that forms from the embryonic radicle.

Primary xylem: Xylem tissue that results from differentiation of derivative cells (procambium).

Procambium: New cells in the central part of the root that will mature to become the vascular tissue (xylem, phloem, and vascular cambium).

Prop root: Adventitious root that arises from the stem, penetrates the soil, and helps support the stem, as in corn.

Prophase: First stage of mitosis; chromatin begins to coil and condense to form chromosomes.

Protandry: When the pollen is shed before the stigma is receptive.

Proteins: Sources of amino acids for production of enzymes and other nitrogen-rich compounds in the seed.

Protoderm: New, primarily epidermis, cells laid down toward the exterior of the root which will mature to become the root dermal tissue.

Protogyny: When the stigma is receptive prior to the pollen shedding.

Pruning: Cutting away dead, overgrown, or unwanted branches or stems to improve safety, aesthetics, or productivity.

Punnett square: Simple database used to visualize the types of zygotes and their expected frequency that form from male and female gametes.

Pure line: True breeding plant produced by inbreeding so it is homozygous at most loci and produces identical plants by seed.

Purine: Consists of the base pairs Adenine and Guanine and contains two rings of carbon atoms.

Pyrimidine: Consists of the base pairs Cytosine and Thymine and contains one ring of carbon atoms.

Qualitative differences: Large differences that can easily be seen or measured in qualitative terms; e.g. fruit color.

Quantitative differences: Small differences that are measured numerically; e.g. yield in kg/ha. They can be influenced by the environment.

Quiescent: When a seed does not germinate until given proper conditions for germination (oxygen, water, temperature, and sometimes light).

Rachis: Stalk of a flower that is situated between the peduncle and the pedicel.

Radicle: Embryonic root that breaks through the seed coat during germination and develops into the seedling's root system.

Randomization: Act of randomly assigning treatments to experimental units using a random number table or computer-generated randomization to help minimize any bias that has not been recognized in advance and controlled for in other ways.

Reaction center: Complex of pigments, proteins, and other factors that execute the primary energy conversion reactions of photosynthesis, primarily where water is split in the LR to form the energy carriers ATP & NADPH.

Receptacle: Base of the flower where the floral parts are attached.

Recessive allele: Allele(s) that are not expressed if a dominant allele is present; they will be expressed if there is no dominant allele.

Reciprocal cross: Matching cross where the pollinator becomes the female and female of the former cross becomes the pollen donor.

Replication: When the same treatment is applied to more than one experimental unit.

Reproductive meristem: Apical meristem that transforms into the reproductive tissues (the inflorescence) of the plant.

Resonance: Energy that is passed from one molecule to the next.

Rhizome: Horizontal stem growing just below the soil surface.

Ribose-Phosphate backbone: Chain of alternating ribose and phosphate that make-up the sides of the DNA structure.

Root: Organ that anchors the plant into the soil, takes up water and nutrients, and stores food.

Root cap: Thimble-shaped mass of cells that covers and protects the root apical meristem from rocks, dirt, and pathogens.

Root hair: Thin, hairlike outgrowth of an epidermal cell of a plant root that absorbs water and minerals from the soil. Root hairs live for only a few weeks, deteriorate, and are then replaced by fresh root hairs.

Rootstock: Portion of a graft that contains the root system.

Rubisco: One of the most abundant proteins on earth; it catalyzes the step in the process where carbon from atmospheric CO2 is incorporated into an organic molecule; full name: Ribulose-1,5-bisphosphate carboxylase/oxygenase.

S stage of interphase: Second stage of interphase where the chromosomes replicate (DNA replicated).

Sand: Largest particle in soil and helps increase aeration.

Sapwood: Younger, lighter xylem in the stem that is resin-free and transports water up the trunk.

Saturated fatty acids: Fatty acids that have no double bonds in the chain with all carbon atoms in the interior of the chain having 2 attached hydrogen atoms.

Scarification: Process used to break a physical seed dormancy (hard seed coat).

Science: Systematic study of the structure and behavior of the physical and natural world through observation and experiment.

Scientific discovery: Process of scientific inquiry; it builds on what is known by testing hypotheses.

Scion: Portion of a graft that contains the shoot system and all above-ground parts.

Sclerenchyma: A cell type with thickened, rigid, secondary walls that are hardened with lignin and provides support for the plant.

Secondary growth: Production of xylem and phloem from cambium cells.

Secondary phloem: New phloem that is formed on the outside and is produced by the fascicular cambium.

Secondary root: Root that forms off of the primary root.

Secondary xylem: New xylem that is formed on the inside and is produced by the fascicular cambium.

Seed: Ripened ovule containing a seed covering, food storage, and an embryo.

Seed coat: Outer layer of the seed.

Seed germination: Activation of metabolic pathways of the embryo leading to the emergence of a new seedling.

seeds: When there are genetic mechanisms that inhibit self-pollination of a flower.

Self-pollination: When the pollen from the plant pollinates the stigma of the same plant.

Sepals: Outermost whorl of the flower that protects the flower and photosynthesizes.

Sessile: When a leaf lacks a petiole; called a sessile leaf.

Sexual propagation: Form of propagation that results in plants with genetics that differ from the parent plants, also called seed propagation.

Shoot: Made up of a central axis (stem) with a repeating pattern of nodes and internodes.

Sieve tube members: Elongated cells that join end to end to form tubes for passage of liquids. The end walls have pores. Unlike xylem cells, these cells are still alive. They have a thin cell membrane containing a layer of living protoplasm that hugs the wall of the cell.

Silt: Intermediate particle size in soil.

Simple fruit: Fruit formed from a flower with one carpel or multiple carpels fused together so that it looks like just one carpel.

Simple inflorescence: Type of inflorescence with a peduncle, rachis, pedicel, and single flower structure.

Simple leaf: Leaf with an uninterrupted blade margin.

Simple sugars: Monosaccharides; examples include glucose and fructose.

Sister chromatids: The two chromosomes that are exact copies that are created during S stage of interphase.

Soil compaction: When the pore spaces between soil aggregates are compressed.

Soil organic matter: Carbon-based plant, animal, and/or microbe tissues that are in the process of breaking down; increasing soil organic matter improves and stabilizes soil aggregation.

Soil structure: The way in which the soil particles and other materials, like the organic matter in the soil, bind together into clumps.

Soil texture: Relative proportion of sand, silt, and clay particles in the soil.

Somatic cells: Cells of flowering plants, other than the reproductive cells; always 2n.

Specific epithet: Uncapitalized Latin adjective or noun that follows a capitalized Genus name in binomial nomenclature and serves to distinguish a species from others in the same genus, as saccharum in Acer saccharum (sugar maple).

Spindle apparatus: Microtubules associated with movement of the chromosomes during division.

Spongy mesophyll: Loosely packed cells with large air spaces in between the cells, which allows movement and exchange of gases, specifically oxygen, carbon dioxide and water vapor. Spongy mesophyll cells also contain chloroplasts.

Sporangia: Structures in the androecium and gynoecium where meiosis takes place and the gametophyte generation develops.

Spore: Haploid single cell produced by meiosis in the sporangium of a diploid sporophyte.

Stamen: Modified leaf and collectively they make up the androecium. A stamen is made-up of the anther and filament.

Staminate flower: When an imperfect flower only contains the androecium.

Starch: Key energy storage compound in plant cells; it is a long glucose chain; it sequesters atmospheric carbon for short-term use.

Statocytes: Specialized cells that help the plant to sense gravity and grow accordingly.

Stem: Supporting and conducting organ usually developed initially from the epicotyl and growing upward and consists of nodes and internodes.

Stigma: Receptive apex of the carpel of a flower, on which pollen is deposited at pollination.

Stipule: Usually a pair of appendages located at the base of a leaf but may be fused into a ring around the stem; variable in size, shape, and texture; serves for protection or to attract pollinators.

Stolon: Stem with long internodes that grows along the surface of the ground.

Stomate/Stoma/Stomata: Gap in the epidermis that allows gas exchange between the atmosphere and internal parts of the leaf.

Storage root: Root that is modified for storage of nutrients, such as carrots and beets.

Stratification: Process used to break a physiological dormancy, such as embryonic or endo/eco-dormancies.

Stroma: Interior of the chloroplast; it is the site of the LIR

Style: Part of the carpel that elevates the stigma to a position for reception of pollen and is a conduit for pollen tube growth.

Suberin: Impermeable (to water and gases), waxy substance present in the cell walls of corky tissues.

Sucrose: Sugar that is transported by the phloem throughout the plant to provide energy and building blocks for other organic molecules like starch and cellulose.

Super weed: Weed produced by crossing with a GMO crop and inherits the GMO trait, like herbicide tolerance.

Superior ovary: When the perianth and androecium are attached below the ovary; also called a hypogynous flower.

Suspensor: Produced by multiple mitotic cell divisions of the embryo's basal cell; the suspensor anchors the apical cell of the embryo to the ovule wall.

Symplast: Interior to the cell membrane, where water and minerals are transported through cells.

Synergid cells: Cells flanking the egg cell in the mature female gametophyte.

T-budding: A type of budding performed using dormant scion buds on actively growing rootstocks; typically done in late summer outdoors.

Tap root: Main root of a plant, usually stouter than the lateral roots and growing straight downward from the stem.

Taxonomy: Science of classifying organisms.

Telophase: Fourth and final stage of mitosis; the nuclear membrane forms around the chromosomes in each of the daughter cells.

Tension: Differential pressure; in plants this occurs as water molecules are pulled through the plant via transpiration.

Tepal: When the sepals and petals are showy and indistinguishable.

Terminal bud: Bud located at the apex of a stem.

Tetrads: Groupings of four sister chromatids.

Thylakoid membrane: Membrane that surrounds the thylakoid.

Thylakoids: Membrane-bound compartments inside chloroplasts and cyanobacteria, and are the site of the light-dependent reactions of photosynthesis.

Tillage: Process of incorporating the residue from the top of the soil into the soil; there are many types of tillage.

Tissue: A group of cells that share a function.

Topworking: A type of grafting performed on established orchard trees.

Totipotent: Ability of a single plant cell to grow into a whole plant.

Tracheids: Elongated and narrower than vessels, connected by overlapping at their ends, are dead at maturity, and contain pits through which water can move.

Transformation: Process of using the gene gun or Agrobacterium tumefaciens to introduce a transgene into a plant.

Transformed: Synonym for GMO, or plant carrying a transgene.

Transgene: Gene introduced into a plant from another organism, not through sexual reproduction.

Transgenic: Synonym for GMO, or plant carrying a transgene.

Translocation: Movement of a substance from one place to another.

Transpiration: Movement of water in the plant from the root to stem to leaf and out through the stomata to the atmosphere; also called evapotranspiration.

Treatments: Administration or application of agents to a plant to prevent disease or facilitate growth.

Trichome: Either unicellular or multicellular hair-like outgrowths arising from the epidermis; found on stems.

Trifoliate leaf: Compound leaf with three leaflets that attach to a rachis.

Triglycerides: Another name for lipids.

Triose phosphate: A 3-carbon sugar (triose) with phosphorus and oxygen atoms (phosphate); G3P is an example.

Triploid: Term used for endosperm that has three sets of chromosomes; abbreviated 3n.

Tube nucleus (or vegetative nucleus): Nucleus in the male gametophyte that is associated with pollen tube growth.

Tuber: Swollen, underground, modified stems that store food.

Tunicate bulb: An underground storage organ formed primarily of modified leaves formed in concentric circles around the active meristem. The bulb is covered with a papery covering.

Turgor pressure: Water pressure inside of cells.

Umbel: Inflorescence with multiple flowers originating from a common point.

Unsaturated fatty acids: Fatty acids that have one or more double bonds between one or more carbon atoms in the chain, lack some hydrogen atoms, and therefore the carbon atoms are not saturated with hydrogen.

Vacuole: An organelle containing various fluids including stored chemical energy like starch and waste products from the cell. The vacuole takes up much of the cell volume and gives shape to the cell.

Vascular bundle: System containing vessels that carry or circulate fluids and dissolved minerals in the plant; composed of xylem, phloem, and bundle sheath cells.

Vascular cambium: Lateral meristem producing vascular tissues.

Vascular tissue: System containing vessels that carry or circulate fluids and dissolved minerals in the plant; composed of xylem, phloem, and bundle sheath cells.

Venation: Pattern of veins on a leaf.

Vessels: Elongated xylem cells that connect end to end to form tubes, are dead at maturity, and have perforated end walls so water can move freely through the holes and flow from cell to cell. Vessels have a relatively large diameter compared to other xylem cells and allow greater movement of water.

Viral coat protein: Protein that surrounds the viral genome, protecting it, and is essential to virus replication.

Whip and tongue graft: A type of graft where both scion and rootstock are dormant and the same diameter; it is much more secure than other types of bench grafts.

Whorl: Node on the receptacle where the four types of modified leaves are attached (four whorls of a flower).

Whorled leaves: Where the leaves are oriented in a whorled formation in which their point of attachment appears to spiral up the stem.

Woody perennial: Plant that lives for more than a year, has hard rather than fleshy stems, and bears buds that survive above ground in winter. Trees, shrubs, many vines, and bamboo are examples of woody perennials.

Xylem: Supporting and water-conducting tissue of vascular plants.

Zone of Differentiation: Area in roots where tissues are formed (expand in width).

Zone of Elongation: Area in roots where recently produced cells grow and elongate prior to differentiation.

INDEX

S

T

V

Z

www.ingramcontent.com/pod-product-compliance
Lightning Source LLC
LaVergne TN
LVHW062217200926
848484LV00006B/1416